Fossil Fuels in the Arab World: Facts and Fiction

Global and Arab Insights of Oil, Natural Gas & Coal

Fossil Fuels in the Arab World: Facts and Fiction

Global and Arab Insights of Oil, Natural Gas & Coal

Basel Nashat Asmar

2050 consulting

First published in the United Kingdom in 2010 by
2050 Consulting
48 Imperial Hall
104-122 City Road
London EC1V 2NR
United Kingdom

Copyright ©Basel Nashat Asmar, 2010

Basel Nashat Asmar has asserted his right under the Copyright, Designs and Patents Act 1988 to be identified as the author of this work.

All rights reserved. No part of this publication may be reproduced, stored in a retrieval system, or transmitted, in any form or by any means, electronic, mechanical, photocopying, recording, or otherwise, without the prior written permission of the publisher.

A CIP for this book is available from the British Library.

ISBN 978-0-9567368-0-2

Author's Notes:
1. This book was written prior to the author joining IHS CERA. The material in the book made no use of any proprietary data owned by IHS, and does not express IHS CERA's opinion. The opinions expressed in this book are the responsibility of the author.

2. The use of particular designations of countries or territories does not imply any judgement by the author as to the legal status of such countries or territories, of their authorities and institutions or of the delimitation of their boundaries.

Printed in the United Kingdom

Cover design by Eman Faidi

to my mum

LIST OF CONTENTS

Acknowledgments	xii
About the Author	xiii
Foreword	xiv
Introduction	xviii

Chapter 1	**Fossil Fuels – Overview and Background**	**1**
1.1	Fossil Fuels	1
1.2	Fossil Fuels Classification	2
1.3	Fossil Fuels Uses	2
Chapter 2	**Petroleum - Basic Knowledge**	**7**
2.1	Oil Classifications	8
2.2	Extra Heavy Oil and Natural Bitumen	13
2.2.1	*Orimulsion*	*14*
2.3	Oil Shale	15
2.4	Natural Asphalt	16
2.5	Arctic and Antarctic Oil	17
2.6	Natural Gas Liquids (NGL)	18
2.7	Refined Petroleum Products	19
2.8	Other Liquid Fossil Fuels	20
2.9	Oil Uses	23
2.10	Oil Transport	24
2.11	Oil Storage	25
2.12	Oil Measurement Units	26
Chapter 3	**Natural Gas – Basic Knowledge**	**28**
3.1	Natural Gas Classifications	30
3.2	Tight Natural Gas	32
3.3	Shale Gas	33
3.4	Coalbed Methane	33
3.5	Methane Hydrates	33
3.6	Gas in Geopressurised and Hydropressurised Zones	34
3.7	Arctic and Antarctic Natural Gas	34
3.8	Secondary Natural Gas	35
3.9	Other Gaseous Fossil Fuels	35
3.10	Natural Gas Uses	37
3.11	Natural Gas Transport	38
3.12	Natural Gas Storage	39

3.13	Natural Gas Measurement Units	41

Chapter 4 Coal – Basic Knowledge — 42

4.1	Coal Classifications	44
4.2	Peat	46
4.3	Arctic and Antarctic Coal	47
4.4	Secondary Coal	48
4.5	Other Solid Fossil Fuels	49
4.6	Coal Uses	49
4.7	Coal Transport	51
4.8	Coal Storage	52
4.9	Coal Measurement Units	53

Chapter 5 Biofuels – Basic Knowledge — 54

5.1	Liquid Biofuels	54
5.2	Gaseous Biofuels (Biogas)	55
5.3	Solid Biofuels	57
5.4	Synthetic Fuels from Pollution	58

Chapter 6 Resources and Reserves: Decoupling the Ambiguity — 59

6.1	Fossil Fuel Resources	60
6.2	Ultimately Recoverable Resources	64
6.3	Fossil Fuel Reserves	64
6.3.1	*Case Studies*	*69*
6.4	Fossil Fuel Contingent Resources	71
6.5	Fossil Fuel Prospective Resources	72
6.6	Recovery Factors	73
6.7	Reserve Growth	74
6.8	Peak Theory	76

Chapter 7 Geographical and Political Contexts — 81

7.1	Arab World	81
7.2	Middle East	82
7.3	MENA Countries	83
7.4	Africa	83
7.5	European Union	84
7.6	OPEC Countries	85
7.7	OAPEC Countries	86
7.8	OECD Countries	86
7.9	GECF Countries	87

Chapter 8 Data Sources — 88

8.1	Classification of Data Sources	89
8.1.1	*Oil, Gas and Mining Companies*	*89*
8.1.2	*Governmental Bodies*	*91*
8.1.3	*Trade and Industry Journals*	*92*

8.1.4	International Organisations	93
8.1.5	Industrial and Commercial Bodies and Organisations	95
8.1.6	Professional Bodies and Organisations	95
8.1.7	Information Service Companies and Consultant Companies	96
8.1.8	Financial Institutions and Investment Banks	97
8.1.9	Think Tanks	98
8.1.10	Scientific and Academic Publications	98
8.1.11	Mass Media and the Internet	99
8.2	Underlying Data Quality Challenges	100

Chapter 9 Oil – Global and Arab Perspectives — **103**

9.1	Conventional Oil Reserves and Resources	104
9.1.1	Conventional Oil Reserves	104
9.1.2	Conventional Oil Resources	112
9.1.3	Reliability of Conventional Oil Reserves and Resources Data	116
9.2	Unconventional Oil Reserves and Resources	124
9.2.1	Extra Heavy Oil and Natural Bitumen	126
9.2.2	Oil Shale	134
9.2.3	Reliability of Unconventional Oil Reserves and Resources Data	138
9.3	Overall Oil Reserves and Resources	140
9.4	Oil Production	141
9.4.1	Converted Oil Production	153
9.4.2	Other Liquid Fossil Fuels Production	154
9.5	Oil Reserve to Production Ratio (R/P)	154
9.6	Refined Petroleum Production and Oil Refining Capacity	157
9.7	Oil Consumption	167
9.8	Oil Trade	177
9.8.1	Oil Transport	189

Chapter 10 Natural Gas – Global and Arab Perspective — **192**

10.1	Conventional Natural Gas Reserves and Resources	193
10.1.1	Conventional Natural Gas Reserves	193
10.1.2	Conventional Natural Gas Resources	200
10.1.3	Reliability of Conventional Natural Gas Reserves and Resources Data	204
10.2	Unconventional Natural Gas Reserves and Resources	208
10.2.1	Tight Natural Gas	210
10.2.2	Shale Gas	216
10.2.3	Coalbed Methane	221
10.2.4	Methane Hydrates	225
10.2.5	Geopressurised and Hydropressurised Gas	229
10.2.6	Reliability of Unconventional Natural Gas Reserves and Resources Data	233
10.3	Overall Natural Gas Reserves and Resources	235

10.4	Natural Gas Production	236
10.4.1	*Converted Natural Gas Production*	*246*
10.4.2	*Other Gaseous Fossil Fuels Production*	*247*
10.5	Natural Gas Reserve to Production Ratio (R/P)	248
10.6	Natural Gas Consumption	249
10.7	Natural Gas Trade	259
10.7.1	*Natural Gas Transport*	*270*

Chapter 11 Coal – Global and Arab Perspective — 274

11.1	Conventional Coal Reserves and Resources	275
11.1.1	*Conventional Coal Reserves*	*275*
11.1.2	*Conventional Coal Resources*	*283*
11.1.3	*Reliability of Conventional Coal Reserves and Resources Data*	*288*
11.2	Unconventional Coal Reserves and Resources	291
11.2.1	*Peat*	*292*
11.2.2	*Reliability of Unconventional Coal Reserves and Resources Data*	*296*
11.3	Overall Coal Reserves and Resources	297
11.4	Coal Production	298
11.4.1	*Converted Coal Production*	*306*
11.4.2	*Other Solid Fossil Fuels Production*	*308*
11.4.3	*Secondary Coal Production*	*308*
11.5	Coal Reserve to Production Ratio (R/P)	310
11.6	Coal Consumption	311
11.7	Coal Trade	319
11.7.1	*Coal Transport*	*329*

Chapter 12 Fossil Fuels - Overall Global and Arab Perspectives — 330

12.1	Interrelation of Fossil Fuels Analysis	330
12.1.1	*Conversion Factors*	*331*
12.2	Overall Conventional Fossil Fuels Reserves and Resources	332
12.2.1	*Overall Conventional Fossil Fuels Reserves*	*332*
12.2.2	*Overall Conventional Fossil Fuels Resources*	*335*
12.3	Overall Unconventional Fossil Fuels Reserves and Resources	337
12.4	Overall Totalised Fossil Fuels Reserves and Resources	339
12.5	Overall Fossil Fuels Production	341
12.6	Overall Fossil Fuels Consumption	344
12.7	Overall Fossil Fuels Trade	349
12.8	The Bottom Line	355

Chapter 13 Fossil Fuels and Alternative Energy — 356

13.1	Alternative Energy Overview	356
13.1.1	*Alternative Energy Types*	*356*
13.1.2	*Alternative Energy Classifications*	*358*

13.2	Can Alternative Energy Replace Fossil Fuels?	361
Chapter 14	**Fossil Fuels and the Arab World – Differentiating Facts from Fiction**	**366**
14.1	Myths Debunked	366
14.1.1	*Myth 1: We are running out of oil, or as some depletionists recently extend it, of all fossil fuels.*	*366*
14.1.2	*Myth 2: Technology does not enhance fossil fuels availability.*	*367*
14.1.3	*Myth 3: The Arab world is home to two thirds of the world's oil reserves, and thus its oil is indispensable to the world.*	*368*
14.1.4	*Myth 4: OPEC and both national and international oil and gas companies control oil price.*	*369*
14.1.5	*Myth 5: The USA is dependent on Arab oil.*	*370*
14.1.6	*Myth 6: Energy independence is possible.*	*371*
14.1.7	*Myth 7: Alternative energy can replace fossil fuels.*	*372*
14.2	Questions For Thought	372
Appendix A: Arab World Statistics		**374**

Acknowledgments

I would like to thank Dr Paul Langston for his insightful comments on the draft chapters, for discussing key ideas and for gauging the opinion of numerous colleagues at Nottingham University, in particular Professor John Patrick.

I am also thankful to my professional colleagues at CB&I Lummus, Jonathan Taylor and Stephen Lamport, for their encouragement and constructive criticism.

I am indebted to my friends and family in London and Amman, who encouraged and supported me during the long process, especially my brothers Ghassan and Hussam.

Special praise is given in particular to Ian Holder, whose patience in listening to my continuous ramblings was phenomenal and who, through his coaching skills, offered frequent guidance that shaped this book.

Completion of this book would have not been possible without the efforts of Karen Hall, who not only carefully edited my manuscript, but also challenged my thinking by raising astute questions and making shrewd comments that steered me in the right direction.

ABOUT THE AUTHOR

Basel Nashat Asmar is an expert in oil and gas fundamentals and a dynamic simulation expert with extensive computer, modelling, and simulation skills. He is currently an Associate Director with IHS CERA, based in London, UK. Dr. Asmar worked previously in several roles with major engineering companies concerned with large liquefied natural gas (LNG) regasification and liquefaction terminals, natural gas compression stations, and offshore oil and gas production platforms. He was as a senior process engineer and dynamic simulation specialist with CB&I, Mott MacDonald, and IMEG, as well as a lead consultant with 2050 Consulting Ltd and Trident Consultants Ltd. Prior to that he worked in academia as a research associate at the University of Nottingham.

Dr. Asmar has published more than 40 articles in international journals, conference proceedings, and newspapers. He is a Chartered Engineer, a member of the Institution of Chemical Engineers (IChemE), a senior member of the American Institute of Chemical Engineers (AIChE) and a member of the Jordanian Engineers Association. He holds a BSc in Chemical Engineering from the University of Jordan, an MSc in Process and Project Engineering and a PhD in Chemical Engineering from the University of Nottingham, and a doctorate in Geoscience from Freie Universität Berlin, Germany.

FOREWORD

By

Philippe Brandt[1]
Minister
Federal Department of Foreign Affairs
Bern, Switzerland

Although it has been some years since I have had personal and professional dealings with the diplomatic world relating to oil producing countries, reading this book has been like a 'crash' revision in the relevant subject matter. We have here a brilliant insider analysis of 'the' hot issue that not only dominates world politics at present, but is also high on the agenda of our daily lives i.e. energy sources, with a specific emphasis on the Arab world.

Like everyone in Western Europe, when I look through information presented by the media about energy, I assume that I know what the challenges of the next decades will look like: fuel shortages, pollution, conflicts of all kinds, but also new techniques and alternative energy sources.

Living in the Northern Hemisphere all of my life has brought to me a sort of security that can be expressed in a few simple words: warmth, travel, convenience, a life style reliant on energy sources, thus daily life and habits that are taken for granted. However, this feeling of security has been somewhat reduced by a diffuse impression from the media suggesting that perhaps, in a future not that far away from us, our sources of energy might be lacking. It seems that our daily routine is threatened in the future by a possible shortage of energy. This could in turn force us to

---

[1] The author is expressing a personal point of view here. The opinions stated in his preface should not be regarded as the official position of Switzerland or of the Swiss Federal Department of Foreign Affairs.

radically change our lives and give up many things that we are used to. But is this true? What are the real challenges?

Have we been living with a myth or with many myths? It is our future and that of future generations which is at stake and just behaving as if we were not concerned would be totally irresponsible in my eyes. Sustainable development – including energy use and global warming – has become the key to the future. As was already mentioned, we have been accustomed to think that our way of life depends very much on energy sources in the Arab world, at least in the past. More recently, this perception has changed, but is it really a twisted reality?

Dr. Asmar starts by examining a commonly held belief, expounded by the media, politicians, pundits and government spokespeople, that our world mainly relies on Arab countries in terms of fossil fuel supply – in particular oil. He discusses the consequences that this widely shared view i.e. energy dependence on Arab energy sources, may have on Western societies. Obviously, he makes the link with fuel and politics in the region, as well as the autocratic regimes and their relationships with other countries all over the world. Finally, he endeavours to debunk the many myths and inaccuracies with an analysis of facts and figures.

Part I sheds light on technical issues and terminology, whereas Part II introduces the reader to the actual analysis of the facts presented in the previous part. Part III, in particular, allows the reader to get a broader picture and put into perspective what all these figures mean, in economical and political terms. Moreover, it helps us to grasp what is happening in other parts of the world where energy also represents a vital resource, e.g. Iran, Caucasus and Central Asia. The major efforts of diversification of the energy mix made by governments e.g. building corridors from the Caspian to Western Europe or a gas pipeline between Russia and Germany across the Baltic Sea confirm the enormity of the challenges ahead of us.

One after another, Dr. Asmar debunks the myths – separating

facts from fiction. Here are a few examples: he demonstrates that our way of life in not about to collapse because of lack of oil resources nor are the Arab world oil reserves indispensable to our functioning. He also gives us a clearer view of one of the most hotly debated topics, i.e. the dependence of the USA on Arab oil; his conclusions are surprising.

Of course, the readers can agree or not with Dr. Asmar's conclusions, but the many pieces of information at his disposal will enable you to judge the facts about fossil fuels for yourselves. He has given us food for thought and opportunities to reflect on our condition. He demonstrates that politics is the key to comprehending energy issues. World politics will tip the balance in one direction or another – and we are not talking about supply and demand! We know that the Arab world is not lacking in political challenges! We are all aware of the bloody conflicts that have plagued this region for decades now, in addition to the more recent conflagrations like the Iraq war. Access to fossil fuels was, some said, one of the reasons to go to war in this part of the globe. In other words, energy issues can lead to war. The link between corruption in Arab governments and dictators is certainly based on many factors, not only on energy sources, but this book clearly shows that political pronouncements can have an impact on the perception that we have of the Arab world as a fossil fuel provider.

It has been a real pleasure to go through this presentation of facts and analysis. Dr. Asmar helps the reader to sort out the available information on a vast topic and has written in such a way as to be accessible to people without a deep knowledge of political, oil and energy matters. Equally, it will be very useful for politicians, diplomats, journalists or commentators dealing with energy and politics, as well as to ordinary citizens wishing to be better informed, by correcting their understanding of commonly misused terms and supplementing their opinions with correct facts and figures.

Even if some readers disagree with the conclusions, I believe it is of the utmost importance to have the knowledge he presents as it gives us a different perspective that enables us a better understanding of a major component of our world and hence make enlightened choices in our lives.

Philippe Brandt
Bern
October 2010

INTRODUCTION

Apart from association with terrorism, oil is the most important word linked to the Arab world. The continuous bombardment by the 'experts' using the terms oil money and Arab money interchangeably has lead the public worldwide, to unequivocally believe that the Arab world is an indispensable energy source without which our civilisation will come to a halt and collapse. This belief is often used as a pretext to justify continuous political and military meddling in the Arab world's affairs, and is even used by democratic western leaders to defend supporting autocratic and authoritarian Arab governments to stay in power. But is the world really this vulnerable? And does our civilisation rely for its energy supplies on what happens in one unstable region in the world? This book tries to answer this question by analysing the facts and endeavouring to separate the facts from the fiction.

In constructing the response to the question above, this book uniquely examines not only oil, but the three types of fossil fuels: oil (petroleum), natural gas, and coal, which are in essence convertible, and thus it is crucial to inspect all three to paint a true picture of mankind's dependence on fossil fuels, particularly oil, and to clearly assess the position of the Arab world in this picture.

This book tackles the fossil fuels topic from two main angles; the first angle is the need to clarify the terminology and the terms used by the media, the politicians, the scientists, and the so-called experts, when discussing oil, natural gas and coal issues, while the second angle is to evaluate the hard facts by analysing the numbers and scrutinise them impartially to come with a definitive quantitative answer. This answer is then subjected to further investigation by viewing it in relation to alternative energy, and from political perspective so that we can draw meaningful conclusions.

The text is arranged in fourteen chapters that are structured into three generalised parts:

- Part I: Overview and Background. It consists of eight short chapters, in which all oil, natural gas, coal, and biofuels terminology and concepts are introduced, as well as a generalised discussion of the notions of reserves and resources, the geographical context, and the data sources and their objectivity and reliability.

- Part II: Analysis. It consists of three detailed chapters, in which sets of data describing the status of oil, natural gas, and coal, in terms of four categories (reserves and resources, production, consumption and trade) are presented and analysed. For each fuel type, the analysis is presented firstly on a global scale to give a worldwide viewpoint, then focusing on examining the situation in the Arab world with its share and contribution are put in perspective.

- Part III: Bigger Picture. It consists of three short chapters, in which an overall fossil fuel situation for the world and the positioning of the Arab world within the full picture is presented. The analysis extends to compare the fossil fuel situation to the much hyped alternative energy to establish the facts exerting a sense of reality to if these alternatives can substitute fossil fuels, quenching our thirst and addiction! Finally, the situation is reviewed considering the current political insight and spin, where a few well publicised myths are exposed and debunked.

Part I is essential reading for readers who have no prior knowledge of the subject, but are interested in developing a basic understanding of fossil fuels concepts and terminology, and can be used as necessary reference. Part II is the analysis that constitutes the bulk of the book. Part III is indispensable reading for all readers as it puts everything in global perspective -

politically, economically, and technically, challenging the readers, to judge for themselves the overall global fossil fuel situation and whether the Arab world's energy endowment is as crucial as we are made to believe.

Chapter 1
FOSSIL FUELS – OVERVIEW AND BACKGROUND

1.1 Fossil Fuels

It is no understatement to claim that the world development in the last two centuries has depended on the carbon based materials widely known as fossil fuels. Most of us are fully aware that these raw materials provide the majority of the necessary energy used by industry, in homes and offices, and for transport, either directly or by means of generated electricity or heat. Yet, most are completely oblivious that these same fossil fuels are also responsible for the significant increase in food production, needed to support the ever increasing world population, by providing the crucial and affordable feedstocks to manufacture fertilizers and pesticides, thus maintaining the current boom in agricultural output.

Furthermore, just look around! It is virtually impossible to spend more than few minutes wherever we are in the world without coming into contact with plastic or similar polymers, which achieved dominant position as the material of choice for everyday appliances. The raw materials to produce plastics are derived from fossil fuels.

Chemically, fossil fuels are energy-rich combustible geologic deposits consisting largely of hydrocarbons, which are compounds composed of hydrogen and carbon, as well as smaller amounts of other compounds such as nitrogen and sulphur. They are, practically speaking, non-renewable resources of energy as the rate of their consumption far exceeds the rate of their formation.

Geologically, two conflicting theories claim to explain the

formation of fossil fuels: the biogenic (organic) and the abiogenic (inorganic) theories. The biogenic theory of fossil fuels formation is almost universally accepted. It states that fossil fuels were formed of organic materials, from decayed plants and animals, which have been converted to natural gas, petroleum (oil), or coal by exposure to heat and pressure in the Earth's crust over hundreds of millions of years. However a few dissident opinions endorse the abiogenic theory, where they believe that some fossil fuels, particularly light hydrocarbons, were formed from deep carbon deposits by abiogenic processes, without any living material involved in their formation, perhaps dating to the formation of the Earth. Some even go further and suggest that the oil and natural gas are merely condensates from the magma in the Earth's core.

1.2   Fossil Fuels Classification
The classification of fossil fuels is not an exact science resulting in numerous classifications being used worldwide. Broadly speaking, fossil fuels are classified into three types, based on the phase, i.e. the state of matter, of the fuel at atmospheric conditions, even though the fuel may exist in a different phase at reservoir or deposit conditions. The three fossil fuel types are:
- Coal – solid state.
- Petroleum (Oil) – liquid state.
- Natural Gas – gaseous state.

These types are discussed in more detail in the subsequent chapters. The boundaries between the three types are not precisely defined, especially when a fuel changes its physical phase between reservoir and atmospheric conditions. Therefore ambiguity exists in some cases, for example in the case of natural gas liquids, which are gaseous at reservoir conditions but turn into liquid at controlled atmospheric pressure or temperature. More detailed sub-classifications have been developed within each type, which take individual or combined characteristics into consideration.

1.3   Fossil Fuels Uses
Energy demand continues to rise due to rapid global

industrialisation and population growth. This is illustrated in Figure 1.1 which shows the world total energy consumption since 1980. Although all sources show similar increasing trends, discrepancies are evident in terms of energy consumption estimated values. The discrepancies in the reported numbers are mainly attributed to the contribution of energy from biomass, which is totally ignored by BP, only considered in the United States by the American Energy Information Administration (EIA), while it is included globally in the International Energy Agency (IEA) numbers. This discrepancy is further confirmed by examining the coal, oil and natural gas reported consumption on the same graph, which demonstrates that the three sources report very close values for the three fossil fuels. Naturally minor differences are always inevitable in the reported numbers since not all countries report their energy consumption with the same rigor, and the exact energy content of each resource varies with quality and according to standards used by the nations defining energy content of diverse resources.

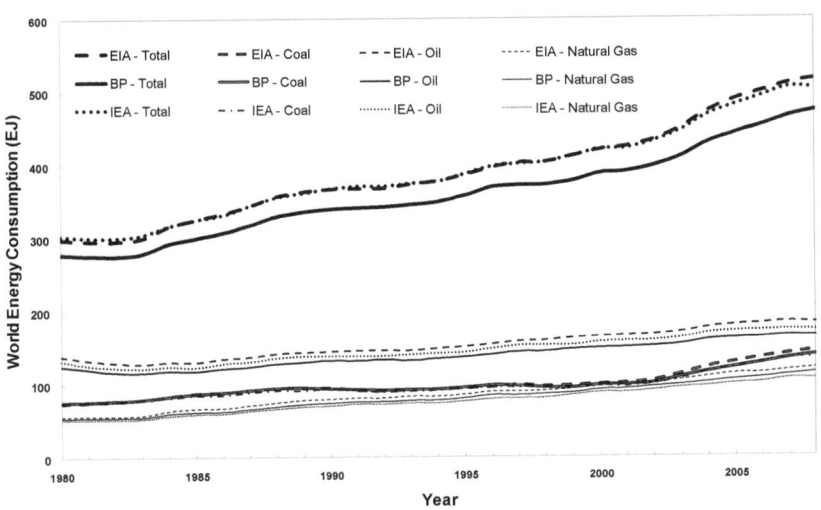

Figure 1.1: World total energy demand (1980-2008)
Source: EIA (http://www.eia.doe.gov/international); IEA (http://www.iea.org/stats/index.asp); BP (Statistical Review of World Energy 2001-2009).

Furthermore, it can be seen by inspecting Figure 1.1, that all sources agree fossil fuels are the primary source of fuel on the planet. This is confirmed in Figure 1.2, which features the fossil fuel contribution as a percentage of the world's total energy consumption since 1980. The figure highlights the world dependence on fossil fuels, showing that they account for over 80% of the total energy demand, even according to the IEA which due to its inclusion of biomass energy credits fossil fuels with lower share of total consumption. Thus it can be seen that despite all the publicity enjoyed by all alternative and renewable sources of energy, the share of fossil fuels supply required to satisfy total demand remains stubbornly high.

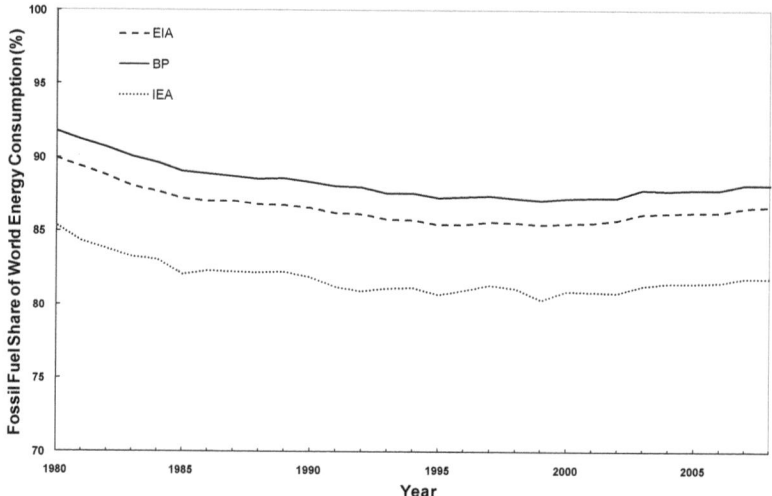

Figure 1.2: Fossil Fuels contribution of the world's total energy (1980-2008)
Source: EIA (http://www.eia.doe.gov/international); IEA (http://www.iea.org/stats/index.asp); BP (Statistical Review of World Energy 2001-2009).

According to the IEA, the amount of energy consumed worldwide was estimated to be 491.7 EJ in 2006. It is estimated that 81.5% of this total consumption was from fossil fuels (35% oil, 26% coal, 20.5% natural gas). Nuclear energy contributed another 6.2%, while the remainder came from biomass and other renewable energy resources.

Fossil fuels are used as the primary energy source in transportation, industrial, residential and commercial sectors, as well as generating electricity and heat in power plants which in turn are used as a secondary energy source in the above four sectors. As a matter of fact, electricity and heat generation are transformative processes where energy is converted from one form to another, then reused. Thus they are not considered final usage sectors; rather they are treated as intermediaries, which contribute to the main usage sectors.

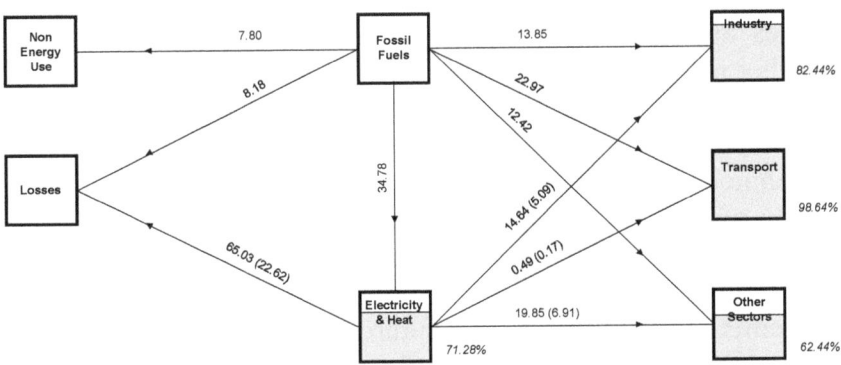

Figure 1.3: Distribution of total fossil fuels global consumption (2006)
Source: IEA (http://www.iea.org/stats/index.asp).

Figure 1.3 details the distribution of the combined fossil fuels global consumption derived from the data reported by the IEA in 2006. Transportation consumes around 23.1% of the total fossil fuels supply, which constitute over 98.6% of the sector's total need; industry consumes approximately 18.9% of the total fossil fuels supply which constitute around 82.4% of the sector's total need; and other sectors including residential, commercial and agricultural consume approximately 19.3% of the total fossil fuels supply which constitute around 62.4% of the sectors' total need. These numbers include the primary consumption and the consumption supplied by means of electricity and heat output from power plants. The figure also illustrates that, if treated as a sector, electricity and heat generation consume almost 34.8% of the total fossil fuels supply which in turn constitutes approximately 71.3% of the generation's total need.

A staggering 65.0% of the total fossil fuel used to generate electricity and heat in power plants is lost or wasted. Thus, only a fraction of all fossil fuels supply translates to energy use. Quantitatively, almost 30.8% of fossil fuels consumption is lost both primarily due to own usage of fuels extraction and transmission, and secondarily due to inefficiencies in generating electricity and heat, and their transmission.

It is necessary to note that approximately 7.8% of fossil fuels available have a non-energy use. They are used as raw material supplies and are processed into plastics, chemicals, lubricants, and other nonfuel products.

Unfortunately, the production and use of fossil fuels cause environmental problems - it is universally accepted that the burning of fossil fuels for energy is a major source of emissions, causing air pollution and contributing to both greenhouse effect (global warming) and acid rain. Efforts to reduce or mitigate these emissions are shaping the future of the energy industry and the transport sector. Additionally, global initiatives toward enhancing energy conservation, as well as the generation of renewable energy, are under way and continue to gain support to help meet increased energy needs.

Unarguably, concern about the security of fossil fuel supplies is considered to be amongst the main causes of regional and global conflicts. The American invasion of Iraq in 2003 is a prime example, finally admitted by George Bush.[2]

---

[2] http://www.boston.com/news/nation/articles/2005/08/31/bush_gives_new_reason_for_iraq_war).

Chapter 2
PETROLEUM - BASIC KNOWLEDGE

Petroleum, which translates from Greek as 'rock oil', is a liquid fossil fuel that consists of a complex mixture of hydrocarbon compounds, mostly alkanes (also referred to as paraffins), cycloalkanes (also referred to as naphthenes), aromatics, and asphaltics. It usually contains other organic compounds containing nitrogen, oxygen, or sulphur, as well as small amounts of various non-organic components such as trace metals. Petroleum is more commonly referred to by its colloquial term 'Oil', and this terminology is used throughout this book. The molecular composition of oil is not exact and differs from oil well to oil well; however the elemental composition is confined to narrow limits which are outlined in Table 2.1.

Table 2.1: Typical composition of oil

Element	Range %
Carbon	83 - 87
Hydrogen	10 - 14
Nitrogen	0.1 - 2
Oxygen	0.1 - 1.5
Sulphur	0.5 - 6
Metals	< 0.1

Source: Standard handbook of petroleum and natural gas engineering By William C. Lyons, Gary J. Plisga, Gulf Professional Publishing, 2005; The chemistry and technology of petroleum, J. G. Speight, CRC Press/Taylor & Francis, 2007.

The majority of oil exists in liquid phase within natural underground reservoirs. It remains liquid at atmospheric pressure after passing through surface separating facilities. However, oil can also exist in gaseous form within natural gas reservoirs, but transforms into liquid at controlled atmospheric pressure or temperature. Furthermore oil also is found in semi-solid state mixed with sand and water.

Oil is commonly extracted from oil and condensate wells onshore or offshore by drilling and pumping. Several methods are applied to optimise the recovery of oil. Primary recovery methods are used to extract oil to the surface utilising the reservoirs' underground pressure, where natural differential pressure drives oil up to the surface. When this pressure is depleted enhanced oil recovery methods (EOR) are employed to extract more oil from the reservoirs. When the reservoirs' pressure reaches the point that prevents natural flow, secondary recovery methods are employed by injecting water or natural gas to repressurise the reservoirs. Afterwards upon exhausting these methods tertiary recovery methods are applied to mainly reduce the oil's viscosity to enhance its flowability and thus facilitate its extraction. These methods include the injection of heat, vapour, solvents, surfactants, or miscible gases such as carbon dioxide. Alternative extraction methods exist to extract oil from other resources such as tar sands, extra heavy oil and oil shale. These methods require separating the oil from inorganic material by mining and in-situ thermal treatment prior to upgrading it. Additionally, oil can be extracted via chemical transformation of natural gas or coal. Oil extraction methods and technology are outside the scope of this book. Interested readers can find a wealth of knowledge elsewhere.[1]

2.1 Oil Classifications
Oil is classified according to several criteria: chemical composition, physical characteristics, geographical locations, political conditions, economic viability and technical achievability. These criteria are not standardised, and differ between countries and organisations. However, since oil has no uniform physical characteristics, its classification varies depending on the particular reference chemical or physical

---

[1] Standard Book of Petroleum and Natural Gas Engineering, W C Lyons and G J Plisga, Gulf Professional Publishing, 2005

characteristic chosen.

In terms of chemical characteristics, oil is classified based on sulphur content, where it is labelled *sour* if it contains substantial amounts of sulphur, and *sweet* if the sulphur content is relatively low. Generally speaking sweet oil commands higher prices as it is less harmful for the environment and is easier to process.

In terms of physical characteristics, oil is classified based on two different criteria: density and viscosity.

In terms of density, oil is generally labelled *heavy* if it has high density or *light* if it has low density. As a rule of thumb, lighter oils are more valuable than heavy oils since they produce higher grades of refined products such as gasoline.

Traditionally, oil density is expressed in terms of API gravity, which is a unit that measures the relative density of petroleum liquid and the density of water at 60°F, and is defined by the American Petroleum Institute using a simple formula, API gravity = $141.5/SG - 131.5$, where SG is the specific gravity of oil.

Thus in simple terms, if the API gravity of oil is greater than 10, it is lighter than and floats on water; if it is less than 10, it is heavier than and sinks in water. Note that API gravity decreases when density increases, thus oil with API equals 10 is denser than oil with API equals 40.

Accordingly oil is divided into four categories based on generally accepted API gravity limits as follows:
- Light Oil, where API gravity is larger than 31.1;
- Medium Oil, where API gravity is between 22.3 and 31.1;
- Heavy Oil, where API gravity is between 10 and 22.3;
- Extra Heavy Oil and Bitumen, where API gravity is less than 10.

Nonetheless API limits are not standard, and different limits are used by different countries, organisations or even companies. For example Canada defines heavy oil as oil with API gravity of less than 20, whereas the UK sets the API gravity limit at 22.

Viscosity characterisation of oil - used as the kinematic viscosity expressed in centipoise (cP) – is not independent but complements the density characterisation. Light, medium and heavy oil all have viscosities less than 10000 cP, and are all flowing freely at both reservoir and atmospheric conditions. Extra heavy oil viscosities can exceed 10000 cP. In these cases the extra heavy oil is termed 'bitumen'. In reservoir and at atmospheric conditions bitumen is extremely dense semi-solid and cannot flow.

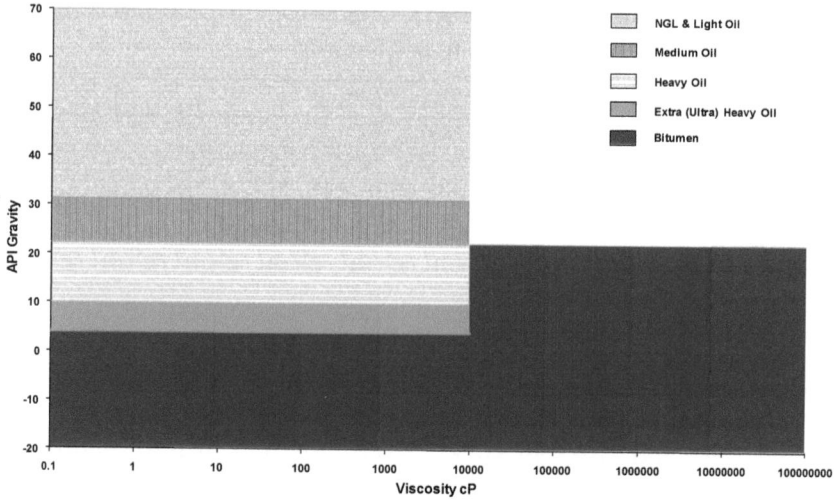

Figure 2.1: Oil classification based on physical characteristics
Source: EIA (http://www.eia.doe.gov/international); OPEC (Annual Statistical Bulletins 1999-2008); BP (Statistical Review of World Energy 2001-2009).

To visualise clearly this classification, Figure 2.1 is a simplified graphical representation that illustrates the physical classification of the oil in terms of density and viscosity. Condensates, NGLs, light oil, medium oil, heavy oil and extra heavy oil are located on left hand side of the figure with viscosity limited to 10000 cP, and API gravity above 4. Bitumen is located at the lower part of the figure, and can share same API gravity with heavy and extra heavy oil.

The oil industry and markets classify oil in terms of geographic origin, since this affects the costs of oil transport from producing to consuming countries. Examples of such classifications are *West Texas Intermediate*, *Brent* or *Dubai-Oman* blends. This classification is generalised as it references oil blends from specific areas even though the characteristics of each shipment may be different. These geographic oil classifications are further defined by physical and chemical characteristics. This classification is used as pricing reference throughout the world, with oil blends referred to as *benchmarks* or *market markers*.

Based on technical and economic criteria including extraction and production methods, geographical location, and costs, crude oil is divided broadly into two groups: *conventional* and *unconventional* (also referred to as *non-conventional*). In the broadest sense, conventional oil is the oil extracted from economically feasible deposits using fully developed practical and easy methods, while unconventional oil is the oil that is more difficult, and less economically feasible to extract, usually because of its location, the physical or chemical nature of the deposit, that the technology to reach it has not been developed fully, or that it is too expensive.

With advancing technical development and changing economic conditions, a significant proportion of unconventional oil is being reclassified as conventional oil. Therefore, what is considered unconventional oil continues to change over time, from deposit to deposit and from location to location. Generally, light, medium and heavy oils fall under conventional oil, whereas extra heavy oil falls under unconventional oil. As a conservative measure and to avoid arguments, this classification is adopted here. However, in recent years several organisations and companies reclassified the tar sands (oil sands/bitumen) of Canada partially, or in totality, as a conventional oil resource. This reclassification created substantial controversy and is not accepted universally, notably it is rejected by OPEC. The same applies to oil deposits in the Arctic region, which were considered unconventional, not because of technical issues, but due to their geographical location. These deposits have recently made the leap and moved into conventional deposits since the rapid increase in the oil price placed them firmly in the economic feasibility window, outweighing the logistical difficulties that hindered their extraction previously.

Furthermore deepwater oil is no longer considered unconventional oil, as developments in technology have led to the depth definition of 'ultra/extra' deepwater oil changing and increasing steadily. Extraction depth capabilities have more than tripled since 1990. Nowadays, oil has been extracted in water deeper than 2800 metres with Shell in 2008 setting the record at 2852 metres below the water's surface in the Silvertip Field at the Perdido Development project in the Gulf of Mexico, and in the process of developing even deeper deposits. While Deepwater Horizon's oil spill in the Gulf of Mexico in 2010 caused a temporary moratorium on deepwater drilling in the USA, the industry is expected to continue pursuing deepwater oil once the media storm is over.

Broadly speaking, there are two main categories of unconventional oil: extra heavy oil including bitumen, and oil shale. It has to be emphasised again that there is no universally agreed definition of what constitutes unconventional oil, and therefore some experts' reports may add, split, combine or remove a category. In this book, unconventional oil categories are extra heavy oil, including natural bitumen (oil sands or tar sands), and oil shale. These categories are discussed below.

Figure 2.2 presents the oil resource triangle which demonstrates oil categories and illustrates the relationship between cost, technology, and resource volume. In simple terms the better grades of oil (i.e. light and medium oil) occupy the smaller upper parts of the triangle. They exist in smaller volumes, are easier to develop, and are more feasible and profitable with current technology. Less valuable grades (i.e. heavy oil, extra heavy oil and natural bitumen) are located lower on the triangle. They exist in larger volumes, although they are more difficult and expensive to develop, and are less profitable with current technology. At the bottom part of the triangle are the worst oil grades (i.e. oil shale) which, although they exist in even higher volumes, are yet to be developed due to being either prohibitively expensive to extract or no current technology exists to extract them. The development frontier which defines the current status of the technically feasible development is influenced by the price of oil, thus the higher the price the further down the triangle the frontier can move, as the increase in price will overcome many obstacles by offsetting additional costs and encouraging more complex technology development and adaptation.

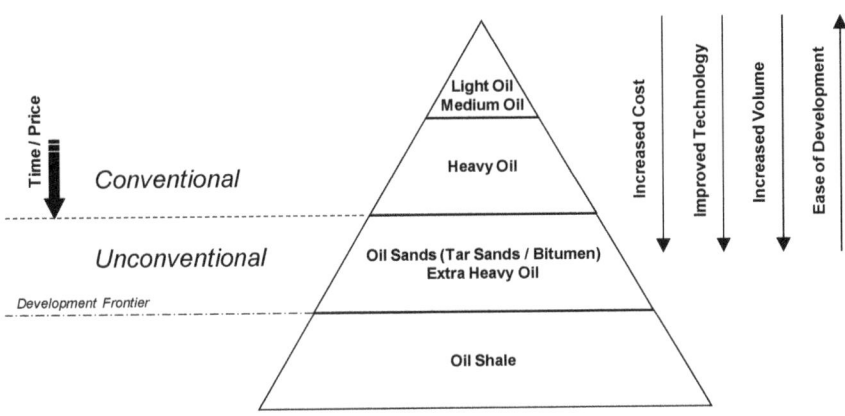

Figure 2.2: Oil resource triangle

2.2 Extra Heavy Oil and Natural Bitumen

Extra heavy oil and natural bitumen are in fact the same substance in terms of composition and density but they differ in terms of viscosity. Extra heavy oil – also known as ultra heavy oil – is commonly defined as oil having an API gravity of less than 10 and a reservoir viscosity of no more than 10000 cP, whereas natural bitumen – also known as oil sand, tar sand, or bituminous sand - is defined as oil having an API gravity of less than 10 with a reservoir viscosity larger than 10000 cP. According to the IEA definition, if viscosity measurements are not available for a reservoir, then extra heavy oil is considered to have a lower API gravity limit of 4, and with API under 4, the oil is labelled as natural bitumen.

Extra heavy oil and natural bitumen differ from higher grade oils by the degree which their original deposits have been degraded resulting in severe loss of light hydrocarbon molecules and an increased concentration of heavier hydrocarbon molecules. Therefore it is essential to note that because of variations in temperature and depth within reservoirs and the different degrees of degradation, many reservoirs can contain extra heavy oil, and (occasionally) heavy oil, in addition to natural bitumen.

As indicated earlier, extracting and processing extra heavy oil or natural bitumen requires additional investment, with the application of specialised technologies. In the case of tar sands, oil exists as part of a sedimentary material mixed with sand, clay and water, thus oil is extracted either by surface mining or in-situ thermal methods to enhance its flowability. The enhanced recovery methods commonly used include steam injection such as cyclic steam stimulation and steam flooding, including the frequently used steam assisted gravity drainage (SAGD) method. In the case of extra heavy oil, oil does not always require in-situ thermal methods to mobilise it and can sometimes be extracted conventionally using 'cold production' methods. If the extracted oil is mixed with water, sand and other inorganic materials, it is then essential to be separated prior to processing. Subsequently, the extracted crude requires further processing such as blending it with lighter oils to make it transportable by pipelines and then upgrading it into higher API gravity oil where effectively the large molecules are split or cracked into smaller molecules, with the resulting oil labelled *synthetic crude* or *syncrude* (not to be mixed with Syncrude Canada Ltd, which is the largest producer of synthetic crude in the world) that is suitable for further refining into normal petroleum products in conventional oil refineries.

2.2.1 *Orimulsion*

Orimulsion is a bitumen-based fuel that was developed by Intevep of PDVSA of Venezuela with earlier involvement from BP. It was made by mixing the bitumen with about 30% fresh water and a small amount of surfactant. The resultant product's physical and chemical characteristics permitted its utilisation in a manner similar to that used for conventional liquid fuels facilitating its transportation, storage and handling. It was used as a commercial boiler fuel in power plants worldwide - notably in China, Japan, Canada, and Italy - where with minor modifications the plants could use it as a substitute to coal or fuel oil.

The Venezuelan government wound down orimulsion's production in 2004 and decided to cease the production of the fuel in 2006 since its profitability margins were low, and with increasing oil prices it became more profitable to sell crude oil blends by mixing the bitumen with lighter crude oil.

2.3 Oil Shale

Oil shale is a misleading term (it is not shale and contains no oil!) that refers to organic-rich fine-grained sedimentary rock. It contains significant amounts of hydrocarbons in the form of *kerogen* (a complex mixture of hydrocarbon compounds of large molecules, containing hydrogen, carbon, oxygen, nitrogen, and sulphur), from which liquid hydrocarbons can be extracted through destructive distillation or exposure to heat into a form of crude oil termed *shale oil*, which subsequently may be refined into normal petroleum products. Oil shale differs from other oil deposits since it does not contain liquid hydrocarbons or petroleum as such, but organic matter derived mainly from aquatic organisms.

As is in the case of extra heavy oil and natural bitumen, extracting and processing oil shale requires additional investment and the application of specialised technologies to convert the kerogen into suitable crude oil substitute. Currently there is limited exploitation of oil shale, and most existing processes involve mining the rock – both using surface open pit and strip mining or underground mining - then transporting it elsewhere, after which it can be burnt directly to generate electricity. In some instances, further processing to extract oil has been undertaken, but this mainly has been done on experimental scale. This processing is usually ex-situ, i.e. it takes place above ground. Recently several newer technologies have been, or are being, developed to perform the extraction and processing of oil shale in-situ, i.e. underground. In both cases, the processing is referred to as destructive distillation, which is simply the chemical process of pyrolysis that uses heat, in the absence of oxygen, to convert the kerogen in the oil shale to condensable synthetic crude oil, shale gas and a solid residue. In-situ processes are preferred since they can potentially extract more oil from a deposit than ex-situ processes due to their ability to access the deposits at greater depths than mining, plus they have less adverse environmental impacts.

Historically oil shale has been used as a possible substitute to coal in power stations where it can be burnt directly to generate electricity and heat. These processes are being phased out worldwide due to adverse environmental consequences; however they are still active in Estonia, Brazil and China. The IEA, EU and the UN classify directly burnt oil shale as coal for statistical purposes (see Section 4.5), however this classification is not adopted here, where oil shale is considered as part of unconventional oil resources and reserves. Its production is considered part of coal production if burnt directly and part of oil production if oil is extracted from it, in agreement with the EIA and the World Energy Council (WEC).

2.4   Natural Asphalt

Asphalt, sometimes termed asphaltum, is a sticky black or brown material that varies in consistency from highly viscous liquid to semi-solid. Asphalt is a heavy oil fraction that consists of compounds of hydrogen and carbon with minor proportions of nitrogen, sulphur, and oxygen.

Natural asphalt exists in deposits as a fossilised natural resource. Natural deposits include asphalt lakes or pits (sometimes referred to wrongly as tar pits), such as the Pitch Lake in Trinidad and Tobago, Bermudez Lake in Venezuela, and La Brea Tar Pits in California, USA; underground mines such as from the Unita Basin in Utah, USA, where it is known as uintaite or uintahite and is marketed under the trademark Gilsonite; and floating asphalt blocks in the Dead Sea.

Asphalt can also be obtained as a by-product of petroleum refining as a residue from distillation (see Section 2.7). Currently, the asphalt from petroleum residue is the major source of asphalt and has surpassed natural asphalt production.

Asphalt should not be confused with *tar*, which is a viscous black liquid produced by the destructive distillation of coal, peat wood or even petroleum, and has different uses. The confusion stems from the misuse of the world 'tarmac' which is colloquially used to refer to asphalt type road or pathway surface materials - historically the material utilised was tar. This also applies to tar sands, which although containing asphalt in large quantities, are higher grade natural bitumen and are considered unconventional oil resource (see Section 2.2).

The principal use of asphalt is in road paving, it is also used extensively in the construction industry for surface lining, for protection against weathering and mechanical damage. Other applications include use in heat-resistant enamels and batteries. Thus, due to its usage as non fuel material and the unfavourable economics of extracting oil from it, natural asphalt is not considered an oil resource, even though it is a fossilised material. However, in cases where oil is extracted from natural asphalt deposits, the extracted oil is treated as part of crude oil production according to the EIA.

2.5 Arctic and Antarctic Oil
It is widely documented that oil exists in the Arctic and Antarctic regions. This oil is usually termed as *polar oil*. Even though some still consider these oil deposits to be unconventional, current technologies has rendered Arctic oil to be considered conventional, as mentioned earlier in Section 2.1. Oil fields in the Arctic are currently being exploited in Russia, Canada and Alaska. The appetite for oil extracting from the Arctic Ocean has increased as a result of the rapid melting of ice in the Arctic, this has prompted the bordering countries to place conflicting territorial claims in the area and in order to add additional reserves to their endowments in due course.

On the other hand, Antarctic oil is not considered available as yet. This is due because of inaccessibility, due to the international Antarctic Treaty System, which freezes territorial sovereignty claims, imposes environmental restrictions, and bans the exploitation of Antarctica's natural resources south of 60°S latitude. It is feared, however, that the Antarctic oil may be used in the future if discovered in large quantities which may subsequently cause political and military conflicts. It is noted as well that the harsh weather of Antarctica will present substantial unprecedented technological challenges that will require the development new technologies to attain feasible oil.

2.6 Natural Gas Liquids (NGL)

Natural gas liquids refer to hydrocarbon components in oil, condensate and gas fields that exist in the natural underground reservoirs in gaseous state but which are recovered as liquids at normal operating pressure and temperature conditions (usually atmospheric) utilising processing facilities incorporating lease separators, field facilities or gas-processing plants. NGLs are separated from the gas as liquids through chemical and physical processes such as absorption, condensation, adsorption, or other methods.

NGLs include (but are not limited to) ethane, propane, butanes, pentanes, and heavier hydrocarbons; they may also include small quantities of non-hydrocarbons. NGLs are distilled with crude oil in refineries, blended with refined petroleum products or used directly depending on their characteristics.

NGL terminology is rather confusing and the term NGL is a generalised term that encompasses what is commonly referred to as *lease or field condensate, plant condensate, natural gasoline,* and *liquefied petroleum gas (LPG)*. **Lease or field condensate** refers primarily to pentanes and heavier hydrocarbons produced from natural gas at lease separators and field facilities, whereas **plant condensate** refers to the same components recovered and separated as liquids at gas inlet separators or scrubbers in processing plants. **Natural gasoline** refers to a mixture of liquid hydrocarbons, mostly pentanes and heavier, extracted from natural gas, that meets specifications for gasoline in terms of vapour pressure, temperature and density. **LPG** consists mainly of propane and butane or a combination of the two. It is gaseous under normal temperature and pressure conditions but is liquefied by moderate compression or cooling to facilitate storage, handling and transportation.

Natural gas plant liquids refer to NGLs that are separated at natural gas processing plants, fractionating and cycling plants, and, in some instances, field facilities. Lease condensate is excluded from the term, but plant condensate, LPG and natural gasoline are included.

Even though NGL exists in gaseous state at reservoir conditions, its reserves and production numbers are reported as part of crude oil reserves and production, and not as part of natural gas reserves and production. This convention is widely adopted. However OPEC only includes lease condensate in its production quotas but exclude other NGL varieties (mainly because the former is produced at the fields, whereas the latter are produced at processing plants, and therefore their production is counted as natural gas at the fields – refer to Sections 9.4 and 9.1.1 for more details). Note that gas, which remains in gaseous state both at oil and condensate reservoirs, and at atmospheric conditions upon processing, is termed 'associated gas' and 'non-associated gas' respectively. These gas reserves and production numbers are included in natural gas reserves and production numbers. For more details refer to Chapter 10.

2.7 Refined Petroleum Products

Crude oil is not generally useful in its raw form; therefore it has to be processed in oil refineries to separate it into components that can be used as fuels, lubricants, and as feedstock in numerous industrial petrochemical processes. Oil can be separated into hundreds of different hydrocarbons that have different boiling points by means of distillation. Since the lighter liquid products are in greater demand, oil refineries are designed to convert heavy hydrocarbons and lighter gaseous elements into the lighter liquid products that command higher prices. Oil refineries are usually designed for specific types of crude oil, using diverse technologies and processes to optimise its output products based on markets needs.

Refined petroleum products are sometimes referred to as *secondary oil*, they are usually grouped into three categories based on the way crude oil is distilled and separated into fractions (called distillates and residuum):
- Light distillates: LPG, gasoline, naphtha.
- Middle distillates: kerosene, jet fuels, diesel.
- Heavy distillates and residuum: fuel oil, lubricating oils, wax, asphalt, tar, petroleum coke.

Details of oil refining technologies and processes, and the characteristics of refined petroleum products are outside the scope of this book.

2.8 Other Liquid Fossil Fuels

Not all liquid fossil fuels are derived from oil resources. Processes exist to derive liquid fuels by liquefying solid or gaseous fossil fuels, utilising coal and natural gas resources. This is usually done for the relative ease of transport and storage of liquid fuels in comparison to gaseous or solid fuels (see Sections 2.10 and 2.11). A derived liquid fuel from natural gas or coal is usually referred to as *synthetic fuel* or *synfuel*. To add further confusion, the term *synthetic fuel* is also used to refer to liquid fuels derived from biomass, which often employ similar liquefaction processes (see Section 5.1). In addition, the term has also been used loosely to refer to oil extracted from extra heavy oil and natural bitumen (see Section 2.2), oil shale (see Section 2.3), or non-organic waste, such as plastics and tyres.

The processes to convert gaseous hydrocarbons to liquid fuel are simply referred to as **Gas-to-Liquid (GTL)** processes. This is where, in a refinery, natural gas or other gaseous hydrocarbons are converted into longer-chain hydrocarbons such as gasoline or diesel fuel. Two main conversion technologies exist: indirect conversion and direct conversion. Indirect conversion comprises intermediate steps by converting the gas first into *synthetic gas* (usually referred to as *syngas*, which is a gas mixture that contains varying amounts of carbon monoxide, hydrogen, and sometimes carbon dioxide), and then chemically reacting the gas, with the help of catalysts, to produce liquid hydrocarbons. Examples of such commercial processes include the Fischer-Tropsch and Mobil processes. Direct conversion eliminates the syngas intermediate step and converts the natural gas directly into synfuel. Several such technologies are being developed, however they are at present energy intensive and difficult to control. To date none have yet been fully commercialised.

GTL processes are gaining popularity as a way in which refineries can convert some of their gaseous waste products into more valuable liquid fuels, and are also seen as an alternative to develop natural gas deposits in remote locations where it is not economical to build a pipeline. Currently Shell and Sasol run commercial facilities to produce liquid fuel from natural gas in Malaysia and South Africa respectively. Both companies are also constructing larger facilities in Qatar.

Similarly, the processes to convert coal to liquid fuel are simply referred to as **Coal-to-Liquid (CTL)** processes, where coal is converted into light liquid hydrocarbons such as gasoline or diesel fuel. Two main conversion technologies exist: indirect conversion and direct conversion. Indirect conversion is similar to GTL processes, but it also comprises intermediate steps, including coal gasification, to produce *syngas*, which then chemically reacting with the gas, with the help of catalysts to produce liquid hydrocarbons. The Fischer-Tropsch process described earlier has been used commercially for over 50 years in CTL processes. The technology is being developed and can be provided by major multinational oil companies including Shell, BP, ExxonMobil, Statoil, as well as Sasol the world leader in this technology.

Direct conversion eliminates the syngas intermediate step and converts the coal directly into synfuel. This can be achieved by hydrogenation or carbonisation. Hydrogenation processes were used in Germany during the First and Second World Wars. This involves mixing the coal with hydrogen at high temperature and pressure, then dissolving the coal in a solvent, with the help of catalysts, to produce light hydrocarbon liquids. Several such technologies have been developed, including the Bergius process and the Solvent Refined Coal (SRC) processes. Carbonisation or pyrolysis processes distil out liquid and gas from coal by heating the coal, in the absence of oxygen, driving off volatile matter from the coal and generating light hydrocarbons by thermal decomposition of coal during the treatment. These processes can be carried out at high or low temperature. Low temperatures optimise the production of liquids richer in lighter hydrocarbons, whereas high temperatures produce heavier liquid, referred to as *coal tar*, as a by-product during production of metallurgical coke. The produced liquid is then further processed into lighter liquid fuels. Several such technologies have been developed including the Karrick process.

CTL processes are technically proven and have been used in the past by Germany to provide the majority of its liquid fuel during the Second World War. Throughout the apartheid period in South Africa, CTL technologies were further developed by Sasol to provide South Africa's oil needs. Currently Sasol runs commercial facilities to produce liquid fuel from coal in South Africa. Additionally several demonstration facilities are under construction or started operation in China. With abundant coal reserves, wider commercialisation of these technologies will depend on the future price of oil and the environmental regulations concerning carbon dioxide emissions since these technologies produce significant amounts of carbon dioxide in the conversion processes.

Similar processes have been developed to convert non-organic waste, such as plastics and tyres, into liquid hydrocarbon fuels. These processes are possible since the non-organic wastes are originally manufactured using hydrocarbons as their raw material. Here, these processes are referred to as **Non-Organic-Waste-to-Liquid (NOWTL)** processes. Such industrial processes involve a form of hydrous pyrolysis usually referred to as 'thermal depolymerisation' in which the long chain polymers in the hydrocarbon material waste decompose (under high temperature and pressure) into shorter chain hydrocarbons. The resultant liquids may require further processing similar to the CTL technologies described above.

2.9 Oil Uses

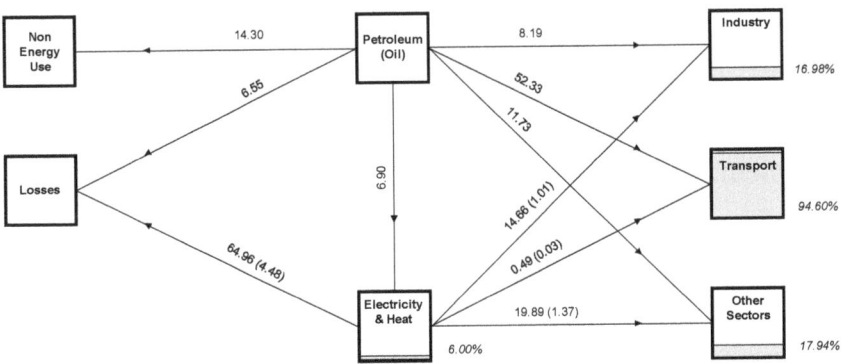

Figure 2.3: Distribution of oil global consumption (2006)
Source: IEA (http://www.iea.org/stats/index.asp).

Oil's primary use is as a fuel, with the remainder used as a chemical feedstock. Details of global oil consumption are shown in Figure 2.3; it also details the distribution of the oil consumption by sector and is derived from the data reported by the IEA in 2006. As can be seen the transportation sector is dominant, consuming just over 52.3% of the total oil supply which constitutes a massive 94.6% of the sector's total need, which demonstrates total dependency on oil; industry consumes a further 8.2% of the total oil supply, which constitutes almost 17.0% of the sector's total need; other sectors including residential, commercial and agricultural, consume around 11.7% of the total oil supply, which constitute just over 17.9% of the sectors' total need. These numbers include the primary consumption and the consumption supplied by means of electricity and heat output from power plants. The figure also demonstrates clearly that, if treated as a sector, electricity and heat generation using oil is heading for extinction; it only consumes 6.9% of the total oil supply, which in turn constitutes a meagre 6.0% of the generation's total need. This cloud however has silver lining since using less oil to generate electricity translates to less losses from oil overall. Thus even though almost 65.0% of the oil used for electricity and heat generation and transmission is lost, it only account for approximately 4.5% of the total oil supply. These losses plus primary losses due to oil consumption in extraction and transmission bring the total losses to over 11.0% of the total supply, which is significantly lower than losses from coal and natural gas. Furthermore a significant proportion of oil supply, standing at around 14.3%, has non-energy use, where it is used as the main raw material for petrochemical and fertiliser industries amongst others.

2.10 Oil Transport

As liquids, oil and refined petroleum products are relatively easy to transport due to their flowability and slightly lower densities compared to water. Crude oil must be moved from production fields to refineries for processing and then the refined products must be transported to consumers. Three main modes of transport are in use:
- Pipelines – which transport crude oil and refined petroleum products. They are considered the most efficient method of transport for crude oil and refined petroleum products. Pipelines are built mainly over land, but some are also built under the sea, connecting offshore facilities to onshore terminals. Pipelines are used to move crude oil from the wellheads to gathering and processing facilities, from there to terminals (to load tankers, trucks and trains destined to oil refineries) or to oil refineries directly. Pipelines are also used to move refined petroleum products from oil refineries and terminals to local distribution facilities. If economically feasible, they can also be used to supply major industrial sites directly. Transport pipelines connect cities, countries and even continents, and include numerous pumping stations to facilitate oil and refined petroleum products movement. Besides their economic benefits, pipelines have a strategic

value that involves international security and are interlinked to international politics. Disputes over their management are known to have caused international political crises, having been used by exporting or transit countries as tools to pressurise and influence the policies of the importing countries. Additionally they are vulnerable and can be targeted by vandalism, sabotage, terrorists or as legitimate targets in military disputes.
- Oil tankers – ships which transport crude oil and refined petroleum products across water. They range in size from small vessels used to transport refined petroleum products from oil refineries to distribution terminals, to huge crude carriers that move crude oil from its points of extraction to refineries and terminals. Specialised tankers are used for LPG transport. The majority of the international oil and refined petroleum products trade uses tankers as the preferred mode of transport due to their relative low cost, easy availability, and the possibility of accessing supplies from a range of sources thus improving security of supply. Despite high profile oil spillages from oil tankers, the safety record of tankers is considered comparatively excellent. Though this record has been affected by military conflicts such as the Iraq-Iran war, and more recently by the new threat of piracy for which oil tankers have been especially vulnerable.
- Tank trucks and trains - which transport crude oil and refined petroleum products over land. They carry oil and products over shorter distances, mainly from refineries to local distribution facilities or directly to consumers. This mode of transport is rarely used for international oil or refined petroleum products transport due to its considerably higher costs compared to oil tankers and pipelines.

2.11 Oil Storage

Crude oil and refined petroleum products are stored to meet variations in demand as well as for strategic and security reasons. They are stored mainly in above-ground or underground tanks for processing into finished products or transported to end users or to other storage locations. These storage facilities are referred to as oil depots or oil terminals.

Oil depots are rather simple facilities comprised of numerous tanks (referred to as tank farms) with no processing or other transformation on site. Tank designs differ and depend on the liquid stored, for example floating roof tanks are used for crude oil, gasoline, and naphtha; fixed roof tanks are used for diesel, kerosene and fuel oil; bullet tanks and spherical tanks are used for LPG.

Oil depots are usually located close to oil refineries or on sites with access to sea or waterways where oil tankers can be loaded or discharged. Some depots can load or discharge through pipelines. Oil depots can also be fed by rail, and by road using dedicated vehicles which can also be used for the discharge and distribution of products. Note, due to their strategic value, many airports have dedicated oil depots.

Additionally, a type of floating tank system that temporarily stores produced oil or gas from offshore platforms until the oil or gas can be offloaded onto waiting tankers or sent through pipelines, is increasingly being used by the offshore industry. This system is referred to as *floating storage and offloading (FSO)* vessels. More advanced vessels also have oil and gas processing capabilities and are referred to as *floating, production, storage and offloading (FPSO)* vessels.

2.12 Oil Measurement Units

Crude oil and refined petroleum products are commonly measured in volumetric quantities Volumetric units usually used are barrels (1 barrel = 42 US gallons), usually shortened to *bbl* or simply *b*, or standard cubic metres (corresponding to 15 °C and 1.01325 bar). Mass units such as metric tons can also be used; however the conversion factor differs from country to country depending on the density of oil used for the conversion. The same units are used to measure NGLs. Note however that LPG is usually measured in metric tons.

To standardise oil and natural gas measurement, units of oil equivalent are becoming increasingly common, and they refer to the amount of energy released by burning a defined quantity of crude oil, either a barrel, where the unit is shortened to *boe* or a metric ton, where it is shortened to *toe* (1 boe = 0.146 toe approximately). The oil equivalent units are used by oil and gas companies, organisations, and researchers as a way of combining oil and natural gas reserves and production, into a single measure.

Chapter 3
NATURAL GAS – BASIC KNOWLEDGE

Natural gas is a gaseous fossil fuel that consists of a mixture of hydrocarbon compounds and a small amount of various non-hydrocarbon components. The prime hydrocarbon compound is methane. Other hydrocarbon compounds that can be found in significant quantities are ethane, propane, butane and pentane. Heavier hydrocarbons can also be found in small quantities. Non-hydrocarbons are primarily carbon dioxide, nitrogen, hydrogen sulphide, helium and trace metals such as mercury. The exact composition of natural gas varies widely between reservoirs. Table 3.1 outlines the natural gas typical makeup.

Table 3.1: Typical composition of natural gas

Component	Range %	
	Non Associated Natural Gas	Associated Natural Gas
Methane	70 - 98	45 - 92
Ethane	1 - 10	4 - 21
Propane	trace - 5	1 - 15
Butanes	trace - 2	0.5 - 7
Pentanes	trace - 1	trace - 3
Hexanes	trace - 0.5	trace - 2
Heptanes and heavier	trace - 0.5	0 - 1.5
Nitrogen	trace - 15	trace - 10
Carbon Dioxide	trace - 5	trace - 4
Hydrogen Sulphide	trace - 3	0 - 6
Helium	0 - 5	0

Source: Standard Book of Petroleum and Natural Gas Engineering, W C Lyons and G J Plisga, Gulf Professional Publishing, 2005.
Note 1: Some natural gas fields have majority carbon dioxide or hydrogen sulphide.

At reservoir conditions, natural gas can exist in gaseous, liquid or even solid forms. In gaseous form, natural gas is found isolated in gas and condensate fields, as a cap above oil fields in oil reservoirs, trapped underground in rock, coal or porous sand formations. Natural gas, as liquid, is found dissolved forming a solution with oil in underground oil fields. As a solid, natural gas is found trapped in crystals in the form of methane hydrates.

If natural gas is produced from a reservoir that does not contain significant quantities of crude oil, it is termed *non-associated* gas. However, if it is produced with crude oil from the same reservoir, it is termed *associated gas*. The latter term is applicable whether the gas is found dissolved in the oil, or if it is free gas above the oil in the reservoir.

Natural gas is commonly extracted from wells onshore or offshore by drilling, and since natural gas is lighter than air, the pressurised gas will rise to the surface with little or no interference. This is usually the case in gas and condensate wells, however extracting associated gas from oil wells requires lifting processes such as pumping (i.e. pumping the oil in which gas is dissolved). Similar to oil extraction, several methods can be applied to optimise the recovery of natural gas.

Primary recovery methods extract natural gas to the surface utilising the reservoirs' underground pressure. When this pressure is depleted enhanced gas recovery methods (EGR) are employed to extract more natural gas from the reservoirs by injecting water or gases, such as carbon dioxide, to repressurise the reservoirs. Stimulation techniques are increasingly being used to extract natural gas from other resources such as tight natural gas, shale gas and from coal beds. These methods include hydraulic fracturing, acidizing, and dewatering. Additionally, natural gas can be extracted via chemical transformation of oil or coal. Natural gas extraction methods and technology are outside the scope of this book. Interested readers can find a wealth of knowledge elsewhere.[1]

---

[1] Standard Book of Petroleum and Natural Gas Engineering, W C Lyons and G J Plisga, Gulf Professional Publishing, 2005

3.1 Natural Gas Classifications

Analogous to oil, natural gas is also classified according to several criteria, which are not standardised, and that differ between countries and organisations. However, unlike oil, natural gas is not classified based on physical characteristics as it has almost uniform physical characteristics and thus differs only slightly in terms of density and viscosity.

In terms of chemical characteristics, natural gas is classified based on three different criteria:
- Hydrocarbon composition, where it is labelled *dry gas* if it is composed from almost pure methane, and *wet gas* if other hydrocarbons are present in the gas mixture. The latter is usually originated in condensate fields which produce raw natural gas as well as natural gas liquids (NGLs).
- Sulphur content, where it is labelled *sour gas* if the gas contains hydrogen sulphide or other sulphur impurities, and *sweet gas* if sulphur content is considerably low.
- Carbon dioxide content, where it is labelled *acid gas* if the gas contains carbon dioxide. The same label applies if the gas contains sulphur dioxide, hydrogen sulphide or similar contaminants.

In terms of technical, economic and geographical criteria, natural gas is classified as either *conventional* or *unconventional* (also referred to as *non-conventional*). In the broadest sense, conventional natural gas is the gas extracted from economically feasible deposits using fully developed practical and easy methods. Unconventional natural gas is the gas that is more difficult, and less economically feasible to extract, usually because of its location or the nature of the deposit, or that the technology to reach it has not been developed fully, or is too expensive. Unlike unconventional oil, which has different physical and chemical properties compared to conventional oil, unconventional natural gas has the same chemical and physical composition as conventional natural gas, the 'unconventional' label is usually given due to the condition of the reservoirs the gas is stored in, which are 'unconventional' reservoirs.

However, as technology and geological knowledge advance, unconventional natural gas deposits are not only being reclassified as conventional deposits, but are also beginning to make up an increasingly large percentage of the gas reserves. Therefore, what is really considered unconventional natural gas is changing over time, and from deposit to deposit. The economics of extraction play a significant role in determining whether or not a particular deposit is unconventional, or simply too expensive to extract. In recent years, deposits of deep natural gas - typically deeper than 5000 metres onshore, or located in deep water fields offshore – are considered conventional. This is due to improved deep drilling, exploration, and extraction techniques resulting in making production of deep gas economical albeit still more expensive than cheaper conventional natural gas. The same applies to natural gas deposits in the Arctic region, which were considered unconventional, not because of technical issues, but due to their geographical location. These deposits have recently made the leap and moved into conventional deposits, since the rapid increase in the gas price placed them firmly in the economic feasibility window and outweighed the logistical difficulties that hindered extraction previously.

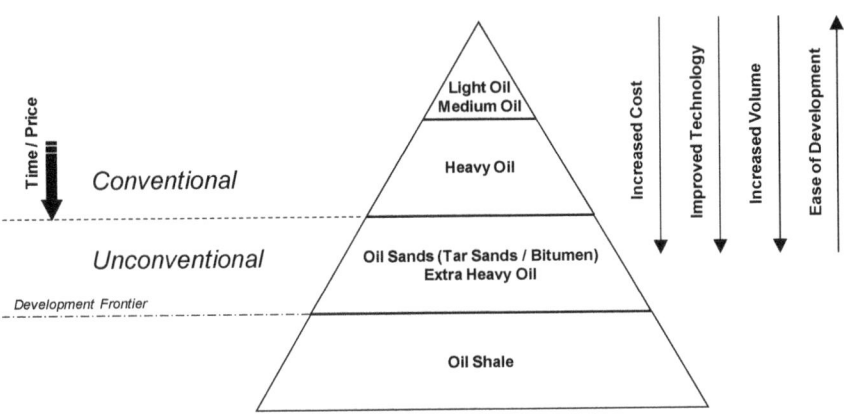

Figure 3.1: Natural gas resource triangle

Broadly speaking, there are five main categories of

unconventional natural gas, even though it has to be noted that there is no universally agreed definition of what constitutes unconventional natural gas, and therefore some experts may add or remove a category. In this work unconventional gas categories are tight gas, shale gas, coalbed methane, gas in geopressurised zones and methane hydrates. Similar to the oil resource triangle, Figure 3.1 presents the natural gas resource triangle which demonstrates natural gas categories and illustrates the relationship between cost, technology, and resource volume. In simple terms the easiest to extract natural gas (i.e. conventional natural gas) occupies the smaller upper parts of the triangle. It exists in smaller volumes, its deposits are easier to develop, are more feasible and profitable with current technology. More difficult deposits (i.e. tight gas, coalbed methane, and shale gas) are located lower on the triangle. They exist in larger volumes, but are more difficult and expensive to develop, and are less profitable with current technology. At the bottom part of the triangle are the most difficult and speculative deposits (i.e. gas in geopressurised zones and methane hydrates) which although they exist in even higher volumes, they are yet to be developed due to being either prohibitively expensive to extract or because there exists no current technology to extract them. As with oil, the natural gas development frontier is defined by the current status of the technically feasible development and is influenced by the price of natural gas, thus the higher the price the further down the triangle the frontier can move as the increase in price will overcome many obstacles by offsetting additional costs and encouraging more complex technology development and adaptation.

3.2 Tight Natural Gas

Tight natural gas is gas that is held in a very tight formation underground where the permeability of the reservoir rocks or sands - that is, their capacity for transmitting fluids - is so low that the gas molecules cannot flow into production wells without help. In a conventional natural gas deposit, once drilled, gas reserves flow unassisted. However, in tight gas reservoirs a great deal more developed technologies and investment are required to allow the gas to be extracted, including fracturing the tight formation and acidizing.

3.3 Shale Gas
Shale gas is gas that exists in shale, which is a very fine-grained soft sedimentary rock. Shale is easily breakable into thin, parallel layers, but does not disintegrate when it becomes wet. Extraction of natural gas from shale formations requires additional investment and more developed technologies, including artificial fractures around well bores and horizontal drilling, than extraction of conventional natural gas.

3.4 Coalbed Methane
Coalbed methane is gas extracted from coal beds. It is also known as *coal seam gas* or *coalbed gas*. The gas is trapped underground and is contained within the coal seams or the surrounding rock. This gas is distinct from conventional gas reservoirs, as the methane is stored within the coal by adsorption, where methane is in a near-liquid state, lining the inside of pores within the coal. Coalbed methane typically contains very little heavier hydrocarbons or hydrogen, and no natural gas condensate. It often contains up to a few percent carbon dioxide. Extraction of coalbed methane requires additional investment and more specialised technologies than for the extraction of conventional natural gas. Specialised technologies include drilling wells into the coal seams and dewatering.

Coalbed methane is generally not released into the atmosphere until coal mining activities unleash it. Previously this gas was considered a nuisance and a safety risk, as when coal was extracted, the methane contained in the seam usually leaked out into the coal mine itself, and had to be vented into the atmosphere.

3.5 Methane Hydrates
'Methane hydrates' is a solid form of water that contains a large amount of methane within its crystal structure. It is also known as *methane clathrate* or *methane ice*. These hydrates look like melting snow and were first discovered in permafrost regions of the Arctic. However, subsequent research has revealed that

deposits seem to be widespread in any continental shelf, as well as in continental rocks trapped in beds of sandstone or siltstone. The USGS estimates that methane hydrates may contain more organic carbon than the world's coal, oil, and conventional natural gas combined. However, no technique to harvesting methane hydrates has been fully developed even though some active research is being conducted.

3.6 Gas in Geopressurised and Hydropressurised Zones

This is natural gas, sometimes referred to as geopressured aquifer gas, which is deposited under very high pressure in porous sand or silt layers. Geopressurised zones are typically located at great depths, usually below 3000 metres below the surface of the earth, or dissolved or dispersed in underground water aquifers, with the amount of methane increasing by increasing depth and salinity of the water aquifer. Both the depth and the high pressure make the extraction of natural gas in geopressurised zones quite complicated. However, experts agree that these deposits have great potential and estimate that geopressurised and hydropressurised zones hold the largest deposits of gas of all of the unconventional sources of natural gas. On the other hand, some sources (academic, governmental or commercial) do not consider this gas to constitute a part of natural gas deposits as the methane content is usually low. These sources consider this gas to be part of geothermal energy resources instead.

3.7 Arctic and Antarctic Natural Gas

Besides methane hydrates, it is widely documented that significant deposits of natural gas exist in the Arctic and Antarctic regions. This gas is usually referred to as *polar natural gas*. Similar to the situation with polar oil, current technologies have rendered Arctic natural gas to be considered conventional, as mentioned earlier in Section 3.1. Natural gas fields in the Arctic are currently being exploited in Russia, Norway, Canada and Alaska, and as mentioned earlier regarding Arctic oil, the appetite for natural gas extracting from the Arctic Ocean has increased as a result of the rapid melting of ice in the Arctic. Refer to Section 2.5 for more details.

The situation is also similar to Antarctic oil regarding Antarctic natural gas, which is not considered available as yet. The discussion in Section 2.5 is equally applicable here.

3.8 Secondary Natural Gas

Unlike crude oil, natural gas is generally useful in its raw form, where it only requires minor treatment prior to its use; therefore it requires no refining and only minor processing. However, as discussed in detail in Sections 3.11 and 3.12, transporting and storing natural gas are considered major challenges. Thus, two main methods have been developed to overcome these challenges and facilitate natural gas usage by reducing the volume of natural gas by either liquefying or compressing it resulting in the following secondary products:

- *Liquefied Natural Gas (LNG)*, which is natural gas that has been converted to liquid form for ease of storage or transport. Natural gas is converted to LNG by cooling it to approximately -163°C at near atmospheric pressures, at which point it becomes a liquid. This process reduces its volume by a factor of more than 600.
- *Compressed Natural Gas (CNG)*, which is natural gas compressed to be used as fuel for transport. Natural gas is compressed to pressures over 200 bar to less than 1% of its standard atmospheric pressure volume.

Details of natural gas liquefaction and compression technologies and processes are outside the scope of this book.

Moreover, strictly speaking, GTL transformation processes result in secondary natural gas products, where natural gas is converted into longer-chain liquid hydrocarbons such as gasoline or diesel fuel. Refer to Section 2.8 for more details.

3.9 Other Gaseous Fossil Fuels

In addition to natural gas, gaseous fossil fuels are commonly derived from oil or coal. Two main processes are used to derive the gas from coal or crude oil. The first process is gasification, which is a process to convert coal or oil into carbon monoxide and hydrogen by reacting the coal or the crude oil at high temperatures with a controlled amount of oxygen and/or steam. The resulting gas mixture, as explained earlier, is called *synthesis gas* or *syngas*, which has been used historically as a fuel. As mentioned previously, gasification is a method for extracting energy from any carbonaceous material, and it is now mostly used as the first step in GTL, CTL or NOWTL processes (see Section 2.8), and in deriving numerous biofuels (see Section 5.2). The second process is destructive distillation or pyrolysis, which, as discussed earlier, is a process to distil out liquid and gas from the coal by heating the coal in the absence of oxygen. The resultant gas is termed *coal gas* or *coke oven gas*, and is a by-product of the coking process. In 1979, methane was produced directly in the UK on an experimental scale using microorganisms that digested coal.[2]

Historically the derived gas from gasification or pyrolysis was marketed as a domestic gas. Its terminology differed from country to country, but usually was referred to as *town gas*, *producer gas*, *water gas* or *manufactured gas*. It was widely used as a domestic fuel for cooking and heating and to power gas turbines. However its usage has been phased out and it has been replaced by natural gas.

Recently, gasification is being promoted in new generation of coal power stations as part of what is termed Integrated Gasification Combined Cycle (IGCC) power plants, where the coal is turned into syngas, and is cleared from impurities before it is combusted. This gas is often used to power a gas turbine whose waste heat is passed to a steam turbine system.

LPG is another fossil fuel, which exists in gaseous phase at processing conditions, but is stored and transported in liquefied state. Therefore, as discussed earlier (see Section 2.6), LPG is not natural gas, but it can be partially considered as a secondary natural gas since some of it is produced during the processing of natural gas (though the majority is produced from oil as one of the main products of the oil refining process), and because it is

---

[2] Deyden, I.J.B., Fuel, Volume 58, Issue 1, January 1979, Pages 2-3.

used in many countries instead of natural gas, domestically for cooking and heating. LPG has widespread use in industry as a refrigerant.

3.10 Natural Gas Uses

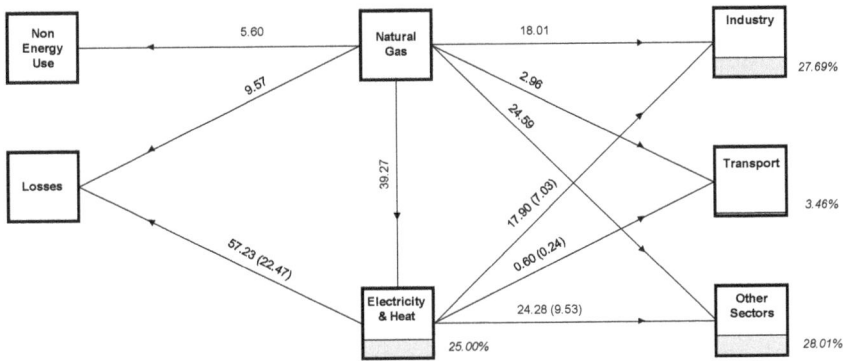

Figure 3.2: Distribution of natural gas global consumption (2006)
Source: IEA (http://www.iea.org/stats/index.asp).

Natural gas' primary use is as a fuel, with the reminder used as a chemical feedstock or in the oil industry as an EOR method, where it is injected to raise the reservoir pressure, though the latter use is not accounted for commercially as part of natural gas production and consumption numbers. Figure 3.2 details the distribution of the global natural gas consumption per sector, and is derived from the data reported by the IEA in 2006. The figure shows that natural gas plays an important role in both industry and 'other' sectors including residential, commercial and agricultural. The figure indicates that industry consumes just over 25.0% of the total natural gas supply which constitutes almost 27.7% of the sector's total need; and other sectors including residential, commercial and agricultural consume around 34.1% of the total natural gas supply which constitute 28.0% of the sectors' total need. On the other hand natural gas use in transportation is still insignificant as it only consumes 3.2% of the total supply which constitutes just under 3.5% of the sector's total need. These numbers include the primary consumption and the consumption supplied by means of electricity and heat output from power plants. The figure also shows that, if treated as a sector, electricity and heat generation is the largest consumer of natural gas as it consumes around 39.3% of the total natural gas supply, which in turn constitutes 25.0% of the generation's total need. However it is noted that, due to generation and transmission inefficiencies, almost 57.2% of the gas used in electricity generation is lost. This loss, as well as using natural gas as fuel in oil and gas extraction and transmission (excluding flaring and re-injection), accounts for over 32.0% of total natural gas consumption. In addition, 5.6% of the total natural gas supply has non-energy use, when it is used as raw material in petrochemical industries where it can be processed into plastics, chemicals, lubricants, and other non-fuel products.

3.11 Natural Gas Transport

Natural gas transport is one of the main difficulties affecting its use. This is mainly due to its low density, which imposes restrictions on the choices of means of transport. Natural gas needs to be transported to consumers either after being processed onsite on production fields or after being moved from production to processing terminals nearer to consumers. As with oil, three main modes of transport are in use:

- Pipelines - which are considered the most efficient and economical method to transport natural gas. Pipelines are built both over land and under the sea, not only connecting offshore facilities to onshore terminals as in the case of oil, but also connecting countries and extending for thousands of kilometres. Pipelines are used to move natural gas from the wellheads to gathering and processing facilities, from there to terminals and subsequently to distribution grids. If economically feasible they can also be used to supply major industrial sites and power stations directly. Natural gas pipelines range from distribution pipelines connecting consumers to longer pipelines connecting cities, countries and even continents. They include numerous compression stations to maintain the gas pressure and facilitate its movement. As is the case of oil pipelines, gas pipelines have not only economic benefits, but also strategic value that involves

international security and are therefore interlinked to international politics. Similar to the oil pipelines, disputes over gas pipelines management are often in the news where they are known to cause international political crises and are currently used by exporting (e.g. Russia) or transit countries (e.g. Belarus) as tools to pressurise and influence the policies of the importing countries. Moreover, similar to oil pipelines, they are also vulnerable and can be targeted by vandalism, sabotage, terrorists or as legitimate targets in military disputes.
- LNG carriers - ships which transport liquefied natural gas (LNG) across water. Transporting LNG requires specialised facilities for liquefaction at the production point, and then gasification at end-use facilities or into a pipeline, therefore using these carriers have a higher cost in comparison to oil tankers. However using this mode of transport allows consumers access to global natural gas supply and thus reduces the dependency on single supplier and improves security of supply.
- Tank trucks - which can carry LNG or CNG on land over shorter distances. They may transport natural gas directly to end-users, or to distribution points such as pipelines for further transport. This mode of transport is rarely used for international natural gas transport due to its considerably higher costs compared to LNG carriers and pipelines.

There have recently been suggestions that natural gas can be transported in solid form as hydrates; however as yet the technology is in its infancy, so it is not feasible.

3.12 Natural Gas Storage

Natural gas is stored, in principle, to stabilise supply and manage variations in demand. It is loaded into storage during periods of low demand and then used at peak demand. Due to the expansion in use of natural gas, its storage has become essential for strategic and security reasons. Unlike oil storage, storing natural gas is difficult due to its low density which means its volume has to be reduced by compression or refrigeration to render storage facilities feasible.

Natural gas is often stored underground in depleted gas reservoirs from previous gas wells, aquifer reservoirs, or in salt caverns. Using depleted gas reservoirs is economically attractive as they are relatively large, need a moderate volume of cushion gas (natural gas to be retained in storage permanently to maintain pressure) and requires less capital investment, as the existing gas production facilities can be re-used with minor modifications to utilise the reservoirs for storage. Salt caverns are considerably smaller than other types of underground storage but require the least amount of cushion gas. On the other hand aquifer reservoirs require the most cushion gas and with increasing gas prices they are considered the least preferred option.

Above ground natural gas is usually stored as LNG in specially designed tanks where it is stored at very low temperatures. LNG tanks do not require cushion gas and can deliver gas at very short notice. However they are more expensive to construct and maintain than underground storage facilities. In addition, natural gas is stored near atmospheric pressure at ambient temperature in tanks usually referred to as holders or gasometers for short term storage to balance distribution. Due to safety regulations overground high pressure gas tanks (bullets) use is discouraged and the existing tanks are gradually coming out of use. Finally, natural gas can be temporarily stored in pipeline systems if needed by increasing the pressure in the pipeline. The latter process is referred to as line packing.

Underground gas storage facilities are dictated by geological constraints, which mean they cannot always be located close to consumers or on sites with access to sea or waterways. Therefore these storage facilities are usually connected to pipelines from which they can load or discharge.

Additionally, a type of floating tank system to store LNG offshore has been developed to overcome local objections to build LNG terminals onshore. This system is called LNG *floating storage and regasification unit (FSRU)* in which the unit receives LNG carriers delivering LNG onboard, where it is regasified and the resulting natural gas is send out through pipelines to shore.

3.13 Natural Gas Measurement Units

Natural gas is measured either in volumetric or in heat quantities, with the latter usually used for trading purposes. Volumetric units usually used are standard cubic feet (corresponding to 60 °F and 14.73 psia), or standard cubic metre (corresponding to 15 °C and 1 atm). Heat units usually used are British Thermal Units (BTU), Joules or oil equivalent (boe or toe), where the conversion factor from volumes of gas to 'boe' or 'toe' depends upon the type of gas. Note that LNG can also be measured in metric tons.

Chapter 4
COAL – BASIC KNOWLEDGE

Coal is a solid fossil fuel composed primarily of carbon. It also contains low percentages of solid, liquid and gaseous hydrocarbons as well as small amounts of other materials such as compounds of nitrogen, sulphur and trace metals. The composition of coal is very complex, and the exact composition varies widely between deposits. Therefore two main approaches are usually used to report coal composition in terms of its *proximate analysis*, which reports the coal composition on a weight basis in terms of four items: fixed carbon, volatile matter, moisture and ash, and its *ultimate analysis*, which reports the coal composition on a weight basis in terms of five elements of the coal's organic fraction. The elements are carbon, hydrogen, oxygen, nitrogen, and sulphur. The composition data is usually reported according to a defined basis such as 'as-received basis', 'dry or moist-free basis', 'ash-free basis', or 'mineral matter-free

Table 4.1: Typical composition of coal

Element	Range %
Carbon	60 - 95
Hydrogen	2 - 6
Nitrogen	0 - 2
Oxygen	2 - 35
Sulphur	0.5 - 3

Source: Chemistry Explained (http://www.chemistryexplained.com/Fe-Ge/Fossil-Fuels.html); Eberhard Lindner; Chemie für Ingenieure; Lindner Verlag Karlsruhe.

basis'. Various combinations of these bases are also used, with the most common being 'dry and ash-free'. Both approaches perform composition analyses according to standards set by institutions such as ASTN (American Society for Testing Materials), DIN (Deutsches Institut für Normung - in English, the

German Institute for Standardization), etc. However although these standards are similar in nature they differ in details, for example at what temperature they measure volatile matter. Figure 4.1 illustrates the different bases of representation, and Table 4.1 shows a typical ultimate analysis coal makeup on dry and ash-free basis.

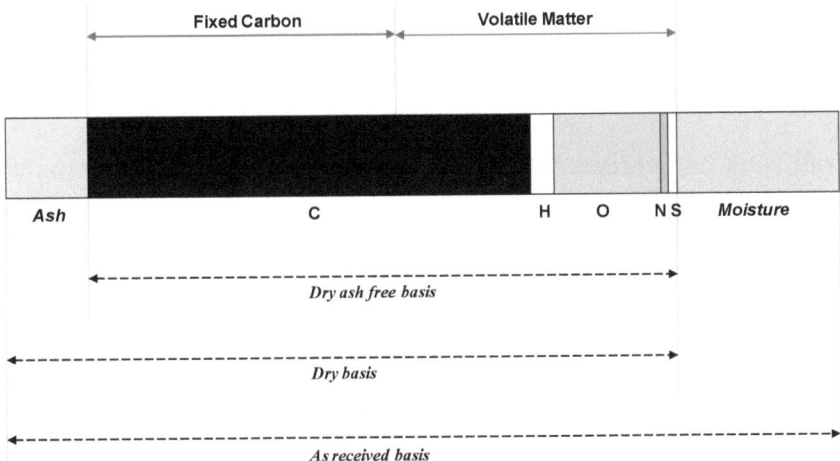

Figure 4.1: Bases of representation of coal

Note that ash and mineral matter are two distinctly different entities. Mineral matter consists of the various minerals contained in the coal; whereas ash is the inorganic solids remaining after the coal is completely combusted. The ash is usually less than the mineral matter because of the weight changes that take place during coal combustion such as the loss of gaseous carbon dioxide from mineral carbonates, loss of water from silica minerals and loss of sulphur (as gaseous sulphur dioxide) from iron pyrites.

Coal is found in reserves in solid form, and is extracted almost entirely by mining either underground or surface mining onshore, though recent offshore mining activities are nowadays active especially in China and Japan. In addition, utilising coal in-situ by gasification is being actively considered by several countries and companies. Coal extraction methods and technology are outside the scope of this book. Interested readers can find a wealth of knowledge elsewhere.[1,2,3]

4.1 Coal Classifications

Coal classification principles differ from those categorizing oil and natural gas, where instead of basing the classification on physical and/or chemical characteristics, or geographical location, coal is classified based on geological history by means of degree of metamorphism – referred to as coal 'rank', or relating to its suitability for use for a particular purpose – referred to as coal 'grade' or 'quality'.

Coal rank categorises the coal relative to other coals in terms of their degree of metamorphism – also referred to as coalification, which is defined as the progressive gradual alteration in the physical and chemical properties of coal in response to temperature, pressure and time. Accordingly, the rank of coal is simply the stage the coal has reached on the coalification path.

Coal grade or quality, on the other hand, are usually used interchangeably as an informal coal classification, where it refers to individual measurements such as the coal's heating value, sulphur content, ash content, moisture content, or the presence of specific trace or heavy elements. Thus, the heating value is the most important measurement, in terms of classifying coal, as a fuel since it defines the fuel content of coal in terms of the amount of potential energy that can be converted into actual heating ability. This value depends on the chemical composition of the coal, and is interlinked with another measurement, namely the ash content, which quantifies the incombustible silicate minerals and other metals in coal. The ash content besides causing disposal problems, also lowers the coal's heating value as the ash materials do not burn. As a result relatively high ash content can exclude a deposit from being classified as coal. For example, the EIA excludes from its coal resource coal estimates

---

[1] Coal Mining: Research, Technology and Safety, Gerald B. Fosdyke, Nova Science Publishers, 2008.
[2] Coal exploration, mine planning, and development, Roy D. Merritt, Noyes Publications, 1986.
[3] Coal preparation, Joseph W. Leonard, Byron C. Hardinge, Society for Mining, Metallurgy, and Exploration, 1991.

that contain more than 33 weight percent ash.

There is no universal agreement or standard on coal ranking. Therefore ranking definitions differ from country to country or even from company to company. In general terms, the higher the rank the coal has, its carbon content, hardness, and heating value increase, while its moisture content, volatile matter content, and ash content decrease.

A widely accepted and adopted ranking system is the system developed by the ASTM, which divides coal into four main ranks. This system is summarised below with coal ranks listed in ascending order from low to high.
- Lignite is the lowest rank of coal with the lowest energy content, the lowest carbon content and highest volatiles content.
- Sub-bituminous coal has a higher heating value, higher coal content and lower volatiles content than lignite.
- Bituminous coal has two to three times the heating value of lignite. It has higher heating content and a lower volatiles content than sub-bituminous coal.
- Anthracite has a heating value slightly lower than bituminous coal. It has the highest carbon content and the lowest volatiles content.

Technically graphite is considered the highest rank coal. However, it is hard to ignite and is not usually used as fuel, but as an industrial raw material. Therefore, it is customary to exclude graphite from coal classification and resource estimates.

Often, to define coal grade, individual quality measurements may be aggregated in various ways to classify and label coal for a particular purpose such as metallurgical, petrochemical or gas usage. The grade classifications usually refer to coal ranks as part of the classification system, since, even though the coal ranks do not have precise chemical and physical characteristics, each coal rank has proximate heating value including having defined ranges of moisture, volatile matter, ash, and fixed carbon. For example, the IEA categorises coal primarily on heating value and secondarily on the volatile matter content of three main types: *hard coal* (above 23865 kJ/kg), *sub-bituminous coal* (between 23865 kJ/kg and 17435 kJ/kg), and *brown coal* (below 17435 kJ/kg). The heating values are set on 'ash-free but moist' basis, and the volatile matter content is set above 31%. Hard coal is further sub-categorised into anthracite and bituminous coal, based on volatile matter content, with the limit set below 10% for the former. Furthermore bituminous coal is further sub-categorised into coking coal and steam coal based on suitability to be used in coke ovens. Brown coal is further sub-categorised into lignite, oil shale or tar sands, produced and combusted directly or consumed in the transformation process (though in this book the latter two are categorised as oil not coal). This IEA categorisation is more-or-less adopted in its entirety by the UN and the EU for their energy statistics.

Examples of other measurements used to grade coal include sulphur content, where coal is graded *low*, *medium* or *high*, depending on the concentration of sulphur and the ease of its removal. For example, sulphur content in the form of pyrite (FeS_2) is easier to remove than organic sulphur.

Even though coal deposits are extracted based on technical and economic criteria similar to oil and natural gas, there is no convention to classify coal deposits as *conventional* and *unconventional*. Thus, unfeasible or technically difficult deposits are usually excluded from resources estimates rather than be classified as unconventional. This is rather an oversight as it can result in underestimating the resources and nowadays, with the development of in-situ gasification, some previously unfeasible deposits that were excluded need to be re-included, re-evaluated, and reconsidered as resources or even reserves.

4.2 Peat
Peat is a soft organic material consisting of partly decayed plant matter together with deposited minerals. Peat forms in wetlands usually referred to as *peatlands*, and its reserves are usually quoted in terms of area. For an area to be qualified as peat reserve

the thickness of the peat layer must be at least 20 cm on drained land, and 30 cm on undrained land. It is difficult to quantify the reserves of peat in energy terms since the energy content of in-situ peat depends on its moisture and ash content. However, dry peat has heating values comparable to lignite. Technically, peat is not coal, but a precursor of coal that will transform to lignite under suitable geological conditions. Peat is brown and has very open structure with high porosity, whereas coal is denser and darker tending towards black.

Peat deposits cover approximately 2% of the world's land, and are found in many places around the world, notably in Russia, Canada, USA, Finland and Sweden. Peat has numerous uses, most of which are non–fuel. It is used mainly in agriculture as growing medium and soil improver. It is also used as a source of organic and chemical products such as activated carbon and waxes, medicinal products such as steroids and antibiotics, and therapeutic applications such as peat baths.

As an energy source, peat is used as a fuel for electricity and heat generation in power plants, and directly as a source of heat for industrial, residential and other purposes. The energy used is almost exclusive to Europe with Ireland, Finland, Sweden, Belarus and Russia account for over 90% of both production and consumption. Thus, due to its majority usage as non fuel material and the unfavourable economics of fuel usage, peat is not considered a coal resource, even though it is considered fossilised material by most governments and organisations. This is the EU official stance, as opposed to Finland, which classifies peat as renewable biomass (see Section 5.3). However, peat's fuel usage is treated as part of coal production and consumption for statistical purposes according to the EIA, EU and the UN.

4.3 Arctic and Antarctic Coal

As with oil and natural gas, it is widely documented that significant deposits of coal exist in the Arctic and Antarctic regions. This coal is usually termed as *polar coal*. Coal fields in the Arctic are currently being exploited in Russia, Norway, Canada and Alaska, and as mentioned earlier regarding Arctic oil and natural gas, the rapid melting of ice in the Arctic is encouraging further mining activities. Refer to Sections 2.5 and 3.7 for more details.

The situation is also similar to Antarctic oil and natural gas regarding Antarctic coal, which is not considered available as yet. The discussion in Sections 2.5 and 3.7 is equally applicable here.

4.4 Secondary Coal

Unlike oil, coal requires minimal processing such as milling (pulverisation), washing, or refining (removing moisture and certain pollutants) prior to using it as fuel. Coal utilised in this way is referred to as *primary coal*. However, this primary coal can be further processed and transformed to enhance its quality, such as improving its heating value, reducing its ash content, and eliminating smoke generation while burning. The resultant transformed manufactured fuels can be solid, liquid or gaseous, and are called *secondary coal*. These transformation processes are somehow analogous to refining oil since their aim is to produce more convenient products to be used as fuel.

As discussed earlier coal can be transformed into liquid (see Section 2.8) or gas (see Section 3.9) with the resultant products considered secondary coal. However, the majority of secondary coal is in the form of solid fuels. Solid secondary coal is produced by transforming coal into three main products:
- Coke, which is a solid carbonaceous material derived from destructive distillation of coal, where the volatile constituents are driven off in coke ovens at high temperatures in an industrial process labelled as coking, resulting in a solid fuel which is low in moisture and volatile matter. Semi-coke, is a solid product obtained from carbonization of coal at low temperature, and is used as a domestic fuel or by the transformation plant itself. The majority of coke is made from bituminous or coking coal, and is labelled *metallurgical coke*. However coke can also be made from lignite and sub-bituminous coal. Coke is used as a fuel and as a chemical agent in the iron and steel industry.

- Briquettes, which are composition fuels manufactured from lignite and sub-bituminous coal, produced by briquetting (where the coal is crushed, dried and molded under high pressure into an even shaped briquette) without the addition of a binding agent. Coal briquettes also include peat briquettes, and dried lignite fines and dust. Additionally peat as a fuel is used in three forms: peat pellets, sod peat or milled peat.
- Patent fuels, which are composition fuels manufactured from bituminous coal fines, with the addition of a binding agent such as pitch. Thus, due to the usage of the binding agent, the amount of patent fuel produced may be slightly higher than the actual amount of coal consumed in the transformation process.

4.5 Other Solid Fossil Fuels

Even though transformation processes are technically possible, there are no commercial processes to specifically convert oil or natural gas into solid fossil fuel, mainly due to the lack of feasible incentives. The only exception is the production of petroleum coke from the residue of the oil refining process. This product has similar uses to metallurgical coke produced from coal.

Additionally, fossilised charcoal-like carbon based solid fuels can be produced as by-products of thermal depolymerisation of non-organic waste such as plastics or tyres.

As explained earlier, several statistics categorise oil shale, burnt directly in power stations, as a solid fuel similar to coal (see Sections 2.3 and 4.2). However in this book oil shale is considered as part of oil resources and is not treated as a solid fossil fuel but regarded as a liquid fuel.

4.6 Coal Uses

Coal's primary use is as a fuel with the reminder used as a chemical feedstock. Lignite is mainly burned at power plants to generate electricity, whereas bituminous coal is used both to generate electricity and to produce coke, which in turn is used as a fuel and a raw material for the steel and iron industries.

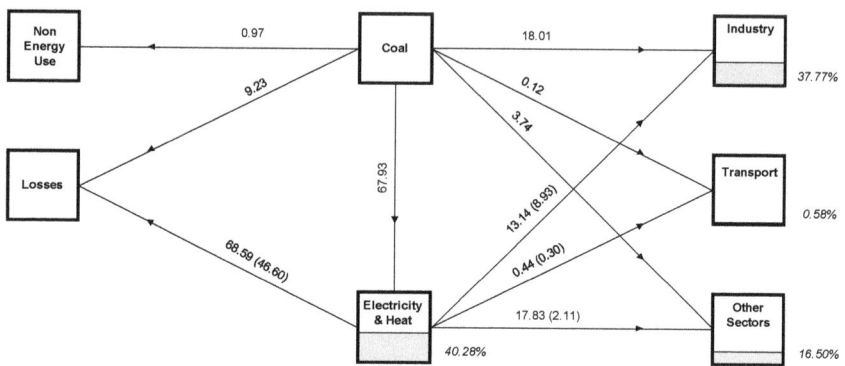

Figure 4.2: Distribution of coal global consumption (2006)
Source: IEA (http://www.iea.org/stats/index.asp).

Figure 4.2 details the distribution of the global coal consumption per sector, and is derived from the data reported by the IEA in 2006. The figure demonstrates that natural coal is still the largest fuel contributor in the industrial sector, it shows that coal's contribution is around 37.8% of the industrial sector's total need, which in turn consumes approximately 26.9% of the coal's total supply. Other sectors including residential, commercial and agricultural consumption of coal, have declined significantly, as they only consume just fewer than 15.9% of the coal's total supply, which constitute 16.5% of those sectors' total need. Even lower than with natural gas, coal use in transportation is trivial as it only consumes over 0.4% of the total supply, which constitutes approximately a tiny 0.6% of the sector's total need. These numbers include both the primary consumption and the consumption supplied by means of electricity and heat output from power plants. As with natural gas, the figure also shows that, if treated as a sector, electricity and heat generation is the largest consumer of coal, as it accounts for almost two thirds of its total supply consuming just over 67.9% which in turn constitutes approximately 40.3% of the generation's total need. However it is noted that due to large generation and transmission inefficiencies, almost 68.6% the coal used in electricity generation is lost. This loss, as well as using coal as fuel in its extraction and transmission processes, accounts for over 55.8% of total coal consumption – that's over half its consumption! In addition, just under 1.0% of the total coal supply has non-energy use, where it is used as raw material in petrochemical industries and other non-fuel products.

4.7 Coal Transport

As bulk solid, coal is technically easy to handle and transport, however the relative high costs of its transport accounts for a significant part of its traded price and thus confines its trade to smaller geographical reach compared to oil or natural gas. In addition its transport requires special care to minimise potential hazards such as risks of fire while being transported due to its susceptibility to spontaneous combustion. Coal must be moved from production mines to delivery points such as ports and railway handling facilities. It is moved either directly to consumers such as power stations or to processing sites to be converted into coke or be pulverised prior to being transported to consumers. Four main modes of transport are in use:

- Land transport by rail containers or trucks - which accounts for the majority of coal transport. Rail is the most common way to transport coal, which although it requires large initial fixed capital investment, is inherently the most efficient due to its appropriateness to handle large volumes for long term and long distance. Road transport is common over shorter distances and is used for moving a significant proportion of the coal mainly from the mines to processing plants or to railway handling facilities. Road transport requires considerably lower capital investment and is more flexible compared to rail transport; however it is hindered by its high running costs, and the limited volumes it can handle. Rail transport is the dominant domestic mode of coal transport, and it is also used by almost all of the coal producing countries to link their producing mines to export ports used for international coal trade; however rail transport is not common to transport coal internationally across borders. Road transport is almost entirely used domestically and is rarely used for international transport.

- Water transport using ships or barges - which carry coal on rivers and canals or across seas. Water transport is used not only for coal export from producing areas to export ports, but also for imported coal distribution from sea ports to end users. The cost of water transport is inexpensive. For inland transport, the cost is competitive with rail transport; however the length of time taken to complete the transport is considerably longer. In addition, the use of water transport on rivers and canals can be constrained by the navigability and the capacity of the waterways. The majority of international coal trade uses sea transport as the preferred mode of transport due to the distance between producing and consuming countries which render land transport unfeasible or impractical.
- Slurry pipelines - which transport coal as a slurry suspended with liquid, mainly water. Slurry pipelines are considered an efficient method to transport coal for long distances and despite having higher initial capital cost than railways their operating costs are considerably lower. However slurry pipelines cannot adapt to changing coal demand and are susceptible to clogging, therefore their usage is still limited and is used only domestically.
- Belt conveyors - which transport coal for short distances. They are used extensively in mines and in loading facilities, but are rarely used as a mode of transport to consumers.

4.8   Coal Storage

Coal is stored mainly to streamline supply where the stored coal acts as a buffer capacity to lessen fluctuations between supply and demand. In addition coal storage is being increasingly used to blend coals from different sources to homogenise feedstock. However, unlike oil or natural gas, coal storage is usually minimised due to its susceptibility to spontaneous combustion, and the deterioration of its properties.

The majority of coal is stored in stockpiles normally designed as open air stores. These storage areas are referred to as stockyards.

Covered storage, for example in bunkers or hoppers, is very expensive and its usage is minimised, it is used only if conditions dictate it, such as adverse weather conditions, or safety concerns due to storage location close to residential areas.

Stockyards are usually located at a number of points along the coal's transport chain. They are found at coal mines, coal preparation plants, importing or exporting terminals, and at the end-user sites, including power stations, coking plants, iron and steel plants, and cement plants.

4.9   Coal Measurement Units
Coal is measured in mass quantities or heat quantities. Mass units usually used are metric or short tons. Heat units usually used are oil equivalent (boe or toe), where the conversion factor from mass of coal to 'boe' or 'toe' depends upon the type of coal.

Chapter 5
BIOFUELS – BASIC KNOWLEDGE

Unlike fossil fuels which are believed to originate from biological materials that died millions of years ago and are non-renewable, biofuels are fuels obtained from living or relatively recently lifeless biological material. They are considered renewable fuels. Accordingly, even though biofuels can have similar chemical composition to fossil fuels the vital difference is one of time scale; therefore, biofuels are not classified as fossil fuels and are excluded from fossil fuel statistics. Analogous to fossil fuels, biofuels are classified into three types based on their state when used as fuels: liquid biofuels; gaseous biofuels or biogas; and solid biofuels.

Biofuels are outside the scope of this work, and therefore are not studied here. However, for the sake of completeness a very brief overview is given below to demonstrate the similarities and differences between biofuels and fossil fuels, to clarify any confusion or ambiguity that may arise due to using similar terminology in literature, and due to the common errors in media reports.

5.1 Liquid Biofuels

Liquid biofuels are simply defined as biofuels in the liquid state. They are usually referred to as 'biofuels' in mass media and have gained widespread publicity. Liquid biofuels production is expanding globally as the most common renewable fuel that can be used, in transportation, to power vehicles engines or domestic sectors for heating and cooking.

Liquid biofuels are derived from agricultural crops, animal waste or other organic material via three main strategies:
- Growing crops high in sugar or starch, such as sugar cane, sugar beet, sweet sorghum or corn, and then produce bio-

alcohols such as bio-ethanol or bio-butanol by yeast fermentation. The resultant alcohol is then used as fuel to run specially designed engines or blended with existing refined petroleum fuels to run traditional petrol engines. Recently, cellulosic alcohol has been developed using non-edible plant parts, waste biomass, and wood, however further technical research is required to render these processes economically feasible.
- Growing plants that contain high amounts of vegetable oil, such as oil palm, soybean, sunflower, algae, or jatropha. The resultant oils can be treated and then burned directly in diesel engines, or chemically processed to produce fuels such as biodiesel.
- Using similar GTL or CTL technologies such as Fisher-Tropsch process or pyrolysis, to convert organic material or biological waste either directly, or from synthetic gas, into light liquid hydrocarbons such as bio-methanol or bio-DME (dimethyl ether). These processes are simply referred to ***Biomass-to-Liquid (BTL)*** processes. Thermal depolymerisation falls into this category, and is simply a hydrous pyrolysis for converting complex organic materials such as animal offal and waste, or bio-degradable plastics into light crude oil.

In the last few years, tremendous advances have been attained developing liquid biofuels. First generation liquid biofuels used feedstocks that are also used by humans and animals as staple food, such as corn, but were criticised for causing food shortages and increasing food prices by diverting the crops away from the human and animal food chains, and by changing the use of agricultural land to produce energy-favoured crops instead of food. This led to active research developing second generation liquid biofuels which either use non-food crops, or produce cellulosic alcohol using non-edible plant parts, waste biomass, and wood. In addition, algae crops are being developed as a high yield energy crop and are dubbed third generation liquid biofuels.

5.2 Gaseous Biofuels (Biogas)
As with liquid biofuels, gaseous biofuels, referred to as biogas,

are simply defined as biofuels in the gaseous state. Biogas production is expanding globally as a low-cost renewable fuel that can be used in to generate electricity or heat in combined heat and power engines, or used directly in the domestic sector for heating and cooking. Biogas can also be compressed, similar to natural gas, and be used in transportation to power vehicles engines.

Biogas is typically produced by three main strategies:
- Biological breakdown of organic matter, in the absence of oxygen, via anaerobic digestion or fermentation of biodegradable materials such as biomass, biodegradable waste including manure or sewage, municipal waste, green waste and energy crops. This type of biogas comprises primarily methane and carbon dioxide. Similarly, naturally occurring uncontrolled anaerobic digestion takes place in landfill sites producing *landfill gas*, which is a less clean form of biogas.
- Gasification of organic matter such as wood, waste or other biomass, which converts the organic matter into carbon monoxide and hydrogen by partial combusting (and sometimes prior drying or pyrolysing) the raw material at high temperatures, with a controlled amount of oxygen and/or steam, resulting in a mixture termed syngas or wood gas. Using this syngas is more efficient than direct combustion of the original organic matter as more of the energy contained in the organic matter is extracted. This type of biogas is comprised primarily of nitrogen, hydrogen, and carbon monoxide, with trace amounts of methane. Furthermore, this gas is not only itself a fuel, but it is also considered the intermediate step in indirect GTL, CTL and BTL processes as discussed earlier (see Sections 2.8, 3.9 and 5.1).
- Thermal depolymerization of waste, which can extract methane as a by-product as part of the process to convert the biomass into light crude oil (see Section 5.1).

The composition of biogas varies depending upon the production method, and the origin of the organic matter. Typical biogas contains 50-75% methane, 25-50% carbon dioxide, with the balance made of nitrogen, oxygen, hydrogen, hydrogen sulphide, and other minor gases. Therefore biogas may require additional treatment to refine it and clean it before using it as a fuel.

Recently, several announcements were made claiming they can produce biological hydrogen directly from water using algae under certain conditions. None of these claims have demonstrated their feasibility.[1,2,3]

5.3 Solid Biofuels
As with liquid biofuels and gaseous biofuels, solid biofuels are simply defined as biofuels in the solid state. They are often referred to, inaccurately, as 'biomass' in the mass media, which technically encompasses all biofuel raw materials. Raw solid biofuels have been used for centuries as readily combustible matter such as wood, domestic waste and dried manure. It can be burnt directly for heating and cooking or to generate steam.

If raw solid biofuels are in an inconvenient form for direct combustion such as sawdust, grass cuttings, or agricultural waste, they can be further processed and shaped into pellets or briquettes which are easier to handle and burn in ovens.

Additionally, raw solid biofuels can be transformed to enhance their quality such as improving the heating value and eliminating smoke generation while burning. Three main processes are used:
- Pyrolysis, which has been used since ancient times to turn wood and wood waste into *charcoal*. Similarly, *biochar* is produced by biomass pyrolysis or as by-product of thermal depolymerisation of organic waste.
- By-products of anaerobic digesters used in producing biogas (see Section 5.2). The solid by-product is called *digestate* and it can be used as a biofuel or a fertilizer.

---

[1] NREL, (http://www.nrel.gov/hydrogen/pdfs/42285.pdf).
[2] NREL, (http://www.nrel.gov/docs/fy04osti/35593.pdf).
[3] Solar Biofuels, (http://www.solarbiofuels.org).

- By-products of low temperature gasification, which while producing syngas produces biochar by-product.

5.4 Synthetic Fuels from Pollution

Several past studies suggested the possibility of producing hydrocarbon synthetic fuels from air or water. The most credible suggestion is the recent conceptual technology developed at Los Alamos National Laboratory in the USA called Green Freedom that demonstrates the production of carbon-neutral hydrocarbon fuel from air and water. The source of the carbon in the manufactured fuel is the carbon dioxide in the atmosphere, which can be captured and recovered. The source of the hydrogen is by splitting water, mainly steam from power stations. The two gases are then converted into synthetic fuel, which can be processed further using technologies similar to GTL or CTL technologies described earlier (see Section 2.8). This concept is still in its infancy, and its feasibility has yet to be proven.[4] Other processes that gained publicity, though are not yet proven, include 'Helioculture', developed by Joule Biotechnologies.[5]

---

[4] Martin, F.J. & Kubic, W.L., Green Freedom – A Concept for Producing Carbon-Neutral Synthetic Fuels and Chemicals, Los Alamos National Laboratory, LA-UR-07-7897, November 2007

[5] Brendan Borrel, http://www.scientificamerican.com/blog/60-second-science/post.cfm?id=joule-biotechnologies-announce-new-2009-07-27.

Chapter 6
RESOURCES AND RESERVES: DECOUPLING THE AMBIGUITY

Nobody knows the exact quantities of fossil fuels that exist under the earth's surface or how much of it was produced, or may be produced in the future. All resource and production numbers are in fact quantities, which are, at best, informed estimates that vary based on numerous assumptions and the quantities' definitions adopted by the estimating bodies.

In this chapter the terms usually used when discussing resources and reserves are explained. In industry and media the use of the two terms is often mixed up, and a lot of confusion exists regarding what does each term means. In the most simplistic terms *resource* is an inclusive term that encompasses *reserve*, so that a reserve constitutes a part of the total resource, which also includes other constituents as will be explained below.

Several guidelines and standards have been published by various governmental agencies, national and international organisations to define and classify fossil fuel resources and reserves. Examples include the classification systems, either in use or proposed, by the US Security and Exchange Commission (SEC), the Norwegian Petroleum Directorate (NPD), the Canadian Security Administrators (CSA), and the United Nations Economic and Social Council (ECOSOC), with the latter classification, known as the United Nations Framework Classification (UNFC), aiming to be applicable to all energy and mineral resources.

Even though the classification systems have a lot in common, they often use different or even conflicting terminology, which frequently does not synchronise or align the different categories exactly, leading to inevitable overlap between categories causing misinterpretation. Thus, it is evident that much of the terminology confusion will be eliminated if the terms are standardised and become universally acceptable. Unfortunately, due to the lack of political will, the UNFC failed to be universally endorsed and was more or less ignored by most agencies.

In 2007, standardisation finally came closer to reality, with the publication of the Petroleum Resources Management System[1], commonly referred to as SPE-PRMS, which is a joint effort by the Society of Petroleum Engineers (SPE), the World Energy Council (WEC), the American Association of Petroleum Engineers (AAPG), and the Society of Petroleum Evaluation Engineers (SPEE). The document is a guideline that provides a consistent approach to estimating fossil fuel quantities and presents the results in comprehensive classification framework. The guideline's intention is not to modify, but rather to complement existing regulatory reporting requirements. In addition, unlike other standards that are applicable to specific fossil fuels and include only partial or no unconventional resources (e.g. SEC classification system does not include natural bitumen, but includes coalbed methane; whereas the CSA classification includes natural bitumen), this guideline applies to all fossil fuels whether currently considered conventional or unconventional. Furthermore it can also apply to any natural resource. The terminology used in the SPE-PRMS is adopted throughout this chapter, with reference made to differing terminologies when necessary.

6.1 Fossil Fuel Resources

Fossil fuel resources are simply defined as the fossil fuels initially-in-place, i.e. the quantity of fossil fuels, in varied sized deposits and reservoirs that exist originally on or under the earth's crust in naturally occurring accumulations. These resources initially in place are referred to by numerous names including *resource base*, *hydrocarbon endowment*, or *original-in-*

---

[1] Society of Petroleum Engineers (SPE), American Association of Petroleum Geologists (AAPG), World Petroleum Council (WPC) & Society of Petroleum Evaluation Engineers (SPEE), Petroleum Resources Management System, 2007.

place.

Several criteria exist to categorise fossil fuels resources. The main criteria are discovery and recovery. In terms of discovery, fossil fuels are either discovered or undiscovered. In terms of recovery they are either recoverable or unrecoverable. The criteria to assess discovery, recovery and commerciality of resources are well defined but vary between standards. For example some standards may require a successful pilot well demonstration or successful process pilot to prove a discovery or a technology. Interested readers can consult specific standards for the details of applied criteria.

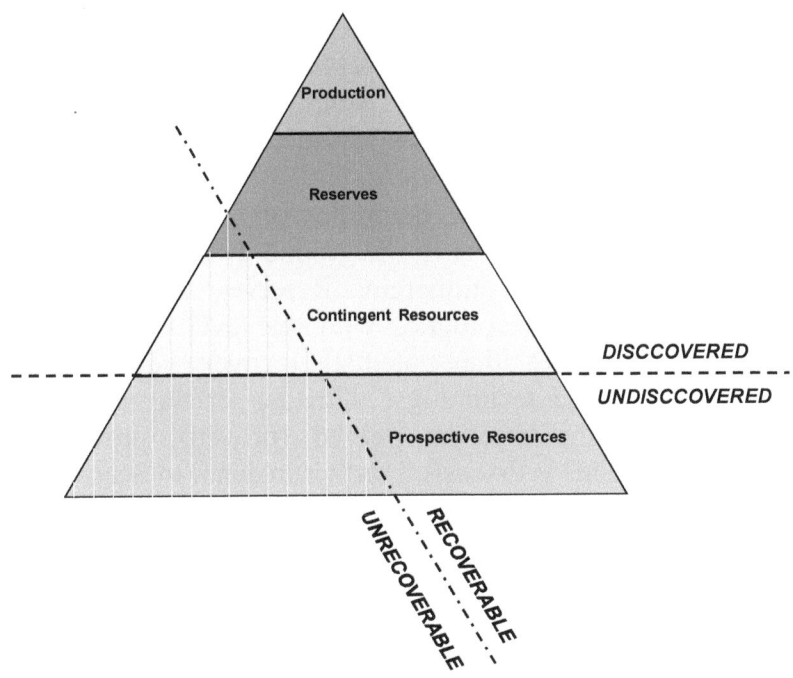

Figure 6.1: Categorisation of fossil fuels resources

The relationship between the above two criteria is demonstrated in Figure 6.1, which positions the reserves within the framework of total resources. The figure also shows that three distinct portions of resources exist: *production, recoverable* resources and *unrecoverable* resources.

Production is the cumulative quantity of a fossil fuel that has been recovered at a given date. This production is the raw production, which includes not only marketed production, but also non-sale production such as flaring and lease fuel (which is the portion of the produced fuel that is consumed in production activities), which may not be measured. In addition, historical production may not all be measured or accounted for. Therefore cumulative production is considered a best estimate quantity.

It follows that the remaining resources are either recoverable or unrecoverable. The remaining recoverable resources are divided into two main categories: discovered and undiscovered. In turn, the discovered resources are further divided into two main categories: commercial and sub-commercial. Amalgamating these definitions together for simplicity, the recoverable resources can be described as divided into three distinct categories:

- **Reserves** - which are quantities of fossil fuels that are discovered, remaining in the ground awaiting production, recoverable and commercial. Reserves are further sub-categorised in accordance with the level of uncertainty associated with the potential recovery due to both feasibility and technology, with any of the two factors rendering the reserves proved (proven), probable or possible. This is discussed further in detail in Section 6.3.
- **Contingent Resources** - which are quantities of fossil fuels that are discovered, remaining in the ground awaiting production, potentially recoverable however they are not yet commercial. Similar to reserves, contingent resources are further sub-categorised in accordance with the level of uncertainty associated with the potential recovery due to both feasibility and technology, with any of the two factors rendering the reserves low, best or high estimates. This is discussed further in detail in Section 6.4.
- **Prospective Resources** - which are quantities of fossil fuels that are undiscovered, remaining in the ground

awaiting production, potentially recoverable and potentially commercial. Also similar to reserves and contingent resources, prospective resources are further sub-categorised in accordance with the level of uncertainty associated with the potential recovery due to both feasibility and technology, with any of the two factors rendering the reserves low, best or high estimates. This is discussed further in detail in Section 6.5.

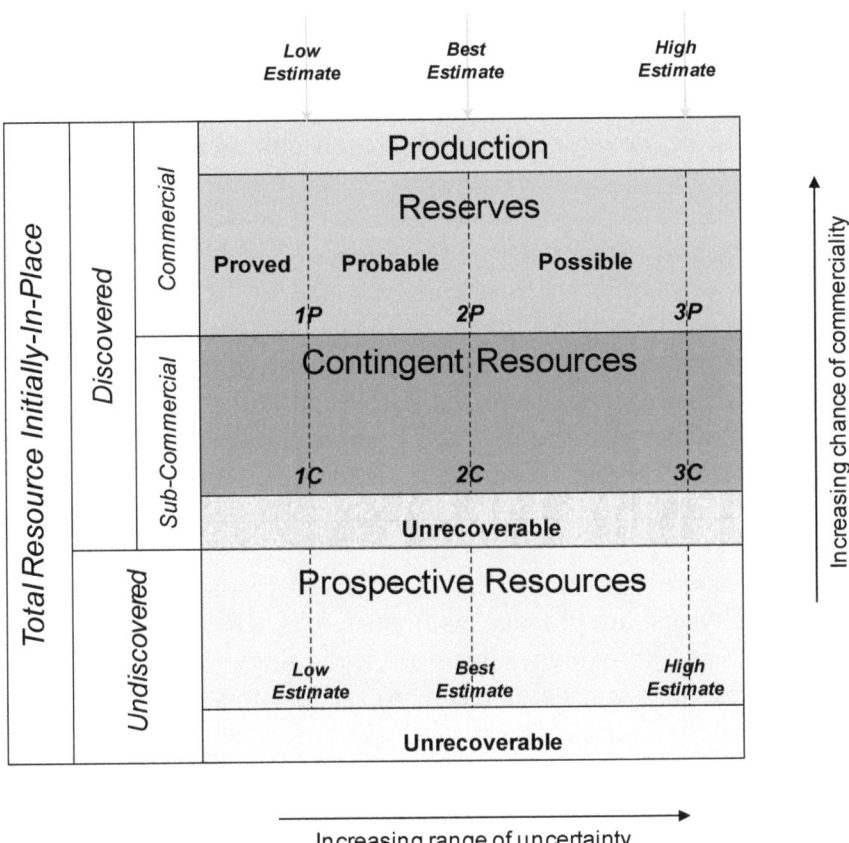

Figure 6.2: Classification framework of fossil fuel resources
Source: Modified from SPE-PRMS.

The unrecoverable resources are the portion of both discovered and undiscovered fossil fuels initially in place, which cannot be recovered at a given date due to geological, commercial accessibility or political constraints. However, a portion of these quantities may become recoverable in the future.

Figure 6.2 above presents the classification framework of fossil fuel resources as defined in the SPE-PRMS. The horizontal axis represents the range of uncertainty, which is a degree of geologic assurance and recovery efficiency, reflecting the range of estimated quantities potentially recoverable from an accumulation. This range has three deterministic classes or scenarios expressed in increasing technical uncertainty as low estimate, best estimate and high estimate. The vertical axis represents the chance of commerciality, i.e. the probability that a project will be developed and can reach commercial producing status.

6.2 Ultimately Recoverable Resources

The term *ultimately recoverable resource* (URR), sometimes referred to as *estimated ultimate recoverable* resource (EUR), is a debatable concept. It is defined as an estimate of the total amount of a fossil fuel that will ever be recovered and produced. Thus, URR is considered to be the sum of four categories: cumulative production, reserves, contingent resources, and prospective resources.

Although some consider URR to be fixed by geology and the laws of physics, in practice estimates of URR continue to be increased as knowledge grows, technology advances and economics change. This leads to redefining some of the unrecoverable discovered or yet-to-find undiscovered reserves into recoverable reserves, and thus extends the fossil fuel endowment that is considered URR. This concept is fundamental in the arguments concerning oil peak theory (see Section 6.8), which is based on the belief that URR is a fixed quantity that cannot grow.

6.3 Fossil Fuel Reserves

It is universally accepted that fossil fuel reserves are based on measurements. However, as indicated earlier, the definition of a reserve is a combined physical-technical-economical-political notion. Hence, taking all these notions into consideration, fossil fuel reserves are categorised into three categories (sometimes referred to as classes or increments):

- **Proved (or proven) reserves**: usually defined as the estimated fossil fuel quantities which geological and engineering data demonstrate, with reasonable certainty, to be recoverable in future years from known reservoirs under current economic and operating conditions, i.e. having a high degree of probability (usually better than 90%, but can range between 80% and 95% depending on the regulatory body) of being produced. These probability 'confidence hurdles' mean that quantities actually recovered will equal or exceed the hurdles limits. It follows that proved reserves are the proportion of fossil fuel in place that is technically and economically recoverable, given current economics and technology. Reservoirs are considered proven if economic production is supported by either actual production or conclusive formation testing. Proven reserves terminology varies between regulatory bodies, and can be referred to as *P90* or *P80*, *1P*, or as *measured* reserves. Proved reserves are further subdivided into Proved Developed (PD) and Proved Undeveloped (PUD).

- **Probable reserves**: these are reserves of fossil fuel-in-place, but which are estimated to have a medium degree of probability (usually better than 50%) of being technically and economically producible. Probable reserves terminology varies between regulatory bodies, and can be referred to as *indicated* reserves. Cumulative quantities of proven and probable reserves are usually referred to as *P50* or *2P*, which is calculated simply by summing the two categories.

- **Possible reserves**: these are reserves of oil-in-place, that, at present, cannot be regarded as probable reserves, but are estimated to have low but still significant degree of probability (usually less than 50% but can be as low as 20%, 10% or 5% depending on the regulatory body) of being technically and economically producible. Possible reserves terminology varies between regulatory bodies, and can be referred to as *inferred* reserves. Cumulative quantities of all three categories are usually referred to as *P10* or *P20*, or *3P*, which is calculated simply by summing the three categories.

To recap the terminology referring to fossil fuel reserves, it is important to remember that the term 'proved' refers to low estimates; probable to best estimates; and possible to high estimates.

Even though the principles of reserves' definitions are common between countries, the exact definitions of reserves vary, with different factors and conditions used in defining specific terms such as commerciality. A study commissioned by the SPE in 2005[2] compared several national and international resource and reserves classification systems, and concluded that although the terminology varies, there is a high degree of communality. Thus it can be generalised that reserves can be categorised into the above three main classes which are differentiated by decreasing technical certainty. It follows that 'proven reserves' refer simply to the *low estimate* of fossil fuel quantities that will be produced from a discovered commercially viable deposit; the *best estimate* constitutes the proven and the probable reserves, while the *high estimate* constitutes the proven, probable and possible reserves.

The existence of so many classification systems makes it clear why arguments rage regarding how much resources and reserves there are, since not only has each country its own classification system, but some countries use more than one classification system with different governmental departments using totally different systems. The prime example is the USA, which uses two main systems: the US Security and Exchange Commission System (SEC-1978), which is used as financial regulatory system for disclosure to investors and thus imposes rules around technical and commercial certainty, and the United States

---

[2] Society of Petroleum Engineers (SPE), Oil and Gas Reserves Committee (OGRC), "Mapping" Subcommittee, Final Report – December 2005, Comparison of Selected Reserves and Resource Classifications and Associated Definitions.

Geological Survey System (USGS-1980), which is used as full resource base assessment and reporting system, and is therefore used by the US government to implement and modify legislation, policies, and strategies on national and global arenas.

To illustrate the confusion, the reporting oil and natural gas reserves around the world is examined briefly.

In the USA, according to SEC rules reserves are defined as oil and natural gas that are reasonably considered to be possible for extraction, in the future, from the known physical resources, with the known techniques and in the present economic conditions. The rules of 'proved reserves' state that oil and gas fields can only be listed as proved if there are data showing actual flows, if the fields will be developed, and if they are commercially viable, which happens only after companies have made a final investment decision to develop a field. Until then the reserves are considered probable or possible.[3] This definition is clearly narrow, very rigid, outdated, and tends to underestimate reserves (see for example the case of Mexico's reserves - Section 6.3.2). Therefore many experts and oil executives argue that the SEC should change these rules, pointing out that a field can simply jump out of the proved category, if the price of oil drops enough that it becomes no longer feasible to develop. Also the criteria regarding the actual flow data is no longer as important, as improvement in seismic images and other technologies, make it possible to estimate field sizes without drilling expensive wells. To elaborate even further, consider the USGS assessment of the USA oil and natural gas reserves, in which they estimate that the USA possesses significantly higher reserves as acknowledged by the SEC. This is due to the less rigid requirements which allow classifying more fields as proved reserves, as well as the inclusion of some probable reserves.

In the rest of the world, reported oil and gas reserves generally correspond to the addition of proved reserves, defined less strictly that dictated by the SEC rules, and a part of other probable or

---

[3] J Carey, C Palmeri & S Reed, Business Week, 26 January 2004, p40.

possible reserves. Many countries estimate reserves using the 2P criteria explained earlier, where reserves are the sum of proved plus probable reserves. This method is adopted in Canada, Australia, Norway, and the UK amongst others. Some authors even include possible reserves and report reserves as the sum of 100% of proved reserves, 50% of probable reserves, and 25% of possible reserves.[4]

Thus, the oil and gas reserves published by the oil and gas companies in the USA correspond to the proved reserves only as dictated by the SEC rules. In the rest of the world most companies use the 2P measure as their best guess of what will be produced, even though they also report proved reserves (1P) separately for different regulatory purposes. Note however that since the proved reserves estimates do not need to comply with the SEC rules, especially regarding the intent of project development, the estimated reserves numbers tend to be larger. These differences in reporting methodologies lead to potential false comparisons between reserves data not only if one compares 1P and 2P values, but also two 1P values, since the reported numbers are not defined according to similar criteria and thus they do not correspond to the same quantities.

It can be concluded then that, since fossil fuel reserves data are obtained from different data sources following different definitions, discrepancies between the data are inevitable and stem mainly from the different definitions and different methods of assessment. However differences may also occur due to the motivation of the data source, which may withhold or exaggerate numbers to serve a certain objective. In reality different countries and companies use different methodologies and the data have varying levels of reliability. Precise comparisons between nations and analyses of time series should be treated with great caution to avoid comparing apples to oranges. Data resources are discussed in detail in Chapter 8.

Taking all above factors into account, it can be deduced that

---

[4] http://www.manicore.com/anglais/documentation_a/oil_reserve.html.

fossil fuel reserves numbers, which are the only numbers to be published, do not designate what remains below or on the surface, but only the fraction of that deposit that we think we will be able to extract in the future, with the techniques available today (or in a near future), and in the current economic conditions (in other words the limit is that the cost of extraction must not be superior to the market price). A reserve is therefore a subjective notion by nature.

Finally, it is important to emphasise that there is a lot of confusion when oil and natural gas reserves are discussed, especially in the mass media. Many authors confuse the definitions and claim that probable and possible reserves mean that the oil and the natural gas are yet to be discovered. This is factually incorrect. Any oil or natural gas labelled as reserve by the oil and gas industry refers to oil and natural gas discovered and known to be in-place; all other undiscovered or uncertain oil or natural gas is labelled as prospective resource. The same applies to coal reserves. A recent example of such confusion is an article published by Newt Gingrich (a former Speaker of the United States House of Representatives) in Newsweek in April 2009 in which he refers to American oil shale resources as reserves and puts them in par with Saudi Arabia's proven oil reserves, stating that the American oil shale reserves are three times the reserves of Saudi Arabia's oil![5] One cannot compare apples to oranges!

6.3.1 *Case Studies*
The following two cases that illustrate how misunderstanding oil reserves terminology leads to exaggerated panic responses and a massive media circus when oil reserves are reported are discussed below.

Case 1: Shell Oil Reserves
The announcement by Shell in 2004, reducing its 'proven' reserves of oil and gas by a whopping one-fifth (from 19.4 to 15.5 billion barrels and then by further 0.25 billion barrels)

---

[5] http://www.newsweek.com/id/192480, printed on April 13th 2009.

generated massive media attention. It caused Shell's share value to collapse, added weight to opinions publicised by peak oil theory advocates (see Section 6.8), it opened a new debate about how accurate are the oil reserves claimed by the major oil producing countries and raised the question whether there is less oil and gas than was originally thought.

The answer to the above question is a simple 'no'. What Shell did was to reclassify or redefine the reserves from 'proven' to 'probable'. In technical terms the oil and gas still exist and are owned by Shell, but the reserves are not close enough to commercialisation for Shell to consider them under US Security and Exchange Commission (SEC) rules as 'proven'.[6] As discussed earlier, 'probable' assets are also very likely to get to market but not as quickly as 'proven' ones.

So as this story shows, the media reported a tornado that turned out to be a storm in a tea cup. Their ignorance of the terminology led to a complete misunderstanding of Shell's announcement and its impact on the market.

Case 2: Mexico Oil Reserves
A big story often cited in the media is the massive increase and subsequent decrease in Mexico's oil reserves estimates. Reports have often claimed that Mexico exaggerated its reserves estimates, claiming huge increase in oil reserves during the 1980s since it needed large reserves estimates to guarantee the loans from the USA and IMF during the Mexican financial crises of 1982 and 1994, and before North American Free Trade Agreement (NAFTA) was signed. Subsequently the argument continued that Mexico decreased the estimates after the financial crises were solved.

Figure 6.3 shows Mexico's oil reserves since 1980. A naïve inspection of the figure appears to support the argument, since one can notice a huge jump in 1982 and a huge cut in 1999. However both changes have a valid explanation differently. The

---

[6] The Economist, 13 March 2004, p71

jump of 1982 was due to the inclusion of the reserves of the massive Cantarell Field – discovered in 1976, while the cut of 1999 was due to the obligation of Mexico, under the NAFTA to report its reserves according to the SEC rules, which means that the reported reserves are only from proven producing fields, thus excluding the giant Chicontepec Field, which, even though discovered since 1926, it has seen minimal development.

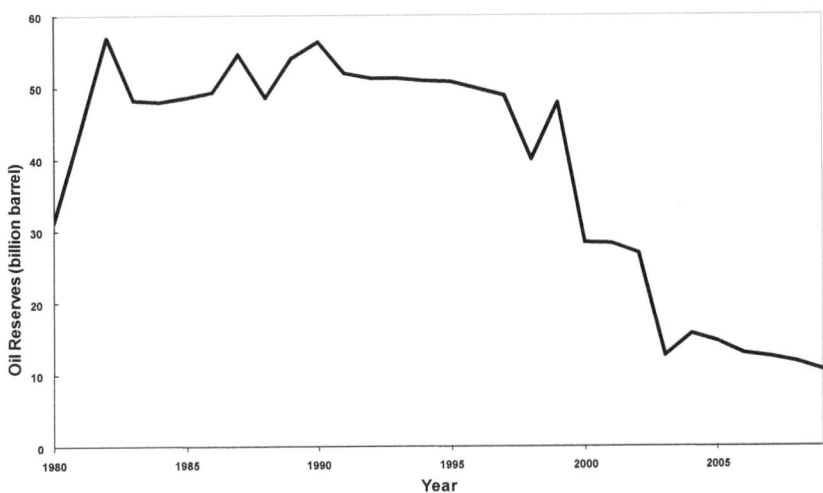

Figure 6.3: Conventional proved oil reserves of Mexico (1980-2009)
Source: EIA (http://www.eia.doe.gov/international).

This case shows that the media was misled by jumping to conclusions without understanding the terminology of oil reserves. Their lack of knowledge once again created sensational baseless stories.

6.4 Fossil Fuel Contingent Resources

Contingent resources are discovered resources. Similar to the reserves categorisation, the contingent resources are categorised into three main classes, which are differentiated by decreasing technical certainty. The SPE-PRMS labels the three cumulative classes as 1C, 2C and 3C. The terminology is analogous to 1P, 2P and 3P used for labelling the reserves. It follows that these categories refer simply to the *low estimate*, *best estimate*, and *high estimate* of fossil fuel quantities that may be produced from a discovered, but still sub-commercial, deposit depending on the economic and technical conditions.

Note that some national standards confusingly use the term 'reserves' to refer to contingent resources, though these standards still treat these so called reserves as a different class of reserves. Examples include the Russian Ministry of Natural Resources classification, which uses the term reserves for all discovered resources and refers to them as contingently profitable and sub-economic reserves, and the Chinese Reserves Office classification, which refers to contingent resources as technically recoverable but not economically recoverable reserves.

Even in the USA, the USGS refers to contingent resources as marginal reserves. Note though that the SPE-PRMS contingent reserves and USGS marginal reserves do not align exactly, and the contingent reserves as defined by SPE-PRMS include also part of what the USGS terms as demonstrated sub-economic resources. Furthermore the USGS introduces a concept called *reserve base*, which encompasses the part of the reserves that have a reasonable potential for becoming economical within the planning time scale, and is defined as the sum of the reserves plus the marginal reserves and a proportion of the demonstrated sub-economic resources.

6.5 Fossil Fuel Prospective Resources

Prospective resources are undiscovered resources. Analogous to the reserves and contingent resources categorisation, the prospective resources are also categorised into three main cumulative classes which are differentiated by decreasing technical certainty. However, unlike reserves and contingent resources, the SPE-PRMS does not label these categories, but simply refers to them as *low estimate*, *best estimate*, and *high estimate*. This is simply the estimates that will potentially be produced from undiscovered deposits without assessing their commerciality.

Note that, unlike SPE-PRMS, some national standards subdivide the prospective resources into different categories. For example, the Russian Ministry of Natural Resources classification splits them into prospects, leads and plays, based on both technical and commercial certainty, whereas the USGS classifies them as hypothetical and speculative based on technical, not commercial, certainty.

6.6 Recovery Factors

The recovery factor, which is also referred to as recovery efficiency, is a technical term that is defined as the proportion of the in-place resource 'discovered', that is recoverable. Thus it designates the proportion of the resource, existing in the reservoir at the beginning of exploitation, that will be possible to extract from this reservoir between the beginning and the end of the exploitation. It follows that understanding the concept of the recovery factor is fundamental to estimating the recoverable fossil fuel resources, since all reserve estimates are heavily dependent on the assigned recovery factors.

Recovery factors are dimensionless quantities, and are often reported in percentage terms or decimal fractions.

Recovery factors vary considerably between reservoirs due to numerous factors including the geology of the reservoirs, their location, the technology used for exploitation, and whether secondary or tertiary recovery methods are use (see introductory sections of Chapters 2, 3, and 4), with the timing of using these methods playing an important role in determining the final recovery factors.

Generally speaking, recovery factors for natural gas reservoirs are reported to be the highest, followed by those for coal reservoirs and lastly for oil reservoirs. Several studies were performed to estimate worldwide recovery factors for oil and natural gas reservoirs. However, predictably, these studies arrived at different estimates. For oil, reported results ranged from 22% to 40%[7], whereas for natural gas the recovery factors ranged

---

[7] Ivan Sandrea and Rafael Sandrea, Global Oil Reserves – Recovery Factors Leave Vast Target for

from 60% to 95%.[8] For coal recovery factors are often reported as 50% for underground mining and 80% for surface mining.[9] Recovery factors for unconventional resources are generally lower, and are discussed in more detail in Chapters 9, 10 and 11.

6.7   Reserve Growth
Reserve growth refers to the notion that fossil fuel reserve estimates are increasing, despite continuous production. Instinctively, this appears to be incorrect, as proven reserves will fall when fossil fuels are produced from a certain reservoir.

However, as explained earlier, reserves are not fixed quantities, but are rather dependant on technical and economic factors, they can and will increase or 'grow' for several reasons such as:

- Successful new exploration and advances in discovery technologies. New resources can be discovered underground, and without changing the technical or economical conditions, this leads to larger reserves. In addition to that better technology such as advanced 3D seismic technologies lead to better knowledge of what earth contains. This leads to an increase of the identified resources, and with same recovery factor assumed, this leads to larger reserves. Indeed, as fossil fuels require millions of years to form, we can consider that the initial endowment is fixed but not all of it is discovered yet. We cannot therefore 'increase' the resources per say, but just discover more resources that we currently do not know of.
- Advances in the recovery and extraction technologies. This leads to an increase of the recovery factors, which, practically, means that we increase the proportion of the fossil fuel enclosed in the reservoir that we are able to extract (or consider that we are able to extract!), and thus label as reserve. If the recovery factor increases enough, it can increase, very substantially, the reserves. Take oil for

---

EOR Technologies, Oil & Gas Journal, Part 1: November 05, 2007 and Part 2: November 12, 2007.
[8] Papay, J., Improved recovery of conventional natural gas. Part 1: Theoretical discussion of recovery methods, Erdöl, Erdgas, Kohle, 1999, vol. 115, no6, pp. 302-308.
[9] EIA, (www.eia.doe.gov/glossary/glossary_r.htm).

example, as discussed earlier, the recovery factor ranges between 22% and 40%, thus assuming a recovery factor value of around 31%, any 1% increase - that is going from 31% to 32% - would increase the reserves by approximately 37 billion barrels according to 2008 reserve numbers[10], which translates to over one year of global consumption of 2008, or an equivalent of discovering a new Nigeria!

- Cost reductions in the technology of extraction and mining. This will improve the economics of fossil fuel production, leading to higher recovery factors and thus increase in reserves.
- Price increase. Economic conditions keep changing so with higher prices more resources move into reserves category. Consider oil price as an example: if the market price of oil is US$20 dollars per barrel; it is meaningless for oil companies to extract oil if it costs US$25 dollars per barrel, even if the quantities that they would be able to extract at this price are potentially huge. If, on the other hand, the oil barrel jumps to US$80 (we experienced US$147 per oil barrel in 2008), then the reservoirs where the extraction cost is US$25 per barrel become economically attractive, and oil resources within these reservoirs are then included in the reserves estimates.

Historically, the first three factors have always materialised, whereas the price has always fluctuated. Therefore, taking all the factors into account this explains why the level of reported reserves has continued to rise over long periods of time. Furthermore, besides reported discoveries, additions and extensions to reserves due to advances in technology and reduction in cost, have exceeded depletion of reserves due to production.

On the other hand, reserves can decrease or 'shrink' for several reasons such as:

- Altering environmental legislation, where tighter laws

---

[10] EIA, (http://www.eia.doe.gov/international).

have an adverse effect and push some reserves back into resources category. This can happen if stricter laws are imposed on greenhouse gases especially carbon dioxide emissions, which imposes restrictions on production and thus render some resources unfeasible.
- Accessibility restrictions to reserves, due to political situations whether legislative or due to security concerns, which again may cause some reserves to be reclassified as resources.
- Price decrease, which has the opposite effect to the price increase as discussed above.

So, contrary to the initial intuition, it is easy to explain why reserves can grow and shrink, i.e. vary quantifiably higher or lower without any change in geological parameters of the reservoirs, but rather by modifying any technical or economical parameters, or merely by defining what is meant by the word 'reasonable' when defining a reserve.

It is important at this point to differentiate between reserve growth and process gain, or refinery gain, at processing facilities and refineries, which is sometimes confused as an apparent reserve growth. This is simply incorrect, since in all processes the mass is always conserved. However, since crude oil and refined products are usually reported in volume units, it appears in some cases that there are more barrels of processed or refined products relative to the feed stock given to the process, which happens due to different densities of different products. Thus these gain numbers can be misleading when overall production or trade numbers are reported.

6.8   Peak Theory
The peak theory hypothesizes that, for the planet as a whole, or for any given geographical area, the rate of a natural resource production tends to follow a logistic distribution bell-shaped curve, which even though it appears similar to normal distribution curve, it has a heavier tail so it is slower in approaching zero.

Figure 6.4 shows a typical logistic peak curve. The curve consists of three main sections:
- The ascending curve (pre-peak), which is the initial curve showing a rapid rise in production rate due to the resource discovery and the addition of production infrastructure. The production rate follows the relatively steep slope of the logistic function. When applied to any natural resource, this part of the curve consists mainly of the actual production data.
- The peak point, which represents the maximum of the mathematical function, and shows the maximum resource production rate.
- The descending curve (post-peak), which is the remaining curve where the function drops from the peak. This is predicted to happen when half of all resources have been consumed, where the production rate will drop following a steep decline logistic function.

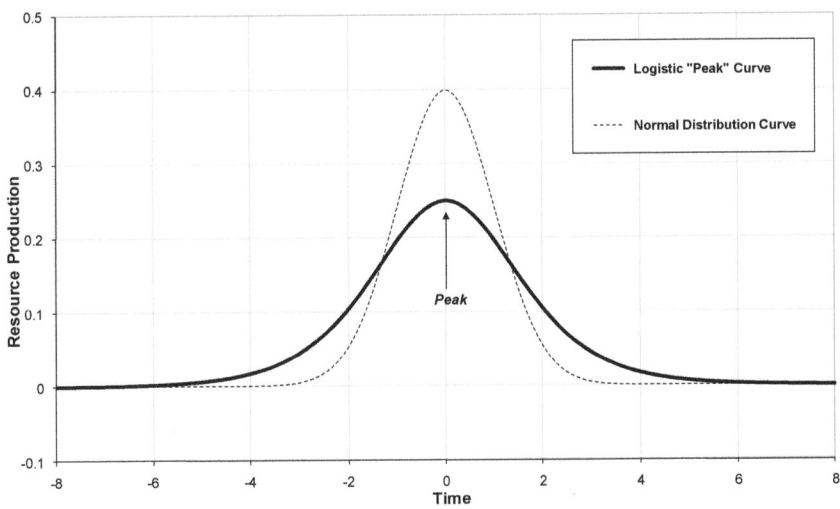

Figure 6.4: Typical logistic peak curve

According to the peak theory model, the amount of any resource on the planet as a whole or in a certain region is finite with an assumed URR; therefore the rate of resource production is determined by the rate of new resource discovery. The rate of discovery which initially increases quickly must reach a maximum and then decline. Production roughly follows the discovery curve after a time lag. The relative steepness of the projected rate of decline of the production curve is the main cause for concern, due to the economic and social impacts caused by resource depletion.

In both scientific and commercial literature, as well as in the mass media, peak theory is often referred to as 'Hubbert peak theory' since it is based on the work of the geoscientist M. King Hubbert in the mid 1950s, who defined the peak curve, usually referred to as Hubbert curve. Hubbert based his calculations on the oil well discovery in the USA, lower 48 states, to predict the rate of oil production over time and to predict the rate of the peak production, which is referred to as Hubbert peak. He accurately predicted that oil production in the USA, lower 48 states, would peak around 1970. Hubbert's theory was initially greeted with scepticism by many in the oil industry, but has since gained many supporters particularly amongst geologists. The Hubbert curve has been used since then to predict the depletion rate of not only fossil fuels, but also various natural resources, and has become popular amongst sectors in the scientific community.

Promoters of peak oil have since generated a lot of publicity, since they argue that a steep drop in the oil production curve implies that global oil production will decline so rapidly that the world will not have enough time to develop sources of energy to replace the energy now generated from it. The advocates of the theory are mostly geologists and retired oil executives. They are usually referred to by their opponents as alarmists, pessimists, doomsters or depletionists. These advocates have published extensively predicting the date of the peak oil and warning of dire consequences of the aftermath. Among the loudest and most distinguished advocates are Colin Campbell, Jean Laherrère, Kenneth Deffeyes and Mathews Simmons. Numerous articles can be found on the Association for the Study of Peak Oil and Gas (ASPO) website. A simple Google search of the term 'peak oil' points to over 1.6 million instances (as of August 2010).

On the other hand, critics of the theory are mostly economists and serving oil executives. They are also referred to by their opponents as optimists or rejectionists. Peak oil critics argue that the theory is over simplistic and is biased towards underestimating oil reserves. They point out that the predicted oil peak has always failed to materialise and they support their arguments by compiling lists showing the false predicted peak oil dates and highlighting that the advocates have had to keep pushing back their predicted peak dates. Furthermore, critics point out that peak oil advocates base their predictions on easy conventional oil resources and ignore unconventional oil resources in their models, even though the availability of these resources is significant and the costs of unconventional oil extraction and processing, while still very high, are falling due to improved technology. Moreover, critics suggest that the reserves of even conventional oil are underestimated in many areas of the world especially in Iraq, Iran and Russia. In addition, critics also emphasise that the recovery rate from existing world oil fields is continuing to increase due to new technology, and thus is adding additional reserves all the time.

Peak gas and peak coal theories have been suggested recently in a similar manner to peak oil theory. However, despite being based on the same theoretical argument, they have yet to gain publicity to feature prominently in the mass media. To date, they have only few active advocates.

At this point in the book I am not advocating or criticising the peak theory. The theory is presented here as an abstract concept, and it is examined further in Parts II and III of this book where fuller discussion and implications of the theory are presented as appropriate.

In conclusion, there is no denying that, although currently abundant, there is a finite supply of fossil fuels. Similar to any renewable resource, fossil fuels usage will come to an end at some time in the future. However, the question that I have to put forward is, whether the world will stop using fossil fuels as a result of them being exhausted. My simple answer is that I do not believe so. After all one has to remember that the Stone Age did not come to an end because the world ran out of stones!

Chapter 7
GEOGRAPHICAL AND POLITICAL CONTEXTS

The geographical perspective of this book is confined to the Arab world. However, throughout the text, references to the Middle East, MENA countries, Africa, the European Union, OPEC countries, OPAPEC countries, and OECD countries are used. Therefore, a clear definition of these terms is presented. It is noticed though that throughout published literature the exact geographical definition of these terms is loose and relies heavily on the political and economical viewpoints taken by authors and organisations.

7.1   Arab World
The 'Arab world' term in this book refers to the 22 members of the Arab League – a regional organisation of countries in Southwest Asia, North and East Africa based in Cairo. The organisation was formed in 1945 by seven countries: Saudi Arabia, Iraq, Syria, Lebanon, Jordan (Transjordan till 1946), Egypt, and Yemen (as North Yemen, South Yemen joined in 1967 and unified with North Yemen in 1990). Membership has increased since then with the following countries joining: Libya (1953); Sudan (1956); Tunisia and Morocco (1958); Kuwait (1961); Algeria (1962); Bahrain, Qatar and Oman (1971); United Arab Emirates (1972); Mauritania (1973); Somalia (1974); Palestine (1976, represented by the Palestine Liberation Organisation PLO); Djibouti (1977) and Comoros (1993).

All 22 countries declare Arabic as the sole or a co-official language. However Arabs constitute only small minorities in Somalia, Djibouti and Comoros. On the other hand, three countries where Arabic enjoys co-official status are not members of the Arab League: Israel, Eritrea and Chad, nor are several Sub-

Saharan African countries, such as Mali, where Arabic is considered a national language.

Throughout the text, the data referred to as the Arab world considers the 22 countries from the year they first joined the Arab League. Egypt's data are incorporated since 1945 and thus cover the period between 1979 and 1989 when Egypt's membership in the organisation was suspended.

Additionally, the following notes apply:
- Generally, data for Palestine are not reported separately. Between 1949 and 1967 the West Bank was considered part of Jordan, while the Gaza Strip was administered by Egypt. Since 1967, Palestine's data are reported incorporated with Israeli data, since the West Bank is effectively occupied by Israel, and the Gaza Strip was under Israeli occupation until 2005, even though both areas are nominally under partial Palestinian administration since 1994.
- The Arab League recognises the Western Sahara as part of Morocco. Accordingly, the data reported for Morocco incorporates data from the territory.
- The Arab League supports Comoros claim of sovereignty over Mayotte, however the data for the Comoros exclude data for Mayotte which is considered French territory.

7.2 Middle East

The Middle East is a quasi-sub-continent region with no defined borders. In this book the term includes the Asian Arab countries (Saudi Arabia, Kuwait, Qatar, Bahrain, UAE, Oman, Yemen, Iraq, Syria, Lebanon, Jordan, Palestine), plus Iran and Israel. This definition is commonly adopted in energy related statistics, but does not agree with the widely accepted definition of the Middle East, which also includes Egypt.

Other authors, organisations and media outlets have extended the definition to include the entire Arab world, Afghanistan, Pakistan, Turkey, Cyprus, and even Central Asia, South Caucasus and the Horn of Africa.

7.3 MENA Countries
The term MENA is an acronym for 'Middle East and North Africa', used regularly in academia, business and media. The term generally covers most Arab world countries – except Somalia, Djibouti and Comoros - as well as Israel and Iran. Some organisations extend the definition to cover the extended definition of the Middle East.

7.4 Africa
In this book Africa refers to:
- The 52 internationally recognised members of the African Union: Algeria, Angola, Benin, Botswana, Burkina Faso, Burundi, Cameroon, Cape Verde, Central African Republic, Chad, Comoros, Congo (Brazzaville), DR Congo (Kinshasa), Côte d'Ivoire (Ivory Coast), Djibouti, Egypt, Equatorial Guinea, Eritrea, Ethiopia, Gabon, Gambia, Ghana, Guinea, Guinea-Bissau, Kenya, Lesotho, Liberia, Libya, Madagascar, Malawi, Mali, Mauritania, Mauritius, Mozambique, Namibia, Niger, Nigeria, Rwanda, São Tomé and Príncipe, Senegal, Seychelles, Sierra Leone, Somalia, South Africa, Sudan, Swaziland, Tanzania, Togo, Tunisia, Uganda, Zambia, and Zimbabwe.
- Western Sahara (admitted as the 53rd member in the African Union but has only limited international recognition).
- Morocco.

The term Africa therefore excludes the following territories which, although geographically located in Africa, are considered parts of non-African nations, even though some of these territories are subject to several conflicting territorial claims from African countries:
- Spanish territories: Canary Islands, and the plazas de soberanía (places of sovereignty) comprising Ceuta, Melilla, Islas Chafarinas, Peñón de Alhucemas, and

Peñón de Vélez de la Gomera.
- Portuguese territories: Madeira Islands.
- French territories: Mayotte, Réunion, and parts of the Territory of the French Southern and Antarctic Lands including Îles Éparses (Scattered Islands in the Indian Ocean) comprising Glorioso Islands, Juan de Nova, Europa, Bassas da India, Tromelin, and Banc du Geyser; Île Amsterdam; Île Saint-Paul; Îles Crozet; and Îles Kerguelen.
- British territories: Saint Helena including Ascension Island and Tristan da Cunha Islands.

Note that the Portuguese territories of Azores Islands and the British Indian Ocean Territory are not geographically part of Africa, even though the latter territory is claimed by two African Nations – Mauritius and Seychelles.

7.5 European Union
The European Union (EU) is a political and economic organisation that consists of 27 member states located mostly in Europe, but with numerous outlying territories elsewhere. The organisation was formed in 1952 as the European Coal and Steel Community by six countries: France, West Germany, Italy, Netherlands, Belgium, and Luxembourg, to create a common market for coal and steel, and as a way to prevent further war between France and Germany. It evolved into the European Economic Community (EEC) in 1957, the European Community (EC) in 1967, and the EU in 1993. The organisation expanded to comprise the following countries: the United Kingdom, Denmark, and Ireland (1973); Greece (1980); Spain and Portugal (1984); East Germany – as part of a unified Germany (1990); Austria, Finland, and Sweden (1995); Poland, Czech Republic, Slovakia, Hungary, Slovenia, Lithuania, Latvia, Estonia, Malta, and Cyprus (2004); Bulgaria and Romania (2007). In addition three countries are recognised now as official candidates for membership: Croatia, Macedonia and Turkey.

Throughout the text, the data referred to the EU considers the 27 countries are from the year they first joined. France's data includes data for Algeria till 1962, and Denmark's data includes data for Greenland till 1985.

The EU includes all Spanish and Portuguese overseas territories, as well as the French overseas territories of French Guiana, Réunion, Martinique, Guadeloupe, Saint Barthélemy, and Saint Martin. It also includes Gibraltar. All other European overseas territories have special relationships with the EU but do not constitute part of it.

7.6 OPEC Countries

The Organisation of the Petroleum Exporting Countries (OPEC) is an intergovernmental organisation, based in Vienna, which aims to coordinate and unify petroleum policies among its 12 member countries, thus it is viewed as a cartel in the Western world. The idea was initially proposed by Venezuela in 1949, who finally managed to enrol Iraq, Iran, Kuwait and Saudi Arabia to formally establish OPEC in Baghdad in 1960. Membership grew with the following countries joining: Qatar (1961); Indonesia (1962, withdrew in 2008); Libya (1962); United Arab Emirates (1967); Algeria (1969); Nigeria (1971); Ecuador (1973, withdrew in 1992 then rejoined in 2007); Gabon (1975, withdrew in 1994) and Angola (2007).

Indonesia announced its withdrawal from OPEC in 2008 – with its membership suspended in 2009, due to it becoming a net importer. However, it stated that it will return to the organisation if it became a net exporter again. Therefore throughout this book the data for OPEC comprise the data for all active members from the year of joining. This includes data for Indonesia until end of 2008.

It is noted that despite fierce criticism from the Western world of the organisation, both the United Kingdom and the United States were effectively members of OPEC, the former in its capacity as responsible for the foreign affairs of Kuwait, Qatar and the UAE prior to their independence, and the latter as the formal occupier of Iraq after 2003 war. Interestingly both countries made no effort to derail OPEC from its agenda.

7.7 OAPEC Countries
The Organization of Arab Petroleum Exporting Countries (OAPEC) is an intergovernmental organisation, based in Kuwait, which aims to coordinate and unify petroleum policies in the Arab nations. It was established in 1968 by Kuwait, Saudi Arabia and Libya. At present it has eleven members with the following Arab nations joining: Algeria, Bahrain, Qatar, United Arab Emirates (1970); Iraq and Syria (1972); Egypt (1973); and Tunisia (1982). Tunisia asked to withdraw in 1986 and its membership is currently suspended.

Throughout the text, when referred to the organisation, the data refers to its members data from the year they first joined. Egypt's data are incorporated since 1973 and thus cover the period between 1979 and 1989 when Egypt's membership in the organisation was suspended. Similarly Tunisia's data are incorporated since 1982.

7.8 OECD Countries
The Organisation for Economic Co-operation and Development (OECD) is an international organisation of thirty countries which coordinates economic, environmental and social policies of its member states. The organisation consists of high income countries which are considered 'developed', and is based in Paris.

The organisation was established in 1961 by 20 countries: Austria, Belgium, Canada, Denmark, France, Germany, Greece, Iceland, Republic of Ireland, Italy, Luxembourg, Netherlands, Norway, Portugal, Spain, Sweden, Switzerland, Turkey, United Kingdom and United States. Membership increased to 32 with the following countries joining: Japan (1964); Finland (1969); Australia (1971); New Zealand (1973); Mexico (1994); Czech Republic (1995); South Korea, Hungary and Poland (1996); Slovakia (2000), Chile and Slovenia (2010).

In 2007, OECD countries agreed also to invite Estonia, Israel and Russia to open discussions for membership, which has been offered in 2010 to the first two. It also offered enhanced engagement, with a view to possible membership, to Brazil, China, India, Indonesia and South Africa.

The International Energy Agency (IEA) was founded by OECD in 1974, in the wake of 1973 oil crisis, to coordinate the members' energy policies to ensure their energy security and to prevent a repeat of the crisis. Thus IEA member countries are required to maintain total oil stock levels equivalent to at least 90 days of net imports. At the end of July 2009 the combined stockpile held was almost 4.3 billion barrels of oil and refined products[1].

Throughout the text, the data referred to the organisation's members from the year they first joined.

7.9 GECF Countries

The Gas Exporting Countries Forum (GECF) is an intergovernmental organisation of some of the world's leading gas producers based in Doha, Qatar, which aims to coordinate and unify natural gas policies among its 16 member countries while promoting their mutual interests. It is in its infancy and is modelling itself on OPEC, and thus it is feared to become another cartel in the Western world. The idea was initially proposed by Russia and endorsed by Iran in 2001. Several ministerial meetings were held since then, and the organisation was born officially, adopting its charter in 2008 at a meeting in Moscow with the following countries listed as members: Algeria, Bolivia, Brunei, Egypt, Equatorial Guinea, Indonesia, Iran, Libya, Malaysia, Nigeria, Qatar, Russia, Trinidad and Tobago, while the United Arab Emirates and Venezuela. Kazakhstan and Norway are observers.

---

[1] http://en.wikipedia.org/wiki/International_Energy_Agency, Accessed August 2010.

Chapter 8
DATA SOURCES

Publishing and collating fossil fuel data is a complex and intricately managed task. The availability, detail, accuracy and certainty of the data can be influenced by several factors principally political and commercial. The motive behind releasing a particular piece of data depends greatly on the target audience for which the data is made available, whether it is the political establishment, the general public, the media outlets, the financial markets, or the scientific and academic communities. As such, data management plays an important role in portraying an image derived carefully from the data in positive or negative manner.

Fossil fuel data come in different shapes and sizes, whether it is describing and quantifying reserves, production (supply), consumption (demand), imports and exports (trade), stocks (inventories), price, refinery capacity, carbon dioxide emissions, etc. Data sources vary significantly, and are detailed in Section 8.1. One has to note that very few sources offer comprehensive coverage of fossil fuel data. Most sources are somehow specialised, and have narrow or limited scope, where they cover particular types of data such as reserves or prices, or cover only one type of fossil fuel such as oil or coal.

The availability of the data is usually characterised by two major factors: accessibility and cost. Accessibility, in this sense, means whether the data is not classified and is available either freely in the public domain or commercially for a fee if requested. This is interlinked with the cost, with the quality and detail of the data improving significantly with higher costs, and we can say, unfortunately, that the free data available in the public domain often lacks detail, is incomplete and usually out-of-date.

Generally resources, reserves and production data are usually released by either oil and gas companies or mining companies (both public and private), as well as governmental agencies periodically, mostly on monthly or annual basis. Alternatively it is collected by scouts or spies working specifically to gather certain information. Either way, the data are usually collated by several interested bodies (for example, international organisations, trade journals, consultants, websites, researchers) who report it either in its raw form or after subjecting it to further manipulation and editing, according to pre-defined criteria and standards. The data is then made available either freely or for a fee depending mostly on the amount of detail required.

Furthermore, different countries use different methodologies to collate and report data, and thus the data have varying levels of reliability. Precise comparisons between nations and analyses of time series should always be treated with great caution.

8.1 Classification of Data Sources
Data used in most published studies, reports, energy outlooks, etc. originate from a few major sources which usually act as primary data sources with other sources simply relaying or collating the data. The data sources can be classified as originating from eleven major categories: oil, gas and mining companies; governmental bodies; trade and industry journals; international organisations; industrial and commercial bodies and organisations; professional bodies and organisations; information service companies and consultants; financial institutions and investment banks; think tanks; scientific and academic publications; and mass media and the internet. These categories are outlined in the following sub-sections.

8.1.1 Oil, Gas and Mining Companies
Oil, gas, and mining companies of all types (public, private or multinational), whether involved in exploration, production, construction or services, are considered to be the primary source of data regarding fossil fuels, particularly reserves and production data. These companies usually release information in various ways including annual reports, press releases, technical papers,

and via their websites. The data available varies in quality, accuracy, detail, and up-to-dateness.

The motivation to disclose certain data, and the quality of the released data, differs from company to company. It usually depends on several factors with the most important being the type of the company's ownership (i.e. public, private, or multinational), and the country where it is registered legally. Both factors determine the regulations governing the frequency, detail, and accessibility of the data a company is obliged to declare to stakeholders, including regulatory bodies, shareholders, and local communities. For example, oil companies listed in the USA, such as ExxonMobil, are obliged to comply with the SEC regulations (see Section 6.3), and disclose their reserves and production data to the regulatory bodies and the stock market. Companies listed in Russia, such as Gazprom, or Saudi Arabia, such as Aramco, disclose fewer and less detailed data, with reserves data, in particular, are considered secretive and only released selectively in a very controlled manner.

In addition to publishing their own data, some companies compile annual reviews of world energy data. The most prestigious compilation is the *BP Annual Statistical Review of World Energy*, published since 1948 and is considered 'the' authority in energy related data. It collates reserve, production and trade numbers for fossil fuels, nuclear and some renewable energy sources, which it draws from numerous sources and employs judgment to edit certain data if required (e.g. estimating Canada's tar sand reserves).

Other notable companies which are often known as significant data sources include Schlumberger and BG Group, who both publish numerous reports especially in the area of unconventional oil and gas resources; as well as Baker Hughes, which publishes detailed active drilling rig counts.

On the other hand, national companies issue only limited amount of data and impose restrictions on numerous details. The data they issue are usually interlinked with the governmental bodies' data, blurring the distinction between the two data sources. This can be explained because these companies are controlled by the governments and hence release data only to satisfy political objectives and to comply with the national policies of their masters (i.e. governments).

8.1.2 Governmental Bodies

An important source of data is the information officially provided by governmental bodies of almost all countries. The data is made available by ministries (including their divisions or departments) directly or by nominally independent specialised agencies affiliated to the ministries. The data published falls into two categories: firstly, original information published directly by the ministries and agencies, such as oil or energy ministries, which in many countries have direct control of the national oil and gas and mining companies; secondly compiled data collected by the ministries and agencies from numerous sources covering a specific fossil fuel sector or energy and natural resources sectors nationally, regionally or globally to assist the governments in policy planning and decision making.

A prime example of this data source is the comprehensive set of data collated and published by the United States Department of Energy via its Energy Information Agency (EIA). The agency collects data for energy resources including reserves, production, trade, price and environmental impacts for the world. It also publishes annual statistics, historical data, and future outlooks. The majority of the data is accessible freely on the agency's website, and thus are amongst the most quoted by all interested parties.

Other notable governmental bodies whose data are often used, especially regarding non-conventional oil and gas resources, include the German Federal Institute for Geosciencies and Natural Resources (Bundesanstalt für Geowissenschaften und Rohstoffe (BGR)) – an affiliate of the German Ministry of Economics and Technology, and the Canadian province of Alberta's Department of Energy.

Interestingly, most governments have no centralised data services, thus many governmental bodies collect their own data and hardly synchronise it with other agencies, which lead to many instances of reporting contradictory data. For example the United States Geological Survey (USGS) – a division of the US Interior Ministry, generates its own estimates for current and future natural resources including oil, gas and coal reserves, and made them available to the public mostly free of charge. The USGS does not utilise data collated by the EIA. Similarly the CIA, through its world factbook, collates its fossil fuel related data from a number of public and private sources including trade and industry journals instead of relying on the EIA data.

8.1.3 Trade and Industry Journals

Trade and Industry journals are used by professionals and decision makers as the prime source of up-to-date fossil fuel data including reserves, production, trade information, exploration activities, new discoveries, new contracts, and new technologies. The journals usually publish periodically selected data in tabular format, as well as news briefs of the latest industry developments, both in print version and on their website which can be obtained or accessed for a fee. These journals compile their data by asking governments and companies for their latest data, they then organise it in a user friendly easy-to-use and easy-to-find format.

The most important trade and industry journals are *Oil and Gas Journal* published by PennWell Publishing since 1902; *World Oil Journal* – first published in 1916 as Oil Weekly - by Gulf Publishing Company; and *Oil and Energy Trends* published since 1976 by Wiley-Blackwell. Specialised journals are also published that cover a narrower aspect of fossil fuel industry, for example *World Coal Magazine*, *LNG Journal* and *Offshore Magazine*. Currently, almost all trade and industry journals have highly ranked websites, which gives them a wider reach and allows them to update their data more frequently.

Besides the online versions of the trade and industry journals, several online-only webzines and portals have established themselves as formidable competitors to printed media. They are used increasingly as trusted data sources, with many independent websites achieving higher rankings than printed journals websites, according to Alexa ranking service. Amongst the most important websites are:

- http://www.rigzone.com
- http://www.gulfoilandgas.com
- http://www.gasandoil.com
- http://www.energycurrent.com
- http://www.oilonline.com
- http://www.energyintel.com
- http://www.oilandgasinternational.com
- http://www.coalportal.com

Despite being universally used, these journals have some sceptics, especially from the peak oil theory advocates, who criticise the journals for publishing data supplied by governments and companies without any verification. They point out that oil reserve data, published by the Oil and Gas Journal for example, has remained constant for several years. In this instance the Oil and Gas Journal defends its stance by stating its oil reserves figures are *"based on survey responses and updates released by individual countries, which in many cases are not released every year-if ever. OGJ changes a particular reserves figure only when it receives not only evidence that a change is necessary but also a reliable, new estimate."*[1]

8.1.4 *International Organisations*

Fossil fuel related data are often published by international organisations, both intergovernmental (referred to as IGOs) and non-governmental (referred to as INGOs or NGOs). The data published by these organisations are usually compiled from primary sources such as companies and governments, they are organised, reformatted, and offered to the public occasionally for free, but mostly as a paid-for service depending on the organisation's policy and objectives. Several organisations conduct further data analysis and research publishing periodical

---

[1] Oil and Gas Journal (Dec 23 2002)

statistical energy and resources reports (mainly annually) presenting historical and current data, as well as future outlooks.

Over one thousand international organisations are identified by the Union of International Associations (UIA) as working in the energy sector, which range from global to regional, or political to economic. All these organisations can be used for data sourcing in one form or another. The most important organisations where data is usually obtained are the United Nations (UN), the European Union (EU), the World Bank, The International Monetary Fund (IMF), the International Energy Agency (IEA), OPEC and the World Energy Council (WEC).

Examples of the published data are:
- The UN, through its Statistics Division (UNSD), maintains a comprehensive database of fossil fuel resources amongst other economic information and publishes an annual energy statistics yearbook. However the data available tends to be a few years old.
- The IEA – affiliated to OECD, publishes comprehensive detailed studies on world energy resources including fossil fuels and alternative energy. Most studies can be purchased, with shortened versions available on-line for free.
- OPEC and OAPEC - each publishes several bulletins and reports, with the most important being an annual statistical report containing energy statistics of their member states, in detail, and the rest of the world in less detailed terms.
- The WEC publishes an array of specialised energy reports, amongst them the widely cited *Survey of Energy Resources*, which provides comprehensive data on not only fossil fuels but also other non-renewable and renewable energies.

In the 1990s several international organisations, namely: UNSD, IEA, OPEC, Eurostat of the EU, Latin American Energy Organisation (Organización Latinoamericana de Energía or

Data Sources

OLADE), Asia-Pacific Economic Cooperation (APEC), and International Energy Forum (IEF) established the Joint Oil Data Initiative (JODI) to share data and enhance transparency in oil markets. However it is still incomplete and requires further efforts to progress.

8.1.5 Industrial and Commercial Bodies and Organisations

Industrial and commercial bodies or organisations are usually made up of companies with common interests and objectives. Bodies and organisations with interests in fossil fuel usually publish a set of data relevant to their members' interests, objectives and needs. They also participate actively in defining standards, accepted practices and methodologies for their relevant industries. As for international organisations, several industrial bodies, commercial bodies and organisations conduct further data analysis and research, then publish periodical statistical energy and resources reports, which they made available to the public, who can access some of it freely, but are expected to pay for the majority of the data. The access policy is set depending on the organisation's policy and objectives.

Examples of such bodies or organisations that compile and publish data include the World Petroleum Council (WPC), which publishes the proceedings from the World Petroleum Congress; International Information Centre for Natural Gas and Gaseous Hydrocarbons (known by its French name Centre International d'Information sur le Gaz naturel et tous Hydrocarbures Gazeux (Cedigaz)), which publishes comprehensive natural gas data; the World Coal Institute (WCI), which publishes numerous studies on coal and carbon dioxide capture and storage; and the World LP Gas Association (WLPGA), which publishes numerous LPG related studies and statistics.

8.1.6 Professional Bodies and Organisations

Professional bodies and organisations are usually made up of individuals who share common interests and objectives, usually work and study related subjects; naturally many exist with an interest in fossil fuels. Similar to industrial and commercial bodies or organisations, professional bodies and organisations with interest in fossil fuel also publish their own compiled set of data, which they perceive relevant to their members' interests, objectives and needs. They also participate actively in defining standards, accepted practices and methodologies for their relevant industries. Similar to international organisations, several professional bodies and organisations carry out further data analysis and specialised research. They also publish periodical statistical energy and resources reports, which they offer partially to the public freely, while the complete set of data will only be available for their subscribed.

Examples of such bodies and organisations that compile and publish data include American Petroleum Institute (API), which compiles and publishes a wide range of energy statistics including drilling wells costs and refinery statistics. Other notable bodies include the American Association of Petroleum Geologists (AAPG), Independent Petroleum Association of America (IPAA), and the Society of Petroleum Engineers (SPE).

8.1.7 *Information Service Companies and Consultant Companies*

Information service companies and consultants own comprehensive databases with which they provide these data for a fee, with only minimal data available free of charge. These companies collect data not only from the primary sources, but also use their own methods including employing scouts. They analyse both public and private data, author market and policy reports, which can be purchased for a fee – relatively expensive.

The most important information service company is IHS, which started out as a specialised database and technical indexes company and technical data distributor growing substantially to a comprehensive information services company. It acquired various well-known consulting companies in the energy sector including the often cited Petroconsultants, Cambridge Energy Research Associates (CERA), Global Insight, World Markets Research Centre, John S. Herold, McCloskey, Lloyds Register Fairplay, and Jane's Information Group. The combined company maintains comprehensive databases and publishes numerous specialised studies and reports. Interestingly the peak oil theory advocates (ASPO community), often use data from Petroconsultants as their starting point for analysing oil data, considering this set of data as the only reliable source, while at the same time dismissing all other data sources (maybe because ASPO leading figures were former employees of Petroconsultants? – I do not comment but only ask the question). Ironically CERA (effectively a sister division of Petroconsultants within IHS) usually arrives at different conclusions using the same database.

Other specialised information data service companies include ODS-Petrodata, which compiles and publishes data on offshore oil and gas activities, and Platts (now a division of McGraw-Hill), which provides information on energy and metals.

The field of energy consultants who publish energy reports is crowded, with almost every consulting firm having an energy division. Besides the divisions of IHS mentioned earlier, other notable consultants whose data is usually referenced include Wood Mackenzie, Energydata, and PFC Energy.

8.1.8 *Financial Institutions and Investment Banks*
Financial institutions and investment banks provide price and trading data for fossil fuels and other commodities, and recently emissions data. They keep track of price history, conducting research and analysis to predict future price, trading volumes and activities. Investment banks also commission market research to advise their customers, particularly oil and gas and mining companies, while pursuing mergers and acquisitions. Brief results of selected research are usually made public via press releases and the banks websites.

Examples of financial institutions that provide price and trade data include New York Mercantile Exchange (NYMEX) – part of Chicago Mercantile Exchange (CME), Intercontinental Exchange (ICE), European Climate Exchange (ECX), and Bloomberg. Examples of investment banks that provide active services in the energy sector include JPMorgan Chase, Goldman Sachs,

Barclays Capital, Deutsche Bank and Simmons & Company.

8.1.9 Think Tanks

Think tanks working in the energy sector, whether policy institutes, organisations, corporations or groups of similarly minded people, conduct their own research in the energy arena. They then communicate their results usually via media outlets and websites to advocate policy changes and enhance the process of decision making. Even though some think tanks are impartial, many – due to their funding resources or political affiliation - are usually biased, as they advocate a certain ideology or promote certain policies, which are endorsed by their funding sources. They are often cited often by lobbying groups when promoting certain policies.

Think tanks are currently vocal promoting environmental issues and climate change policies, and as such are used by many – including oil companies – to support or dismiss certain environmental policies such as transport emissions or opening access to oil exploration in protected areas. Often, these think tanks have fancy names, which give them perceived authority.

Amongst the most vocal think tanks in the energy sector is the Association for the Study of Peak Oil and Gas (ASPO), which is a network of researchers at institutions and universities worldwide, with a common interest in publicising peak oil theory (see Section 6.8). Other famous think tanks active in the energy sector include the Centre for Global Energy Studies (CGES); the Fraser Institute, which criticises some aspects of scientific opinion on climate change; the Council of Foreign Relations (CFR), which publishes periodical global oil trends reports; and the Brookings Institutions, which is ranked as the world's top think tank by Foreign Policy Magazine.

8.1.10 Scientific and Academic Publications

Scientific and academic publications such as journals, proceedings of conferences, symposia or workshops, monograms, or books are used primarily by researchers pursuing studies in fossil fuels and energy disciplines. The publications are usually specialised and concentrate on a very narrow area. The data obtained from these publications is considered trustworthy as it is usually subjected to peer review and scrutiny prior to being made available. This, however, results in time delays, which means that the data is not up-to-date relative to the data obtained from trade and industry journals. Scientific and academic publications are nonetheless used by professionals and decision makers as the prime source of up-to-date, new and future technologies. The journals are usually published periodically, with most contributions offered for free, but the content is only available to subscribers for relatively hefty fees, being available both in print and electronic versions. In addition, several databases are available online free or for a fee, which allows users to search for relevant articles.

Although thousands of scientific and academic publications are available, with a few hundred of them dealing with energy research, a few publishers are dominant, namely Reed Elsevier, Wiley-Blackwell, Springer, and McGraw-Hill.

8.1.11 Mass Media and the Internet
Mass media including newspapers, magazines, broadcast media, and websites are a major source of data on fossil fuels. However the quality and accuracy of data vary significantly from factual to fanciful, with many relaying data from other sources. Some journals such as the Economist or Scientific American verify their data prior to publishing; others, such as tabloid newspapers rely on sensational headlines and often publish stories with no scientific or factual basis.

Data is available on television and radio stations in different forms including news bulletins, and specialised documentaries, with hundreds of channels accessible to the public.

Similarly, data is now widely available on internet websites such as forums, blogs, or specialised websites. The quality data on websites associated with the printed or broadcast media reflect the quality of the data of the parent media company. On the other hand there is no verification for most data on forums, blogs and many independent websites, thus the quality of the data varies enormously from the very accurate (e.g. Wikipedia) to the blatantly manufactured. Therefore extreme caution must be exercised prior to using such data.

8.2 Underlying Data Quality Challenges

Three important factors affect the quality of fossil fuel data in general, they are amplified regarding resources and reserves estimates in particular. These factors are the reliability, accuracy and trustworthiness of the data and its sources. Any deficiency in any factor causes many arguments and often leads to controversial viewpoints.

These three factors are often influenced by additional aspects, which further complicate the evaluation of the quality of the data. The key aspects are discussed below:

- The lack of standardisation; where different definitions are used by different countries to report 'identically called' data. For example – as explained earlier in Section 6.3 – proven reserves in the USA are defined differently from proven reserves in Canada or in Saudi Arabia
- Political aspects; where data reporting is affected by political views and thus data is often released 'manipulated or decorated' to suit a certain political viewpoint or achieve a particular goal rather than being simply technically accurate. For example, many argue that OPEC oil reserves estimates were influenced by the introduction of the organisation's production quotas in the 1980s, where several OPEC members raced each other to increase their official proved oil reserves so that they can be allocated higher quotas. This argument is untrue, as the quotas are not linearly correlated to the reserves, but are determined by a certain formula. This is discussed in detail in Section 9.1.3. Another example is the fixation of many American and other Western media outlets, with the 'drastic' reduction of Venezuela's oil production, while dismissing all official Venezuelan numbers, this was only highlighted when the Chavez government took controlling stakes in all major oil producing projects.

- Economic aspects; where data is manipulated to obtain economic benefits. An example frequently mentioned is the alleged manipulation of the Mexican oil reserve estimates in the 1980s. It is often argued that due to the severe economic crisis, the country suddenly inflated its proven oil reserves and used the new figure as a guarantee to obtain loans from the World Bank and other financial institutes. Once the crisis had subsided the reserves number was cut back. Once again this can be disputed, as explained in Section 6.3.1 earlier.
- Regulatory aspects; where data is manipulated to comply with certain regulatory instructions. For example as part of its obligations after joining NAFTA Mexico had to revise its oil proven reserve number downwards to comply with the USA SEC's definition (See Section 6.3.1).

It is evident that no impartial unbiased source for fossil fuel data exists and that, due to the lack of standardisation, any value can have several interpretations influenced by political, economic or regulatory aspects.

Criticism of most data sources, especially by Oil Peak Theory advocates, is centred around one point, which is that they accuse all major data sources, such as oil companies, governmental bodies, and trade journals, of having ulterior motives, and that the reporting is often politically biased. Even though this may be true, these critics, however, ignore the fact that data obtained from information service companies and consultants can be also biased, since the data has been compiled for commercial reasons and therefore it may be skewed to fulfil the agendas of these consultants, and their customers, who are their paymasters.

To illustrate how data can be very different when taken from various data sources, consider the oil reserves of Iraq as an example. The following paragraph is taken from a study in 2003 entitled "How Much Oil Does Iraq Have?" authored by G. Luft of the influential Brookings Institution, which advocated the

American invasion of Iraq.[2]

It reads: *"Over the past several months, news organizations and experts have regularly cited Department of Energy (DOE) Energy Information Administration (EIA) figures claiming that the territory of Iraq contains over 112 billion barrels (bbl) of proven reserves - oil that has been definitively discovered and is expected to be economically producible. In addition, since Iraq is the least explored of the oil-rich countries, there have been numerous claims of huge undiscovered reserves there as well— oil thought to exist, and expected to become economically recoverable - to the tune of hundreds of billions of barrels. The respected Petroleum Economist Magazine estimates that there may be as many as 200 bbl of oil in Iraq; the Federation of American Scientists estimates 215 bbl; a study by the Council on Foreign Relations and the James A. Baker III Institute at Rice University claimed that Iraq has 220 bbl of undiscovered oil; and another study by the Center for Global Energy Studies and Petrolog & Associates offered an even more optimistic estimate of 300 bbl - a number that would give Iraq reserves greater even than those of Saudi Arabia. In a Guardian interview before the war, Taha Hmud Moussa, Saddam's deputy oil minister, said that all of Iraq's oil reserves will exceed 300bbl".*

Yes... read the paragraph again and notice the huge differences in estimating the oil reserves! As can be seen, the differences are startling, and thus it is evident that authors can always find data to support their arguments. Therefore to analyse objectively, one has to rely on consistent sources of data and cross reference several sources to verify the data prior to using it.

So, in this study the main sources of data used are the EIA, BP, IEA and OPEC. Other sources are used to complement these data sources, and are highlighted if they are significantly different.

---

[2] G. Luft, How Much Oil Does Iraq Have?, (http://www.brookings.edu/papers/2003/0512globalenvironment_luft.aspx).

Chapter 9
OIL – GLOBAL AND ARAB PERSPECTIVES

As a finite commodity, there is no doubt that oil and other fossil fuels will run out eventually. However, as already discussed in Chapter 6 (see Section 6.8) debate is raging between peak oil theory advocates and opponents whether the world is or will be running out of oil soon, and if so, on the timing of the alleged peak, with the majority of geologists and economists positioning themselves firmly in the opposing camps of the debate respectively. Peak oil theory advocates claim that the world has reached a peak in oil reserves, and has reached or about to reach a peak in oil production after which the production will decline sharply leading to astronomic oil prices. On the other hand peak oil theory opponents point out that increasing oil price will lead to increased oil reserves since a substantial amount of oil will be reclassified as 'reserves' moving from 'resources' category as explained in Chapter 6, and this will lead to increased production.

The above debate shows no signs of subsiding, and may rage forever! We will know the true answer only after the peak is reached (if ever). So, if by reading this chapter you think you will get the definite answer I'm afraid I have to disappoint you, as you won't find it here. This chapter aims to present a brief quantitative analysis of the oil globally and in the Arab world specifically, in terms of four main aspects: reserves and resources; production; consumption and trade, as well as explaining briefly the interlink with the oil refining industry.

The chapter is structured systematically with each section starting by presenting the global standpoint of a certain oil aspect, it is then followed by assessing and quantifying the contribution of the Arab world, and lastly discussing the importance of the Arab world in relation to that specific aspect. An overall analysis of the significance of the Arab world's oil, along with its positioning into the overall energy picture globally, as well as any implications of this positioning, is discussed in Chapter 12.

In a nutshell this chapter endeavours to answer three main questions:
- How much oil reserves and resources are there in the Arab world?
- How long will these oil reserves last?
- Will the Arab world sustain its significance as the major oil supplier of the world?

After reading I encourage the readers to draw their own conclusions – okay with some hints thrown along the way!

9.1 Conventional Oil Reserves and Resources

9.1.1 Conventional Oil Reserves

Contrary to vocal media misconception and a considerable amount of misinformation on the internet, according to reports by all the major data sources, the world's oil reserves are continuing to increase. This upward trend is evident in Figure 9.1, which shows the conventional oil reserves since 1980 reported by the EIA, BP, and OPEC. All three sources report a substantial increase in proved reserves, in the range of between 80% and 100%, between 1980 and 2008. The data reported are for conventional crude oil reserves (as defined in Section 2.1) and thus includes extra light, light and medium crude oil reserves, as well as heavy oil reserves lighter than 10 API gravity. It also includes condensates, deepwater oil and Arctic oil reserves. The data however is inconsistent in its treatment of natural gas liquids (NGLs), which are included by most sources such as the EIA, BP and OPEC, but excluded by others such as the USGS, which reports the latter separately. It is observed that the reserves data for individual countries are often similar when reported in most data sources, which suggests that most sources include NGL numbers in total crude oil reserves, even though some of these sources do not state this explicitly.

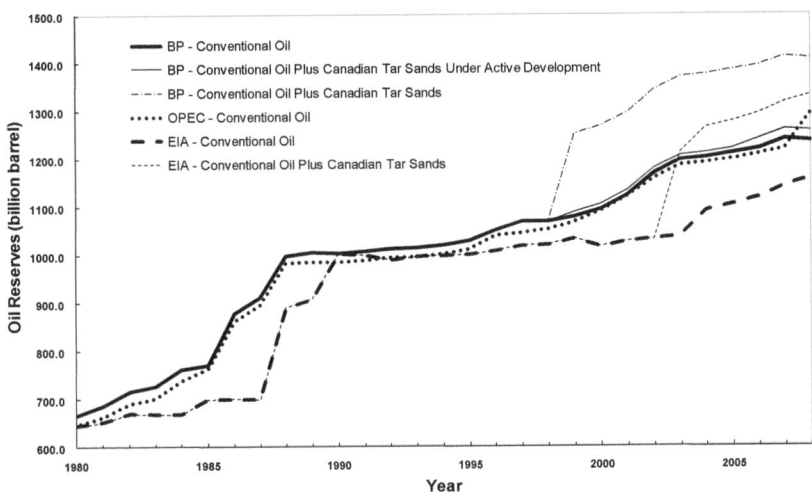

Figure 9.1: World conventional proved oil reserves (1980-2008)

Source: EIA (http://www.eia.doe.gov/international); OPEC (Annual Statistical Bulletins 1999-2008); BP (Statistical Review of World Energy 2001-2009).

The data in Figure 9.1 is presented in a way that identifies the Canadian natural bitumen (tar sands) reserves explicitly, since there is no universal agreement of including them into the conventional reserves as yet, the practice to include them has only been recently adopted by some of the major data sources, though each source uses different definition and includes varying proportions of these oil reserves. This practice was first adopted by the Oil and Gas Journal, which reclassified part of the Canadian natural bitumen oil resources as conventional oil reserves in 2002, and was followed promptly by the EIA. BP followed suit in 2004, however it backdated the data in its statistical review to partially include these natural bitumen resources from 1999, although it divided them into two categories: under active development, which were included in the reserves numbers; and other resources, which were reported separately. OPEC has yet to change its stance, which until now it has not classified any natural bitumen resources as part of the conventional oil reserves. It is observed that, excluding the Canadian natural bitumen reserves, all major data sources estimates are in broad agreement and differ insignificantly.

Unavoidably, the small differences observed are due to different practices in data reporting. According to the EIA, the world's total proved oil reserves in 2009 stand at 1169.5 billion barrels.

On the other hand, until 2008, there was a consensus amongst almost all data sources to exclude the Venezuelan extra heavy oil reserves from the reported data of conventional oil. Despite this, the Venezuelan government managed to effectively include part of these resources into its conventional oil reserves, by adding certified oil reserves from some blocks that contain extra heavy oil, into its conventional oil reserves gradually, and then supplying the revised estimates as official conventional oil reserves estimates to all major data sources, who seem to have accepted the revised values and adjusted their values accordingly. In 2009 OPEC broke rank and almost doubled Venezuela's conventional proved oil reserves, by adding a substantial amount of the extra heavy oil reserves, though it did not state this inclusion explicitly, instead, in its latest statistical bulletin, added a footnote to the new reserves number stating it includes some extra heavy oil developments without explicitly labelling them so. Note however, that numerous reports cited in the media and the internet claim, erroneously, that extra heavy oil reserves are already included within Venezuela's original numbers without substantiating their claims.

Table 9.1 lists the conventional oil reserves in the leading countries in the world. These countries account for almost 84.6% of the global proved reserves. As expected, the majority of the places in the table are occupied by OPEC members, who occupy eight positions including the top six. The data in this instance excludes the Canadian natural bitumen oil reserves, which if included, will place Canada at number two in the top ten list, with estimated reserves of 178.1 billion barrels. It also excludes the Venezuelan uncertified extra heavy oil reserves claimed by the Venezuelan government, but are still not endorsed universally, and which if included, will place Venezuela at the top of the top ten list with estimated reserves of 272 billion barrels, or in the second place, according to OPEC latest numbers, with 172.3 billion barrels.

Table 9.1: World conventional proved oil reserves – top ten countries (2009)

Rank	Country	Crude Oil Reserves billion barrel	Share %
1	Saudi Arabia	266.71	22.81
2	Iran	136.15	11.64
3	Iraq	115.00	9.83
4	Kuwait	104.00	8.89
5	Venezuela	99.38	8.50
6	UAE	97.80	8.36
7	Russia	60.00	5.13
8	Libya	43.66	3.73
9	Nigeria	36.22	3.10
10	Kazakhstan	30.00	2.57
	TOTAL	988.92	84.56
	WORLD	1169.51	

Source: EIA (http://www.eia.doe.gov/international).
Note 1: Totals may not add up due to rounding.

The Arab world's overall conventional oil reserves and their share to the world's total since 1980 are shown in Figure 9.2. The data shows a continuous increase in Arab proved oil reserves,

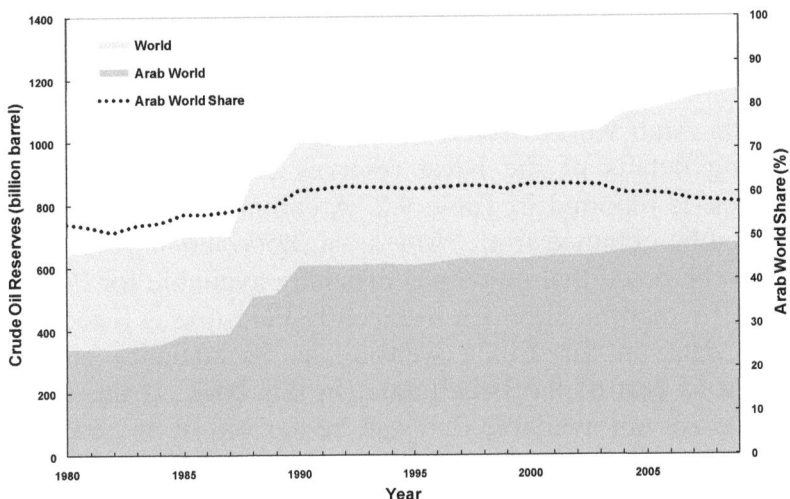

Figure 9.2: Arab world conventional proved oil reserves and its share to the world's total (1980-2009)

Source: EIA (http://www.eia.doe.gov/international).

which are estimated at 674.9 billion barrels in 2009, so if taken as one entity, the Arab world will be placed at the top of the top ten list in terms of oil reserves with five times more reserves than Iran, the second placed country. The data reveals the sheer domination of the Arab world in terms of conventional oil reserves, where its share of the world's total remains relatively constant and hovers highly at around 60%, standing at approximately 57.7% in 2009. The slight share decline observed in the last few years is due to recent significant additions to conventional oil reserves in Iran, Venezuela and Nigeria, all of which are OPEC members.

A first glance at the numbers suggests that with such massive proved oil reserves, the Arab world is in a position to totally dominate the oil markets and name its price for oil. Alas this is wishful thinking! As, besides reserves, there are several other parts to the jigsaw that form the oil market, with oil reserves being only one of them. Other parts include oil actual and spare production capacity, oil consumption, export facilities, global economic conditions, and political conditions. So the Arab world's weight on the other factors does not mirror its huge reserves, therefore its influence is not determined solely by the reserves numbers. The complete picture will become clearer in the next few sections, so read on!

The total Arab world overall conventional oil reserves in 2009, including details of the latest reserves estimates for all Arab countries, is reported in Table 9.2. A complete set of data from 1980 can be obtained from *"http://www.2050consulting.com/books"*. It has to be noted that no data is officially available for Palestine in the EIA reports, although in recently Palestine is listed in the tabular data on the EIA's website, as its statistics are often included as part of the Israeli data. In this book, if the data for Palestine are not available they can be derived by reviewing the data for Israel. This is applied throughout the book. The table above shows that proved oil reserves exist in 17 Arab countries. We can classify countries into five tiers in terms of total proved conventional oil reserves: countries with major, significant, medium, minor, or no reserves. The Arab countries fall into this

Table 9.2: Conventional proved oil reserves in the Arab countries (2009)

Country	Crude Oil Reserves billion barrel	Rank	Share %
Bahrain	0.12	14	0.01
Iraq	115.00	2	9.83
Jordan	0.00	16	0.00
Kuwait	104.00	3	8.89
Lebanon	0.00		
Oman	5.50	8	0.47
Palestine	0.00		
Qatar	15.21	6	1.30
Saudi Arabia	266.71	1	22.81
Syria	2.50	12	0.21
UAE	97.80	4	8.36
Yemen	3.00	11	0.26
Algeria	12.20	7	1.04
Comoros	0.00		
Djibouti	0.00		
Egypt	3.70	10	0.32
Libya	43.66	5	3.73
Mauritania	0.10	15	0.01
Morocco	0.00	17	0.00
Somalia	0.00		
Sudan	5.00	9	0.43
Tunisia	0.43	13	0.04
ARAB WORLD	674.93		57.71

Source: EIA (http://www.eia.doe.gov/international).
Note 1: Totals may not add up due to rounding.

classification as follows:
- Four countries belong to the first tier with proved oil reserves exceeding 50 billion barrels each and a combined share of approximately 49.9% of the world's proved oil reserves, which is just under half the global proved conventional endowment. These countries in descending order in terms of proved oil reserves are: Saudi Arabia, Iraq, Kuwait, and UAE, with all four also ranked in the top ten list globally, including the top spot (see Table 9.1). All these countries are members of OPEC all have a long history with oil reserves discoveries, with the first commercial oil discovered as early as 1927 in Iraq, 1938

in Saudi Arabia and Kuwait, and 1958 in UAE. The reported proved oil reserves numbers of all four countries continue to increase despite continuous oil production. This apparent unexplained reserve growth is often cited by peak oil theory advocates as the main proof that these reserves numbers are manufactured. However, there are plausible explanations that vindicate these reserves numbers and are discussed in detail in Section 9.1.3.

- Three countries belong to the second tier, with proved oil reserves exceeding 10 billion barrels but less than 50 billion barrels each and a combined share of approximately 6.1% of the world's proved oil reserves. These countries in descending order in terms of proved oil reserves are: Libya, Algeria, and Qatar, with the Libya also ranked in the top ten list globally (see Table 9.1). All these countries are also members of OPEC, all have a long history with oil reserves discoveries, with the first commercial oil discovered in 1940 in Qatar. First discoveries were made later in Algeria in 1956 and Libya in 1959. The reported proved oil reserves numbers of all three countries also continue to increase despite continuous oil production. This situation is similar to the first tier countries above. Note that recently, Qatar almost doubled its proved oil reserves according to OPEC data, however these new reserves are yet to be endorsed by other sources.
- Five countries belong to the third tier, with proved oil reserves exceeding 1 billion barrels but less than 10 billion barrels each and a combined share of approximately 1.7% of the world's proved oil reserves. These countries in descending order in terms of their reserves are: Oman, Sudan, Egypt, Yemen, and Syria. None of these countries are members of OPEC and their commercial oil reserves history varies widely with the first commercial discovery in Egypt being the first in the Arab world, going back as far as 1869. Discoveries in other countries are relatively recent, starting in 1956 in Syria, followed by Oman in 1964, Sudan in 1974, and most recently Yemen in 1984. The reported reserves

numbers of these countries do not show the same upward trend. Only in Sudan have substantial additional recent reserves discoveries increased its endowment, while at the same time declines in reserves due to production were recorded in the other four countries.

- Five countries belong to the fourth tier, with minor proved oil reserves less than 1 billion barrels each and a combined share of less than 0.1% of the world's proved oil reserves. These countries in descending order in terms of their reserves are: Tunisia, Bahrain, Mauritania, Morocco, and Jordan. The first two countries have a relatively long history of oil discovery, with Bahrain recording the first commercial oil discovery in the Arab Gulf countries in 1932, while oil was discovered in Tunisia shortly after Libya and Algeria in 1964. The story is not yet clear in Morocco and Jordan, where several announcements of major oil discoveries hit the media front pages from time to time, but to date, no substantial proved reserves have been confirmed, despite several exploration concessions being awarded. However optimism has increased in Morocco recently, especially regarding offshore potential, after the first commercial discoveries in Mauritania in 2001.
- Finally five countries have no proved oil reserves. These countries are Lebanon, Palestine, Somalia, Djibouti, and Comoros. Currently several companies are exploring the Somali waters, with potential oil reserves expected on the horizon, though this has been hampered by the ongoing war in Somalia and the increasing pirate activity off its coasts. No exploration activities are being perused in the other four countries.

The Middle East Perspective

The Middle East region is home to approximately 63.8% of the world's total conventional proved oil reserves, 81.8% of which are in its Arab countries. The region's countries occupy five places in the top ten list of leading conventional oil reserves countries including the top four. This confirms the famous quote, that the Middle East contains two thirds of the world's oil reserves, though this quote needs quantifying, as in reality, it only refers to conventional oil reserves and excludes unconventional oil.

Source: EIA, (http://www.eia.doe.gov/international), based on 2009 data.

OPEC Perspective

OPEC member states are home to approximately 80.7% of the world's total conventional proved oil reserves, 69.3% of which are in its Arab member states. The organisation's members occupy eight places in the top ten list of conventional oil reserves countries including the top six. This illustrates the potential dominant position the organisation enjoys in the oil markets, and reaffirms the power it can extract in shaping global energy policies and economics.

Source: EIA, (http://www.eia.doe.gov/international), based on 2009 data.

9.1.2 Conventional Oil Resources

Unlike oil reserves data, oil resources data are not readily available. Therefore these data need to be estimated using the latest available information. To achieve this, a simplified method is implemented to attain a rough estimate of these resources on a global level. Obviously, this estimation is a conservative approximation that gives an idea of the resources.

The methodology employed is straightforward, but effective and uses the available data from different sources to calculate resources quantities. It starts by estimating the original oil in place (OOIP) on a global level using average recovery factors to extract oil as reported by Sandrea and Sandrea[1] for current and ultimate oil recovery. This calculation assumes a uniform recovery factor for all global conventional oil resources, which is not true, as each field has unique recovery factor; however its usage is justified to achieve approximate values in the absence of detailed data for every single oil field in the world.

The BGR publishes what it labels resources estimates on a country-by-country basis. These resources are in fact remaining

---

[1] Ivan Sandrea and Rafael Sandrea, Global Oil Reserves – Recovery Factors Leave Vast Target for EOR Technologies, Oil & Gas Journal, Part 1: November 05, 2007 & Part 2: November 12, 2007.

recoverable resources, excluding proved reserves and cumulative production. Thus the sum of these resources plus the proved reserves and the cumulative production leads to a total recoverable resource estimate, which is considered to be the ultimate oil recoverable resources (UORR). Using an ultimate recovery factor of 70%[2], the OOIP based on the BGR data can be calculated by dividing the UORR by the ultimate recovery factor.

Other data sources provide only proved reserves and production data, thus using the sum of the reserves and the cumulative production data, the OOIP can be calculated, using recovery factors ranging between 22% and 40%[3], by dividing the sum of the reserves and the cumulative production over an average recovery factor, which in this instance is the harmonic average of the recovery factor range above, calculated to be 28.4%. The proved reserves data are obtained from the latest EIA data, while the cumulative production data are obtained from the BGR[4] till 2007, then corrected to 2008 using the EIA latest production data.

The two calculated OOIP quantities are compared, while the recovery factors for all countries, based on BGR reserves and resources data, are back calculated. So if the calculated recovery factor for a country is found to be above the average value of 28.4%, it suggests the BGR resources value, and thus the OOIP for that country is underestimated, while the OOIP, based on the EIA reserves data, is used instead for that country. After reconciling the data for all countries, a global OOIP value is derived and is shown in Table 9.3. This derived OOIP value is then used to back calculate all other quantities.

One point to stress is that the estimated OOIP numbers are calculated not only considering proved reserves numbers, but also considering oil remaining in place in oil fields that ceased production, while still containing a significant amount of oil that can be produced in the future, if conditions change, such as higher price or advances in technology. The inclusion of the latter

---

[2] Ibid.
[3] Ibid.
[4] BGR, Energierohstoffe 2009 – Reserven, Ressourcen, Verfügbarkeit.

is important to achieve more accurate oil resources estimation and is unfortunately often overlooked by many estimators. Furthermore, even though ideally, all the above values must include only conventional oil, implicitly they include a few quantities of unconventional oil that were not reported separately, either as reserves or as production.

Table 9.3: World and Arab world conventional oil in place and oil resources (2009)

Quantity (billion barrel)	World	Arab World	Share
	Estimated	Estimated	%
Original Oil in Place (OOIP)	7994	3406	42.6
Remaining Oil in Place (ROIP)	6855	3099	45.2
Ultimate Oil Recoverable Resources (UORR)	5596	2384	42.6
Remaining Oil Recoverable Resources (RORR)	4457	2077	46.6
Remaining Oil Resources Excluding Proved Reserves	3287	1402	42.7
	Reported	Reported	
Remaining Oil Proved Reserves	1170	675	57.7
Cumulative Production	1139	307	27.0

Source: EIA (http://www.eia.doe.gov/international), BGR (Energierohstoffe 2009 – Reserven, Ressourcen, Verfügbarkeit).

Based on the reconciled OOIP of 7994 billion barrels, the UORR is calculated by multiplying the OGIP estimates by 70% - the ultimate recovery factor. Thus, performing a simple calculation gives a global UORR estimate of 5596 billion barrels. All the above values include conventional crude oil, condensates and NGLs. The above OOIP estimates are lower than the conventional oil and NGL estimates, provided by the EIA, of 10200 billion barrels[5], and are thus on the conservative side.

By deducting the cumulative production, the remaining oil-in-place (ROIP) can be calculated as 6855 billion barrels. By subtracting the cumulative production, the remaining oil recoverable resources (RORR) can be calculated, and are shown to be 4457 billion barrels. These values are calculated using total

---
[5] EIA, International Energy Outlook 2008.

conventional world proved reserves of 1169.5 billion barrels, and total world oil cumulative production of approximately 1138.7 billion barrels.

The above remaining resources estimations are in line with the conventional oil estimates provided by Schlumberger of 4005 to 5850 billion barrels[6] agreeing with Holditch and Chianelli's[7] estimate of 4500 billion barrels and with the upper estimated numbers published by the USGS of 3919 billion barrels.[8] However, they are considerably higher than the lower estimated numbers published by the USGS of 2269 billion barrels[9]; although it is noticeable that the lower estimate of the USGS is very conservative.

The same methodology, described above, is used to estimate the OOIP, ROIP, UORR and RORR in the Arab world. These values are calculated using total conventional world proved reserves of 674.9 billion barrels, and total world oil cumulative production of approximately 307.1 billion barrels. The results are shown in Tables 9.3 and 9.4.

It is clear that, the Arab world's share of RORR decreases to approximately 46.6%, which, even though lower than proved reserves share of 57.7% remains very high and cements the position of the Arab world as the dominant force in terms of conventional oil resources. The reason for this decline is that more oil to date has been produced from non-Arab oil fields, which even though considered exhausted now have the potential to be reproducing if technical and economic conditions change. As expected though the ranking of the Arab countries in terms of conventional oil; resources is identical to their ranking in terms of conventional oil reserves, though obviously the oil quantities differ.

---

[6] Schlumberger (http://www.heavyoilinfo.com/blog-posts/billion_bbls_6_uk.pdf).
[7] Stephen A Holditch and Russel R Chianelli, Factors that will influence oil and gas supply and demand in the 21st century, MRS Bulletin, Volume 33, April 2008.
[8] USGS, Digital Data Series 60, 2000.
[9] Ibid.

Table 9.4: Conventional oil in place and oil resources in the Arab countries (2009)

Country	Original Oil in Place (OOIP)	Remaining Oil in Place (ROIP)	Ultimate Oil Recoverable Resources (UORR)	Remaining Oil Recoverable Resources (RORR)	Remaining Oil Resources Excluding Proved Reserves	Remaining Oil Proved Reserves	Cumulative Production
				billion barrel			
Bahrain	5.25	3.87	4.72	3.34	3.22	0.12	1.38
Iraq	513.41	481.80	462.07	430.46	315.46	115.00	31.61
Jordan	0.05	0.05	0.05	0.05	0.05	0.00	0.00
Kuwait	495.37	455.25	445.83	405.71	301.71	104.00	40.12
Lebanon	0.00	0.00	0.00	0.00	0.00	0.00	0.00
Oman	49.32	40.62	44.39	35.68	30.18	5.50	8.71
Palestine	0.00	0.00	0.00	0.00	0.00	0.00	0.00
Qatar	83.75	74.75	75.38	66.38	51.17	15.21	9.00
Saudi Arabia	1350.22	1227.23	1215.20	1092.21	825.50	266.71	122.99
Syria	25.83	20.84	23.25	18.25	15.75	2.50	5.00
UAE	438.95	411.03	395.05	367.14	269.34	97.80	27.92
Yemen	18.27	15.78	16.45	13.95	10.95	3.00	2.50
Algeria	107.11	88.09	96.40	77.38	65.18	12.20	19.02
Comoros	0.00	0.00	0.00	0.00	0.00	0.00	0.00
Djibouti	0.00	0.00	0.00	0.00	0.00	0.00	0.00
Egypt	49.76	39.37	44.78	34.40	30.70	3.70	10.39
Libya	235.40	209.36	211.86	185.82	142.16	43.66	26.04
Mauritania	0.44	0.41	0.39	0.37	0.27	0.10	0.03
Morocco	0.34	0.33	0.30	0.29	0.29	0.00	0.01
Somalia	0.21	0.21	0.19	0.19	0.19	0.00	0.00
Sudan	25.58	24.53	23.02	21.97	16.97	5.00	1.05
Tunisia	6.83	5.46	6.14	4.78	4.35	0.43	1.37
ARAB WORLD	3406.08	3098.97	3065.48	2758.36	2083.43	674.93	307.12

Source: EIA (http://www.eia.doe.gov/international); BGR (Energierohstoffe 2009 – Reserven, Ressourcen, Verfügbarkeit).

9.1.3 Reliability of Conventional Oil Reserves and Resources Data

We scratched the surface regarding the issue of data sources in Chapter 8. Here more detailed examination is carried out concerning the oil resources and reserves estimated quantities. Using reliable data to obtain correct estimates is essential in determining current and future energy policies as well as impacts on other oil aspects such as production, trade and refining.

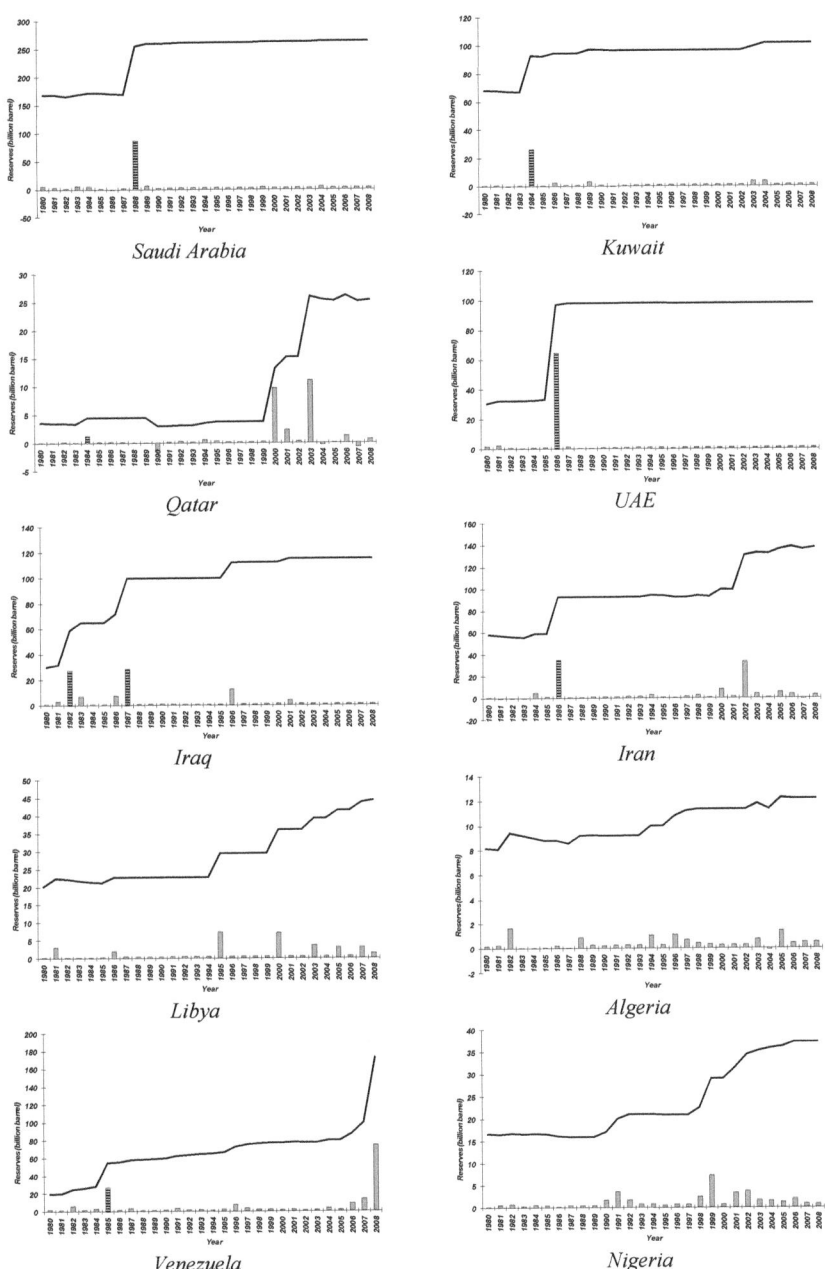

Figure 9.3: Conventional proved oil reserves and reserves growth in the selected OPEC countries (1980-2008)

Source: OPEC (Annual Statistical Bulletins 1999-2008).

The reliability of conventional oil reserves data has been questioned continuously, especially by peak oil theory advocates, who have authored numerous books, articles and blogs, thus succeeding in creating media frenzy. These authors usually point out that most data reported has no independent auditing procedures and that the data seem to stay constant for years, before having a sudden jump. Furthermore, they claim that a considerable proportion of the reserves are inflated and exaggerated. They usually highlight a period of a few years, in the 1980s, when several OPEC countries increased their reserves estimates significantly. The story they often tell is that during 1980s, OPEC had restricted its members' oil production to quotas, related to the size of their reserves estimates, in response OPEC countries increased and exaggerated their reserves artificially to gain enlarged production quotas. The accused countries are Saudi Arabia, Kuwait, Iraq, UAE, Iran and Venezuela, with endless numbers of reports and articles containing tabular data highlighting the 'suspicious' sudden increases in these countries oil reserves estimates.

Figure 9.3 shows the official conventional oil reserves for ten OPEC countries, including the six countries suspected of inflating their reserves, as well as the reserves data for Qatar, Nigeria, Libya, and Algeria. These numbers are obtained from OPEC official statistical bulletins. The oil reserve growth and the percentage change in oil reserves are also shown. Oil reserve growth columns striped in Figure 9.3 are the contested values. Note that these oil reserves data exclude the NGL reserves (as is the normal practice by OPEC) and the unconventional oil reserves (with the exception of Venezuela's latest numbers).

At first, the suspicions raised regarding oil reserves estimate exaggeration initially appear to be plausible. The six alleged countries almost doubled their conventional oil reserves estimates without apparent explanation. However, a thorough analysis rebukes these suspicions when the following considerations are taken into account:
    1. Prior to the nationalisation of the oil industry by OPEC member states, the bulk of OPEC conventional oil

reserves' rights were owned by multinational oil and gas companies. The royalty fees and taxes paid on these reserves to the local governments were proportional to the size of the declared oil reserves. Therefore it was in the interest of these multinational companies to underestimate the size of their reserves to minimise their fees and tax bills.

2. The multinational oil and gas companies deliberately withheld data regarding part of the discovered oil reserves so that they could announce to the financial markets the 'good news' of additional reserves when needed especially if these companies were facing difficulties or when market conditions were tough, as these added reserves could enhance significantly their financial positions. So although the share value for these publicly listed oil companies is always proportional to the amount of reserves they declare they already possess in the ground, the declaration of a massive new oil reserve find can lift the share price significantly. And if the companies can prove to the financial markets a trend of continuous reserve growth, rather than one-off jumps, the share price can be maintained at a higher rate.

3. As explained in Chapter 6, it is easy to prove that the oil reserves estimates can vary, higher or lower, without the modification of any technical or physical parameters, just by changing economic factors, because this will mean that the definition of what is 'reasonable' has changed.

4. OPEC countries had no incentive to update the conventional oil reserves data before the introduction of the production quota system which was imposed in 1986. However, even though these production quotas were never proportional to the conventional oil reserve estimates published[10,11,12], and were governed by a complicated formula based on several factors either oil

---

[10] John Gault, Charles Spiererb, Jean-Luc Bertholetb and Bahman, How does OPEC allocate quotas?, Journal of Energy Finance & Development, Volume 4, Issue 2, Autumn 1999, Pages 137-148.
[11] Mahmoud Al-Osaimy and Aziz Yahyai, The importance of weighted variables to OPEC's production quota allocation, OPEC Review, Volume 27 Issue 2, Pages 129 – 141.
[12] R Sandrea, R Sandrea, Oil and Gas Journal, Vol 109, 2003, Oil and Gas Journal, Vol 109, 2003.

related (e.g. reserves, production capacity, production cost, historical production share, domestic oil consumption) or socio-economical (e.g. population, dependence on oil exports, external debt), OPEC countries found themselves obliged to update the conventional oil reserves data for the fear of future quota formulae becoming proportional to conventional oil reserves estimates. As a proof, even though the official oil reserve estimates of the UAE, Kuwait and Iran were relatively of equal magnitude, their OPEC oil production quotas always differed significantly. Interestingly, during the same period, NGL reserves also increased even though they were, and are still, excluded from OPEC oil production quotas.

5. The increase in conventional oil reserves was not only observed in the six accused countries, but also by Qatar (who announced significant discoveries in the 1980s), and by Nigeria, whose oil industry is controlled by multinational oil and gas companies in partnership with a governmental national oil company. The remaining OPEC countries did not increase their conventional oil reserves estimates. This triggers an obvious question, which is, if as the peak oil theory advocates claim, the increase in conventional oil reserves was to enhance the quotas, what stopped the remaining countries from following suit and maintaining the exact ratio of reserves? I invite the reader to draw his or her conclusions.

6. The OPEC countries had no incentives to invest in oil exploration activities. This was due, not only to the huge conventional oil reserves they already possessed but also, to the collapse of the oil prices in the 1980s. This collapse meant that OPEC countries found it difficult to justify investing any capital in looking for additional cheap oil that they did not need to produce especially as there was no market for it. The financial strains they were under restricted future expansion in the oil industry for years to come.

7. The interest, currently shown by all international oil companies, to develop Iraq's remaining oil reserves

indicates that they believe in the massive reserves estimates (based on the figures provided in the 1980s) declared by Iraqi officials, and vindicates the numbers.

To add further to the suspicion and mistrust of the reported data, the peak oil theory advocates conjecture that, since the oil reserves have increased in spite of an ever-growing consumption, many new oilfields should have been discovered, also that these discoveries should be large enough to more than compensate for oil consumption. They point out that this is not true, and that almost no giant oilfields (such as Ghawar in Saudi Arabia, Burgan in Kuwait or Kirkuk in Iraq) - those fields which 'make a difference' (as they call them) for the world total of conventional oil reserves, have been discovered since 1980. The argument continues by drawing attention to the fact that most conventional oil reserves in the Middle East - which totals almost two thirds of the world's conventional oil reserves, and almost one third of the world's production (see Section 9.4) - come from oilfields discovered a very long time ago, and that these oilfields have either reached or are about to reach their peak production. Furthermore peak oil theory advocates not only discredit existing numbers of oil reserves in known oilfields, they also claim that major oil field discoveries are now pretty ancient, and even relatively big discoveries are now rare.

The argument regarding the non-discovery of new major fields is naïve. It implies that the only major reason for increasing oil reserves is discovery of new oil fields. This has been rebuked already in Chapter 6, where it has been shown that enhanced oil recovery methods (EOR), increased recovery factors, better feasibility, and improved technology are all very important factors in increasing oil reserves estimates. The USGS estimated, in its 2000 report, that reserves growth from currently known reserves would be responsible for 43% of total reserves added to proved reserves and cumulative production, and thus would constitute 22.8% of the UORR.[13] Another argument which refutes the peak oil theory advocates position is if one considers

---

[13] USGS, Digital Data Series 60, 2000.

the proved oil reserves of the USA shown in Figure 9.4, which despite continuous production and no major discoveries, declined very slowly for nearly 30 years.

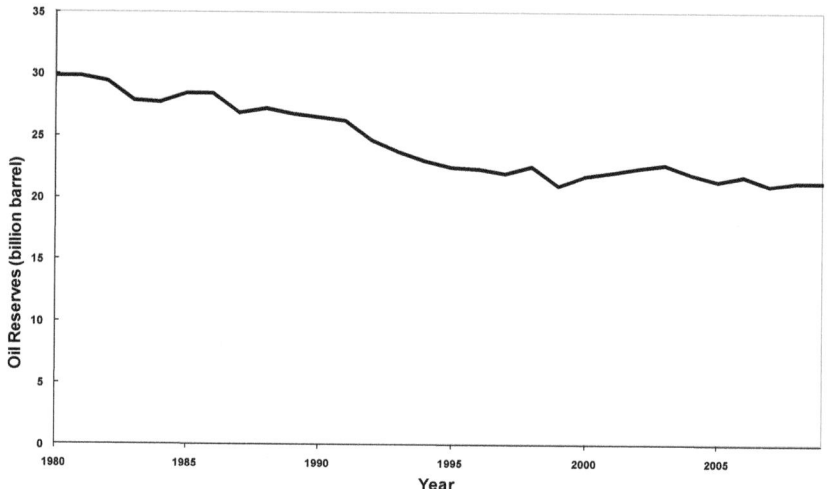

Figure 9.4: Conventional proved oil reserves of the USA (1980-2009)
Source: EIA (http://www.eia.doe.gov/international).

Furthermore - even though not as big as Ghawar and Burgan oilfields, several giant oilfields have been discovered in recent years containing oil reserve estimates of over 10 billion barrels each of recoverable conventional oil. These include Tupi offshore of Brazil, Azadegan in Iran, and Kashagan in Kazakhstan. A simple search on Google can identify a long list. Two excellent references can be consulted for giant oilfield discoveries since 1990.[14,15]

On a more general note, many peak oil theory advocates (especially ASPO affiliates) have the tendency to discredit almost all data sources, using only oil reserves data that they had back-calculated themselves, and which contradict all major data sources. They usually use these unverified data to support their claims that oil peak has already been reached or is just around the

---

[14] AAPG Special Volumes, Volume AAPG Memoir 78, Giant oil and gas fields of the decade 1990-1999
[15] Fredrik Robelius, Giant Oil Fields of the world, Presentation AIM Industrial Contact Day, Uppsala Universitet, 2003.

corner. The data they use originate from ancient Petroconsultants database (see Section 8.1.7), which they had access to while they were working with Petroconsultants – note that they do not use the latest database! Ironically, the same set of data is now owned by IHS, and is used by CERA as a basis for their studies and reports, which, incidentally, dismiss the peak oil notion. Interestingly it seems therefore that using the same set of data does not necessarily guarantee similar outcomes, but on the contrary, can lead to totally opposite conclusions. So who is right and who is wrong? Readers can examine the data and conclude.

Nevertheless, despite appearing to be dismissive of the majority of peak oil theory advocates' arguments, it is imperative to admit that some of their criticisms of the reliability of conventional oil reserves data are valid. The most important of their criticism is that many data sources, including the EIA, BP and OPEC, use the data of the Oil & Gas Journal, which is based on survey responses and updates released by the oil countries, without any independent verification. For example in 2002, 67 countries out of 105, did not change oil reserve values even though they produced to full capacity[16], whereas other organisations, such as the USGS, use computer programs based on the most comprehensive statistics of oil industry maintained by consultants and databanks, therefore the different approaches, i.e. verified data versus surveys, lead to inevitable data discrepancies and confusion.

In addition, several notable odd and unexplainable discrepancies in conventional oil reserves data persist, often being very obvious and clear. For example all major data sources report that Timor-Leste and Uganda have no oil reserves, but the former already produces oil (as published by the same data sources who credit it with zero reserves) and the latter has several fields under development, which are changing hands for billions of dollars in the stock market and are about to start commercial production imminently. Another example that brings published reserves data into question is the reserves of Kuwait, where one has to

---

[16] G Luft, 2003, (www.brookings.edu/views/op-ed/fellows/luft20030512.htm).

remember that during the Iraqi invasion and occupation of Kuwait in 1990-91, it is estimated that up to 6 billion barrels of oil were burnt out from Kuwaiti fields in fires initiated by the retreating Iraqi troops. This lost oil is not accounted for in any data! This prompts a simple query questioning the apparent failure in the reported reserves data deducting this lost oil. Similarly estimates of Iraq conventional oil reserves vary massively depending on data sources and whether the reported reserves estimates originate from organisations or companies that support or oppose the Anglo-American invasion of Iraq. This example has been presented already in detail in Section 8.2, which interested readers can revisit for a quick reminder.

9.2 Unconventional Oil Reserves and Resources

Estimating the reserves and resources of unconventional oil is more an art than a science! It is controversial, far from straight forward, open to misinterpretation and conflicting view points. It is understandable since, what is considered to be unconventional oil keeps changing, where with improving technologies, and fluctuating (but edging higher) oil prices a considerable proportion of what was traditionally deemed unconventional oil is making the gradual shift into being reclassified as conventional oil. The case of reclassifying part of Canada's natural bitumen resources into conventional reserves is a prime example and has already been brought up in Section 9.1.

In this book, two categories of unconventional oil are considered: extra heavy oil, including natural bitumen, and oil shale. The two categories have already been introduced in Sections 2.2 and 2.3, but it is worth reemphasising here that what set these two categories apart from conventional oil are the added technical difficulties in extracting oil, which lead to inevitable lower economic feasibility. Therefore resolving some of these difficulties is paving the way for these deposits to be partially reclassified as conventional. Once this is acknowledged universally by organisations and governments, some of these deposits, such as Canada's natural bitumen reserves and Venezuela's extra heavy oil reserves, will inevitably be recognised as conventional by all sources.

History teaches us that this reclassification is unavoidable, based on the lessons learned from the reclassification of both deepwater oil and heavy oil, as it is commonly accepted now, that these two categories of oil are included within conventional oil. It is unfortunate though, that to date, a great amount of confusion still lingers as many data sources not only disagree on the setting of the upper limit on the definition of heavy oil, but also still list heavy oil resources and reserves separately from conventional oil resources and reserves. Some even go to the extreme of amalgamating heavy oil with extra heavy oil and natural bitumen deposits. The latter practice is strictly outdated, as the technology used to extract heavy oil does not differ from the technology to extract light or medium oil and the feasibility of the extraction is only slightly lower, resulting in numerous heavy oilfields producing oil profitably. An example of such confusion is the extensively cited heavy oil map produced by Schlumberger[17]. In this map, heavy oil is reported as a distinct category from conventional oil in the pie chart, whereas at the same time the colour coding of the map amalgamates heavy oil with extra heavy oil and natural bitumen when reporting total heavy oil resources. Therefore the map lists five countries (namely Canada, Venezuela, Russia, Iraq, and Kuwait) each to have in excess of 350 billion barrels of heavy oil resources, and each exceeding the resources of Saudi Arabia, which is credited with between 50 and 350 billion barrels of heavy oil resources. This is obviously misleading, as the heavy oil of Kuwait or Iraq for example is easier to extract than the oil from the tar sands of Canada, whose heavy oil is actually natural bitumen. The map also gives the false impression that five countries have more oil resources than Saudi Arabia. This is untrue, as, if Saudi Arabia's light and medium oil resources are included in the map, it will certainly overtake these five countries.

Other unconventional resources exist, though they are not considered in the oil endowment as they exist either in small concentrations, or are very scattered so that producing feasible oil

---

[17] Schlumberger (http://www.heavyoilinfo.com/blog-posts/billion_bbls_6_uk.pdf).

from them is considered next to impossible in the foreseeable future. A major example of an unconventional resource is the oil source rocks that did not develop enough to be classified as any other unconventional oil category. The amount of oil in these rocks is estimated to exceed oil shale resources.[18]

9.2.1 Extra Heavy Oil and Natural Bitumen

Despite being located and identified for years, until recently most major data sources did not report data on unconventional oil reserves, as they did not consider them to constitute part of the proved reserves. As stated earlier, this practice started to change in 2002, when Oil and Gas Journal decided to consider part of Canada's natural bitumen reserves are 'conventional' reserves and include them in its estimates of proved reserves. Even though this practice was not followed up immediately by other data sources, it nevertheless initiated a change in attitude. This stance started being adopted by many other data sources, such as BP, which began including a fraction of Canada's natural bitumen (tar sands) reserves in the conventional reserves estimates since 2004 in its Statistical Review, albeit at considerably lower volumes than Oil and Gas Journal. As expected, the different approaches used by different data sources led to massive discrepancies in the reserves estimates. In Canada's case, these discrepancies range from a maximum of 300 billion barrels as reported by Holditch[19] and thus placing Canada on the top spot of the world's top ten of conventional reserves displacing Saudi Arabia, to 178.1 billion barrels as reported by Oil and Gas Journal based on data supplied by the official authorities in Alberta and adopted by the EIA, thus placing Canada on the second spot of the world's conventional reserves, to a minimum of 4.9 billion barrels as reported by OPEC, who so far refuse to include the natural bitumen reserves as part of the conventional oil reserves. In Venezuela's case, the estimates range from a maximum of 272 billion barrels, as reported by Venezuela's official statistics and Holditch[20], thus placing Venezuela on the

---

[18] EIA, International Energy Outlook 2008.
[19] Stephen Holditch, Manpower and Technology Issues in the Oil and Gas Industry, 2007 (www.holditch.com/Portals/66/Wells%20Fargo_%20July%2007.pdf)
[20] Ibid.

top spot of the world's leading bearers of conventional reserves, displacing Saudi Arabia, to a medium estimate of 172.3 billion barrel recently adopted by OPEC, placing the country in the second spot, to a minimum of only 99.4 billion barrels by BP and the EIA. In this book, to stay on the conservative side for the time being, part of Canada's natural bitumen and Venezuela's extra heavy oil reserves are acknowledged as reserves, but are reported as 'unconventional reserves'. The same stand is applied to other unconventional oil resources, with a fraction of them reported as unconventional reserves, calculated on conservative recovery factors values.

Table 9.5: World extra heavy oil and natural bitumen reserves and resources

Data Source	Publication year	Reserves	Resources	Oil in Place
			billion barrel	
Natural Bitumen				
WEC	2007	246	*998*	3327
BGR	2009	287	*894*	2908
EIA	2008	*240*	*720*	2400
Schlumberger	2006	*330*	*990*	*3300*
Swindell	2008	*300*	*900*	3000
Extra Heavy Oil				
WEC	2007	60	*746*	2486
BGR	2009	42	*301*	1550
EIA	2008	*230*	*690*	2300
Schlumberger	2006	*275*	*825*	*2750*
Swindell	2008	*300*	*900*	3000

Source: EIA (http://www.eia.doe.gov/international); BGR (Energierohstoffe 2009 – Reserven, Ressourcen, Verfügbarkeit); World Energy Council, 2007 Survey of Energy Resources; Schlumberger (http://www.heavyoilinfo.com/blog-posts/billion_bbls_6_uk.pdf), Swindell (http://gswindell.com).
Note 1: Numbers in italic are calculated by the author.
Note 2: Schlumberger reference to "resources" actually means oil in place rather than recoverable resources. Numbers are calculated based on a mean value of 11 trillion barrels.

Despite the change in attitude since 2002, reserves estimates data of extra heavy oil and natural bitumen remain patchy and are reported erratically rather than annually, with no updated annual compilation of data available in the public domain. Table 9.5 summarises the latest world extra heavy oil and natural bitumen resources, as reported by different sources, and the estimated reserves calculated based on conservative recovery factors. The data presented covers both natural bitumen (tar sands) and extra heavy oil, but excludes heavy oil (lighter than 10 API), which is considered as part of conventional oil resources. The data in the above table are calculated using the published data of original oil in place (OOIP) and cumulative oil production from these deposits. The reserves are estimated using a conservative recovery factor of 10%, while the resources are estimated using an ultimate recovery factor of 30%, Sandrea and Sandrea[21] It is observed that all major sources estimates are in general agreement and do not differ significantly. Note that while even peak oil theory advocates accept the enormous resources estimates, they dismiss their impact on increasing oil production in the future, crediting them with an insignificant projected oil production, far less than projected by the EIA or the IEA for example.

The data for extra heavy oil and natural bitumen is not amalgamated, so it is reported separately. This is due to the fact that even though the two resources are geologically similar (and that most extra heavy oil resources will inevitably contain some natural bitumen as part of the resource), the production processes needed to extract oil from them differ significantly both technically and economically, with extra heavy oil deposits being simpler, cheaper and have lower energy demand. These differences lead to different recovery factors assigned to both resources, resulting in significantly varying reserves estimates. These recovery factors do not depend only on available technology and economic conditions, but also include political conditions, legal and environmental restrictions, public opinion, and labour availability to name but a few. These different factors explain the reasons why many data sources assign Canadian natural bitumen resources higher recovery factors than Venezuelan extra heavy oil resources, even though that the latter are technically easier and economically more feasible to extract. Currently commercial production from these unconventional

---

[21] Ivan Sandrea and Rafael Sandrea, Global Oil Reserves – Recovery Factors Leave Vast Target for EOR Technologies, Oil & Gas Journal, Part 1: November 05, 2007 & Part 2: November 12, 2007.

deposits occur only in Canada and Venezuela, with planned commercial exploitation already underway in other countries such as China, DR Conge, and Madagascar.

Only the BGR and the WEC publish detailed lists of both extra heavy oil and natural bitumen reserves and resources. In most cases the data they report are in agreement. However it is noticeable that they disagree in estimations for Venezuela, Canada, and Russia, where the BGR tends to report lower estimates than the WEC for the first two, it reports higher estimates for the latter. When the reported data is examined, one notices that the data underestimates the total reserves and resources; this is because some countries only report the existence of deposits without quantification.

Table 9.6: World natural bitumen reserves and resources – top ten countries (2007)

Rank	Country	Reserves billion barrel	Share %	Resources billion barrel	Share %	Oil in Place billion barrel	Share %
1	Canada	240.12	65.85	730.35	66.15	2446.17	66.16
2	Russia	71.60	19.64	214.84	19.46	716.15	19.37
3	Kazakhstan	42.07	11.54	126.21	11.43	420.69	11.38
4	USA	5.32	1.46	16.02	1.45	53.46	1.45
5	Nigeria	3.83	1.05	11.50	1.04	38.32	1.04
6	Angola	0.46	0.13	1.39	0.13	4.65	0.13
7	Madagascar	0.22	0.06	0.66	0.06	2.21	0.06
8	Italy	0.21	0.06	0.63	0.06	2.10	0.06
9	China	0.16	0.04	0.48	0.04	1.59	0.04
10	Columbia	0.09	0.03	0.28	0.03	0.94	0.03
	TOTAL	364.10	99.85	1102.36	99.85	3686.29	99.70
	WORLD	364.63		1104.02		3697.51	

Source: World Energy Council, 2007 Survey of Energy Resources; BGR (Energierohstoffe 2009 – Reserven, Ressourcen, Verfügbarkeit).
Note 1: Totals may not add up due rounding.

Based on the latest data published by these two data sources, but supplemented with other latest available data from different sources for completion, the total OOIP is 3698 billion barrels. Table 9.6 lists the top ten countries in the world in terms of natural bitumen resources and reserves estimates. These countries account for almost 99.9% of the global estimated reserves and resources respectively. The data in the above table are calculated using the published data of original oil in place (OOIP) and cumulative oil production from these deposits. The reserves are estimated using identical recovery factors described above. Note that since a significant proportion of Canada's natural bitumen deposits are under active development, its reserves estimates are verified by comparing them to actual and somehow more accurate recovery factors, whereas for other countries, the recovery factors are conservative guesstimates. One striking observation from the table is, that unlike conventional oil, it is notable that one country, Canada, dominates the reserves with approximately 65.9% of the world's total reserves, and over 66.2% of the world's total resources. Note that two of the top ten countries, in terms of natural bitumen resources or reserves, are members of OPEC, though they only account for 1.2% of the total reserves or resources. This is peculiar and can be attributed to either pure geology, or because OPEC countries are making no effort or spending any money exploring natural bitumen deposits, since they are already well endowed with better quality and more profitable oil deposits.

Similarly, and using identical recovery factors as above, the total OOIP is 2486 billion barrels, Table 9.7 lists the top ten countries in the world in terms of extra heavy oil reserves and resources estimates. These countries account for over 99.9% of both the global estimated reserves and resources respectively. It can be seen that Venezuela, dominates the table, with its resources accounting for approximately 98.8% of the world's total resources, and over 99.3% of the world's total reserves. Note that this number ignores the fact that the same geological formation that hosts the extra heavy oil in Venezuela extends to Ecuador and Columbia, and some suggest even as far as the Falklands, which means that the share of Venezuela is actually less than the number reported above. Note that, here also similar to Canada's natural bitumen, since significant proportion of Venezuela's extra heavy oil deposits are under active or perspective development, its reserves estimates are verified by comparing them to actual and more accurate recovery factors, whereas for other countries, the recovery factors are conservative guesstimates. Note that here

OPEC dominates the resources and reserves even though it is represented by two countries in the top ten countries list, the list is dominated overwhelmingly by Venezuela. Here again, the absence of other OPEC countries from the list is odd, and suggests that many do not bother with reporting these deposits.

Table 9.7: World extra heavy oil reserves and resources – top ten countries (2007)

Rank	Country	Reserves billion barrel	Share %	Resources billion barrel	Share %	Oil in Place billion barrel	Share %
1	Venezuela	230.69	99.37	720.23	98.75	2433.63	97.89
2	China	0.75	0.32	2.53	0.35	8.74	0.35
3	UK	0.18	0.08	2.55	0.35	10.84	0.44
4	Azerbaijan	0.13	0.05	1.89	0.26	8.08	0.33
5	Italy	0.09	0.04	0.63	0.09	2.51	0.10
6	Egypt	0.05	0.02	0.15	0.02	0.50	0.02
7	Cuba	0.05	0.02	0.14	0.02	0.48	0.02
8	USA	0.05	0.02	0.57	0.08	2.42	0.10
9	Ecuador	0.04	0.02	0.23	0.03	0.87	0.03
10	Columbia	0.03	0.01	0.11	0.01	0.37	0.01
	TOTAL	232.05	99.96	729.03	99.95	2468.44	99.29
	WORLD	232.15		729.37		2486.09	

Source: World Energy Council, 2007 Survey of Energy Resources; BGR (Energierohstoffe 2009 – Reserven, Ressourcen, Verfügbarkeit).
Note 1: Totals may not add up due rounding.

It has to be noted that since no data source offers a complete global set of natural bitumen and extra heavy oil resources or reserves data, the individual countries shares of the total resources or reserves are at best approximates and are sometimes no better than informed guesses. To be conservative, this book performs the calculations to estimate the countries shares, based on the most complete set of data, while at the same time taking into consideration a close review of the other data sources if possible, thus it attempts to synchronise the available data using the most up-to-date accurate set of data available.

Table 9.8 shows the natural bitumen and the extra heavy oil reported resources and the estimated reserves in the Arab world, highlighting the Arab world's share to the world total resources

Fossil Fuels in the Arab World: Facts and Fiction

Table 9.8: *Natural bitumen and extra heavy oil and reserves and resources in the Arab countries (2007)*

Country	Reserves billion barrel	Rank	Share %	Resources billion barrel	Share %	Oil in Place billion barrel	Share %	Notes
Natural Bitumen								
Bahrain	0.00			0.00		0.00		
Iraq	0.00			0.00		0.00		
Jordan	0.00			0.00		0.00		Note 1
Kuwait	0.00			0.00		0.00		
Lebanon	0.00			0.00		0.00		
Oman	0.00			0.00		0.00		
Palestine	0.00			0.00		0.00		Note 1
Qatar	0.00			0.00		0.00		
Saudi Arabia	0.00			0.00		0.00		
Syria	0.00			0.00		0.00		Note 1
UAE	0.00			0.00		0.00		
Yemen	0.00			0.00		0.00		
Algeria	0.00			0.00		0.00		
Comoros	0.00			0.00		0.00		
Djibouti	0.00			0.00		0.00		
Egypt	0.00			0.00		0.00		
Libya	0.00			0.00		0.00		
Mauritania	0.00			0.00		0.00		
Morocco	0.00			0.00		0.00		
Somalia	0.00			0.00		0.00		
Sudan	0.00			0.00		0.00		
Tunisia	0.00			0.00		0.00		
ARAB WORLD	0.00			0.00		0.00		
Extra Heavy Oil								
Bahrain	0.00			0.00		0.00		
Iraq	0.00			0.00		0.00		Note 2
Jordan	0.00			0.00		0.00		
Kuwait	0.00			0.00		0.00		
Lebanon	0.00			0.00		0.00		
Oman	0.00			0.00		0.00		
Palestine	0.00			0.00		0.00		
Qatar	0.00			0.00		0.00		
Saudi Arabia	0.00			0.00		0.00		
Syria	0.00			0.00		0.00		
UAE	0.00			0.00		0.00		
Yemen	0.00			0.00		0.00		
Algeria	0.00			0.00		0.00		
Comoros	0.00			0.00		0.00		
Djibouti	0.00			0.00		0.00	0.00	
Egypt	0.05	1	0.02	0.15	0.02	0.50	0.02	

Libya	0.00		0.00		0.00	
Mauritania	0.00		0.00		0.00	
Morocco	0.00		0.00		0.00	
Somalia	0.00		0.00		0.00	
Sudan	0.00		0.00		0.00	
Tunisia	0.00		0.00		0.00	
ARAB WORLD	0.05	0.02	0.15	0.02	0.50	0.02

Source: World Energy Council, 2007 Survey of Energy Resources; BGR (Energierohstoffe 2009 – Reserven, Ressourcen, Verfügbarkeit).
Note 1: Natural bitumen resources identified but not quantified.
Note 2: Extra heavy oil resources identified but not quantified.
Note 3: Totals may not add up due rounding.

and reserves. The table also details the data for each Arab country. It is interesting to note that the resources of both extra heavy oil and bitumen are insignificant in the Arab world, where its share is a mere 0.02% of the extra heavy oil reserves and resources respectively. It does not even register even 0.01% in terms of natural bitumen reserves and resources respectively. This is in stark contrast to the conventional oil reserves and resources, where the Arab world is dominant. It is obvious that further development of these unconventional resources and reserves globally will dent the Arab world share, and in time, render it less significant to global oil markets.

Revisiting Table 9.8 and following the same tier classification of the conventional oil (see Section 9.1), the Arab world countries fall into the classification as follows:
- None of the countries belong to the first three tiers, with extra heavy oil and natural bitumen oil reserves exceeding 1 billion barrels each.
- Four countries belong to the fourth tier, with minor extra heavy oil and natural bitumen reserves of less than 1 billion barrels each, and a combined share of less than 0.02% of the world's reserves. These countries in descending order, in terms of their reserves, are: Egypt, which is the only Arab country that quantifies its reserves and resources, followed by Iraq, Syria and Jordan, which only reports the existence of some deposits, but no estimated data were ever released regarding them.
- Finally 18 countries have no extra heavy oil and natural

bitumen reserves. These countries are Bahrain, Kuwait, Lebanon, Oman, Palestine, Qatar, Saudi Arabia, UAE, Yemen, Algeria, Comoros, Djibouti, Libya, Mauritania, Morocco, Somalia, Sudan, and Tunisia. Note though that some of these countries have unofficially reported the existence of number of deposits. Generally though, no exploration activities of extra heavy oil or natural bitumen resources are being pursued in most countries.

9.2.2 Oil Shale
Unlike natural bitumen and extra heavy oil, there are no commercial processes to date that extract oil from oil shale. Pilot extraction has been tried, or is being planned, in the USA, Australia, Jordan, Estonia, China and Brazil, with the last three burning small quantities of oil shale directly in power stations. Historically the latter activity was previously also practiced by France and the UK. Therefore all deposits of oil shale are treated as resources and are not included as reserves by any data source, even if the technology costs fall in the near future to render the processes feasible, the recovery factors will remain very low, thus triggering a minor reclassification of only a small percentage of the oil shale resources into reserves.

However, similar to resources data for extra heavy oil and natural bitumen, resources data for oil shale remain inconsistent, poor, badly compiled and reported irregularly, with no updated annual compilation available in the public domain. Table 9.9 summarises the latest world oil shale resources as reported by different sources and the estimated reserves, calculated based on conservative recovery factors. The data in the above table are calculated using the published data of original oil in place (OOIP) of 2823 billion barrels while ignoring cumulative oil production from these deposits, since no commercial oil production is active from oil shale yet. The reserves are estimated using a conservative recovery factor of 1%, while the resources were estimated using an ultimate recovery factor of 30% Sandrea and Sandrea.[22] In this instance unexplained high ultimate recovery

---

[22] Ivan Sandrea and Rafael Sandrea, Global Oil Reserves – Recovery Factors Leave Vast Target

factors for some countries like Jordan and Thailand reported by some data sources, are ignored as it is not substantiated. It is observed that generally all major sources estimates are in broad agreement regarding overall resources and do not differ significantly, though they differ in ranking the leading countries, sometimes mysteriously dropping one or two countries from their data. Note that here also, peak oil theory advocates do not dispute the enormous resources estimates, however they correctly dismiss their impact on increasing oil production in the future, crediting them with insignificant projected oil production and in this instance they agree with projections provided by the EIA or the IEA for example. Furthermore it has to be pointed out that extracting oil from coal yields more oil per ton compared to oil shale, therefore these resources need only to be considered for development after all coal resources are evaluated as oil resources![23]

Table 9.9: World oil shale reserves and resources

Data Source	Publication year	Reserves	Resources	Oil in Place
			billion barrel	
WEC	2007	28	*847*	2823
BGR	2009	26	779	2598
EIA	2008	28	*840*	2800
Schlumberger	2006	Note 3	Note 3	Note 3
Swindell	2008	*100*	3000	*10000*

Source: EIA (http://www.eia.doe.gov/international); BGR (Energierohstoffe 2009 – Reserven, Ressourcen, Verfügbarkeit); World Energy Council, 2007 Survey of Energy Resources; Schlumberger (http://www.heavyoilinfo.com/blog-posts/billion_bbls_6_uk.pdf), Swindell (http://gswindell.com).
Note 1: Numbers in italic are calculated by the author.
Note 2: Schlumberger reference to "resources" actually means oil in place rather than recoverable resources. Numbers are calculated based on a mean value of 11 trillion barrels.
Note 3: Schlumberger does not report oil shale as part of total world oil resources in the reference above.

It has to be stated though that if oil shale is burnt directly, it is classified as coal reserves and not oil reserves, and its production

---

for EOR Technologies, Oil & Gas Journal, Part 1: November 05, 2007 & Part 2: November 12, 2007.
[23] Jean Laherrère, Review on oil shale data (http://www.hubbertpeak.com/laherrere/OilShaleReview200509.pdf). He is citing an article in a 1972 publication by the journal Pétrole Informations (ISSN 0755-561X).

is reported as part of coal rather than oil production. This classification is widely accepted and is endorsed by the EIA (see Sections 2.3 and 4.5).

Table 9.10: World oil shale reserves and resources – top ten countries (2007)

Rank	Country	Reserves billion barrel	Share %	Resources billion barrel	Share %	Oil in Place billion barrel	Share %
1	USA	20.85	73.88	625.57	73.88	2085.23	73.88
2	Russia	2.48	8.78	74.36	8.78	247.88	8.78
3	Congo DR	1.00	3.54	30.00	3.54	100.00	3.54
4	Brazil	0.82	2.91	24.60	2.91	82.00	2.91
5	Italy	0.73	2.59	21.90	2.59	73.00	2.59
6	Morocco	0.53	1.89	16.01	1.89	53.38	1.89
7	Jordan	0.34	1.21	10.25	1.21	34.17	1.21
8	Australia	0.32	1.12	9.52	1.12	31.73	1.12
9	Estonia	0.16	0.58	4.89	0.58	16.29	0.58
10	China	0.16	0.57	4.80	0.57	16.00	0.57
	TOTAL	27.40	97.06	821.90	97.06	2739.68	97.06
	WORLD	28.23		846.76		2822.54	

Source: World Energy Council, 2007 Survey of Energy Resources; BGR (Energierohstoffe 2009 – Reserven, Ressourcen, Verfügbarkeit).
Note 1: Totals may not add up due rounding.

Table 9.10 lists the top ten countries in the world in terms of oil shale oil reserves and resources estimates. These countries account for almost 97% of the global estimated reserves and resources respectively. Here again, similar to the situation regarding extra heavy oil and natural bitumen, but in contrast to conventional oil situation, it is notable that one country, the USA, dominates the resources and reserves with approximately 73.9% of the world's total reserves and resources respectively. It is noted that none of the leading countries is a member of OPEC, and hence, once again, the question arises as to whether this is due to pure geology or lack of exploration.

Table 9.11 shows the oil shale reported resources and estimated reserves in the Arab world, highlighting the Arab world's share to the world total reserves and resources. The table also details the data for each Arab country. It is interesting to note that the resources of oil shale are insignificant in the Arab world, where its share is a mere 3.3% of the reserves and resources respectively. Also the resources of oil shale are insignificant in the Arab countries that are well endowed with conventional oil resources, while they are significant in two of the Arab countries with no conventional oil resources, namely Morocco and Jordan, with both appearing in the top ten countries in terms of oil shale resources. This again raises the simple and speculative question of whether the countries rich in conventional oil ever made any effort to look for oil shale or report any findings. This is further supported by the fact that the geology of the areas rich in oil shale, in Jordan and Israel, and the officially oil shale free areas in Saudi Arabia, across the borders, are very similar.

Table 9.11: Oil shale reserves and resources in the Arab countries (2007)

Country	Reserves billion barrel	Rank	Share %	Resources billion barrel	Share %	Oil in Place billion barrel	Share %	Notes
Bahrain	0.00			0.00		0.00		
Iraq	0.00			0.00		0.00		Note 1
Jordan	0.34	2	1.21	10.25	1.21	34.17	1.21	
Kuwait	0.00			0.00		0.00		
Lebanon	0.00			0.00		0.00		
Oman	0.00			0.00		0.00		
Palestine	0.00			0.00		0.00		
Qatar	0.00			0.00		0.00		
Saudi Arabia	0.00			0.00		0.00		
Syria	0.00			0.00		0.00		
UAE	0.00			0.00		0.00		
Yemen	0.00			0.00		0.00		
Algeria	0.00			0.00		0.00		
Comoros	0.00			0.00		0.00		
Djibouti	0.00			0.00		0.00		
Egypt	0.06	3	0.20	1.71	0.20	5.70	0.20	
Libya	0.00			0.00		0.00		
Mauritania	0.00			0.00		0.00		
Morocco	0.53	1	1.89	16.01	1.89	53.38	1.89	
Somalia	0.00			0.00		0.00		
Sudan	0.00			0.00		0.00		
Tunisia	0.00			0.00		0.00		
ARAB WORLD	0.93		3.30	27.98	3.30	93.25	3.30	

Source: World Energy Council, 2007 Survey of Energy Resources; BGR (Energierohstoffe 2009 – Reserven, Ressourcen, Verfügbarkeit).
Note 1: Oil shale resources identified but not quantified.
Note 2: Totals may not add up due rounding.

Revisiting Table 9.11 and following the same tier classification of the conventional oil (see Section 9.1), the Arab world countries fall into the classification as follows:
- None of the countries belonging to the first three tiers, with oil shale reserves exceeding 1 billion barrels each.
- Three countries belong to the fourth tier, with minor oil shale reserves of less than 1 billion barrels each, and a combined share of approximately 3.3% of the world's reserves. These countries in descending order in terms of their reserves are: Morocco, Jordan and Egypt.
- Finally 19 countries have no reported oil shale reserves. These countries are Bahrain, Iraq, Kuwait, Lebanon, Oman, Palestine, Qatar, Saudi Arabia, Syria, UAE, Yemen, Algeria, Comoros, Djibouti, Libya, Mauritania, Somalia, Sudan, and Tunisia. Note that some of these countries have reported the existence of a number of deposits, but no estimated data were ever released regarding them. Generally no exploration activities of oil shale resources are being pursued in most countries.

9.2.3 *Reliability of Unconventional Oil Reserves and Resources Data*

There are no standards on reporting unconventional oil resources and reserves data. Not only is there no agreed definition of what is 'unconventional', but, unlike conventional oil resources and reserves data, most official numbers published by governments ignore unconventional oil. As seen in the previous section most data were obtained from unofficial data sources such as the WEC and the BGR, which are based on surveys from member countries or from scientific publications, both of which are not endorsed by any official organisations. None of the data is complete, as not all countries report their relevant data additionally no independent verification or audit has ever been done on a complete set of data. Moreover, many countries have never conducted a full assessment of their unconventional oil potential, especially in the Middle East and Russia, where the countries see no compelling reason to do so while their conventional oil resources are in abundance. Therefore, it is possible that there may be far more unconventional resources yet to be identified. The numbers reported to date only give a partial and incomplete picture but with time, the blur in the picture will disappear as more data becomes available.

A major cause of unreliability in the unconventional data is the persistent mix-up between resources and reserves. Refer to Section 6.3 for a comprehensive discussion, where examples were presented that clearly highlight the confusion and the wide spread mixing between the two terms.

The wide range of uncertainty in recovery factors used, plays a large role in the reliability of the estimates. For example, a 1% increase in the Canadian natural bitumen deposits recovery factor from 10% to 11% will increase the Canadian reserves by 24 billion barrels. This author's practice has been to always use conservative data, and thus adopt the recovery factors that arrive at conservative estimates of reserves and resources.

Generally, political conditions have a significant effect on data reliability, as has been explained in Section 8.2. These conditions especially apply to unconventional resources data. Important examples include the refusal of OPEC to endorse Canada's natural bitumen resources as reserves since this will affect OPEC's dominant share in the conventional oil reserves. Another is the underestimation of the Venezuelan extra heavy oil reserves by the USA, where the acrimonious relations between the two countries lead to wide variations regarding the reporting of Venezuela's extra heavy oil reserves. Since the start of Hugo Chavez presidency the American estimates of the recovery factors of the Venezuelan extra heavy oil were reduced significantly and are currently reported to be even lower than the recovery factors of Canada's natural bitumen resources, despite the fact that the Venezuelan extra heavy oil resources are easier to extract.[24] This reduction in recovery factor estimates contradicts the trend in the industry, where the recovery factors

---

[24] Richard F Meyer and Emil D Attanasi, USGS, Fact Sheet 70-03, 2003.

have always been improving. If the USA were to adopt realistic recovery factors for Venezuela's extra heavy oil resources, this would lead to more reserves estimates in Venezuela, thus increasing its significance. This situation would obviously not be favoured by the USA!

Another significant source of data confusion is that some data sources amalgamate tar sands and extra heavy oil as one resource. This is inaccurate as, even though the recovery factors from the two resources can be similar, their economics differ significantly. So, even though this practice does not affect resources data, it has a determinant effect on estimating actual reserves data, which leads to inconsistent reserves estimates.

Finally, a major confusion discussed earlier, is the inclusion of some conventional oil resources, such as heavy oil or deepwater oil, in the unconventional oil data. As already stated these reserves are treated as conventional oil reserves, and thus are included in conventional reserves quantities.

9.3 Overall Oil Reserves and Resources

Based on the data presented in the previous sections, the overall oil reserves and resources in the Arab world, and their share to the world's total, can be estimated. The results of the estimates are shown in Table 9.12.

It can be seen that the share of the Arab world, in terms of overall reserves and resources, decreases significantly and stands at approximately 37.7% and 29.5% of overall remaining reserves and resources respectively. As shown previously in Section 9.2, the Arab world is extremely poor in terms of extra heavy oil and natural bitumen resources and is relatively poor in terms of oil shale resources. However, the numbers of the reserves illustrates that the Arab world remains very significant as, even though its share of the total oil reserves decreases when unconventional reserves are taken into consideration, it remains high, accounting for over a third of all reserves, even more significantly the easier and more feasible types. One has to note though, that this percentage will inevitably change (most probably decrease) in the future, with better recovery factors for both conventional and unconventional resources and a continuous exhaustion of conventional reserves. It is anticipated that the share of unconventional reserves will grow, leading to the eventual decline in the share of the Arab world of total oil reserves, though it will remain significantly higher than the Arab world's share of the total oil resources thus cementing the position of the Arab world as the largest bearer of recoverable oil in the world. This may prompt the Arab countries to explore more unconventional deposits to defend their share, which would lead to an increase in the Arab world share if more unconventional reserves were discovered.

Table 9.12: Overall oil reserves and resources in the Arab world and their share to the world's total

		Conventional Oil	Natural Bitumen	Extra Heavy Oil	Oil Shale	Total
		billion barrel				
Remaining Reserves	Arab world	675	0.00	0.05	0.93	676
	World	1170	364.63	232.15	28.23	1795
	Arab world Share (%)	*57.71*	*0.00*	*0.02*	*3.30*	*37.67*
Remaining Resources	Arab world	2077	0.00	0.15	27.98	2105
	World	4457	1104.02	729.37	846.76	7137
	Arab world Share (%)	*46.61*	*0.00*	*0.02*	*3.30*	*29.50*
Original Oil in Place	Arab world	3406	0.00	0.50	93.25	3500
	World	7994	3697.51	2486.09	2822.54	17001
	Arab world Share (%)	*42.61*	*0.00*	*0.02*	*3.30*	*20.59*

Source: Calculated based on data in Sections 9.1 and 9.2.

9.4 Oil Production

Despite peak oil theorists well publicised prediction that there is an imminent severe decline in oil production, this shortage has failed to materialise. In fact oil production continues to demonstrate an upward trend rather than a dip or a plateau. There have been occasional brief declines but these have been due to a global economic downturn or recession, but they have certainly not been 'the peak' warned of. Actual production values, which by their nature tend to be volatile, fluctuating with changing economic and political conditions, can give the illusion that an 'oil production peak' has been reached. They do not indicate a change in real oil production magnitude or capacity available, with the latter being the most important factor that determines if the producing countries have the capability to fulfil the world's needs. Looking at the bigger picture over a longer timescale, the brief 'peaks' level out as the growth in oil production continues its ascending trend. It becomes apparent that the peak oil theorists are incorrect and that the upward trend is maintained. Indeed, it is projected to grow in all major forecasts (as detailed below). Historically, while there have been brief declines in actual production due to economic factors, political factors or technical and environmental incidents, there has been no recorded fall in the global oil production capacity.

To date, this indisputable upward trend is supported by all major data sources, who report that the world's oil production is continuing to increase. Figure 9.5 demonstrates this and shows the reported total oil production since 1980 as published by the EIA, BP, and OPEC, with all sources reporting an increase of approximately 21% to 31% in total oil production between 1980 and 2008. Total oil production numbers include crude oil (both conventional and unconventional) and NGL production, as well as process gain from refining processes, but exclude other liquids production such as biofuels, GTL and CTL. It can be seen that the difference in the reported data is insignificant and that the major data sources not only agree on the upward trend but also on the quantitative total oil production. Unavoidably, there are slight differences but these differences can be attributed to different practices in data reporting (e.g. what exactly is included). Note that the data takes into account production from the natural bitumen deposits of Canada as well as from the extra heavy oil reserves in Venezuela. According to the EIA, the total oil production in 2008 is estimated to total 83.8 million bpd. This total production is the summation of three major constituents: crude oil – both conventional and unconventional – including lease condensates; NGLs; and refinery process gain. The crude oil constituent contributes the lion's share accounting for 88% of the total oil production; a further 9.5% comes from NGL production and a mere 2.5% from refinery process gain.[25]

A close examination of Figure 9.5 shows the difference in numbers of total oil production reported by OPEC and BP. Even though the difference may raise suspicions, in truth it is fully explainable and is attributed to NGL production data which are excluded by the former, but included by the latter. As proof, the data from the EIA are plotted on the same chart with and without NGL, and it can be seen that they agree closely with OPEC numbers when NGL is excluded and with BP numbers when NGL is included.

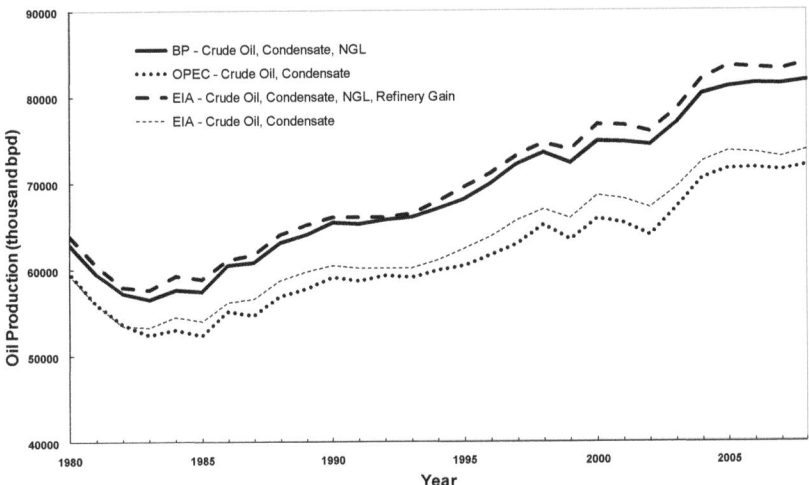

Figure 9.5: World total oil production (1980-2008)
Source: EIA (http://www.eia.doe.gov/international); OPEC (Annual Statistical Bulletins 1999-2008); BP (Statistical Review of World Energy 2001-2009).

Furthermore, the predicted oil production for 2030 is reported to be adequate to satisfy projected demand, which is predicted to reach 105, 106, and 106.6 million bpd by the IEA, OPEC, and EIA respectively.[26,27,28] This contradicts peak oil theory advocates who estimate the world production to decline to around

---

[25] EIA, (http://www.eia.doe.gov/international).
[26] IEA, World Energy Outlook 2009.
[27] OPEC, World Oil Outlook 2009.
[28] EIA, International Energy Outlook 2009.

50 million bpd by 2030.[29] An interesting little fact, which often is rarely reported in popular mass media, is that to date, most peak oil theory advocates predictions of oil production peaks and peak dates have failed spectacularly, with the forecasters of doom and gloom having to keep reforecasting! They have had to justify their changes in pushing the dates of the predicted peaks forward, often resorting to citing several 'unforeseen' reasons for the failure of their predictions. For a list of these failed predictions refer to BTRE[30], Hirsh[31] and Bentley and Boyle.[32]

The majority of oil production comes from conventional oil. Estimates from unconventional oil production constitute a mere 1.6% in 2008 and even though they are projected to increase substantially, their share will not exceed 8.6% by 2030.[33] Currently most unconventional oil production projects are in Canada, exploiting tar sand, or in Venezuela, exploiting extra heavy oil (note that historically, extra heavy oil was used to produce orimulsion in Venezuela, however this production has now been terminated – see Section 2.2.1 for more details), with other countries beginning similar exploitative projects including China, Malaysia, Madagascar and DR Congo. To date, no commercial oil production is taking place from oil shale, though several pilot plants are active around the world. Historically some oil shale is burnt directly as an energy source, though this usage is accounted for as part of coal production.

Generally speaking, oil production data from different sources are more reliable and consistent than oil reserves and resources data. This is largely due to the fact that the reported data reflects actual tangible oil production, which is measurable and accounted for both physically and financially, although it differs

---

[29] Jean Laherrère, Uncertainty of data and forecasts for fossil fuels, Universidad de Castilla-La Mancha, 2007.
[30] Bureau of Transport and Regional Economics, Department of Transport and Regional Services. Australian Government, Is the world running out of oil? A review of the debate. Working Paper 61, 2005.
[31] Robert L Hirsh, Peaking of world oil production: Recent forecasts, World Oil, Vol 228 No. 4, 2007.
[32] Roger Bentley and Godfrey Boyle, Global oil production: forecasts and methodologies, Environmental and Planning B: Planning and Design, volume 35, p 609-626, 2008.
[33] IEA, World Energy Outlook 2009.

from estimates of reserves and resources, which sometimes reflect fanciful mathematical models rather than concrete physical reality. However, as demonstrated in earlier chapters, it is clear that oil production data reporting is not transparent. On the contrary, it is worryingly confusing and without standards – though the confusion exists to a lesser extent than its counterpart in reserves and recourses data. As an exercise, try picking a few oil production reports and compare the definition of oil production reported, undoubtedly you will come across oil production numbers that refer to crude oil only, or to crude oil plus NGL, or both including or excluding unconventional oil, or any combination of the previous categories with or without including biofuels and other liquids – are you confused yet? Furthermore, oil production is reported in different units (e.g. bpd, tonnes/day, boe, Btu, etc.). In this book, when referring to the term total oil production, it includes conventional and unconventional crude oil (including condensates), NGL, and process gain. It excludes other liquid production such as biofuels, GTL and CTL, with the latter two accounted for in coal and natural gas production respectively to avoid double counting.

A revisit to Figure 9.5 prompts a new question: why does OPEC exclude NGL from its total oil production data even though it reports NGL production (admittedly separately) in its statistical bulletins? There is no official rational explanation. However, it is widely speculated that it allows more leeway to its members to relax compliance criteria regarding the oil production quota system, since NGL production is excluded from this system. Therefore its production is reported separately.

On the other hand, OPEC is ambiguous to clarify its position regarding production from unconventional oil reserves, despite strong indications (and no evidence to the contrary) that oil production from the unconventional reserves is included in the total production numbers reported by its members. However, OPEC's position is clearer regarding non members oil production, when reported in OPEC publications, where for example in Canada's oil production, the unconventional oil is included in the total oil production numbers. It remains though that OPEC has no declared official position and made no effort to identify distinct deposits from which the oil has been produced. So officially at least, OPEC members do not report separately production from unconventional oil. The subject as a whole is mostly ignored and brushed aside. This can lead to the assumption that this oil production is included in the total production numbers. However, note the API degree used when reporting unconventional oil in the volumetric unit of barrels, since reporting the original production at lower API or after upgrading to higher API can lead to differences amounting to over 10% of the production quantities. This foggy area often causes arguments in reporting Venezuelan oil production.

Currently the majority of crude oil is produced onshore with a share of approximately 67%, though offshore production is quickly catching up and accounts for 33% in 2008.[34]

Table 9.13: World total oil production – top ten countries (2008)

Rank	Country	Total Oil Production	Share
		thousand bpd	%
1	Saudi Arabia	10701.12	12.76
2	Russia	9789.76	11.68
3	USA	7707.24	9.19
4	Iran	4149.30	4.95
5	China	3973.13	4.74
6	Canada	3353.09	4.00
7	Mexico	3185.64	3.80
8	UAE	3046.47	3.63
9	Kuwait	2741.38	3.27
10	Venezuela	2642.90	3.15
	TOTAL	51290.04	61.18
	WORLD	83835.09	

Source: EIA (http://www.eia.doe.gov/international).
Note 1: Totals may not add up due to rounding.

Table 9.13 lists the leading ten countries in the world in terms of total oil production. These countries account for over 61.2% of

---

[34] Rafael Sandrea, Future offshore/onshore crude oil production capacities, Penn Energy. 2009 (www.pennenergy.com/.../rafael-sandrea/future-offshore-onshore-crude-oil-production-capacities.html).

the global production. Note that four of these countries are not in the top ten countries in terms of oil reserves (refer to Table 9.1), which - contrary to the generally perceived wisdom – illustrates that the level of oil production is not determined solely by the proved oil reserves, but depends on several factors including economics, politics and strategic considerations.

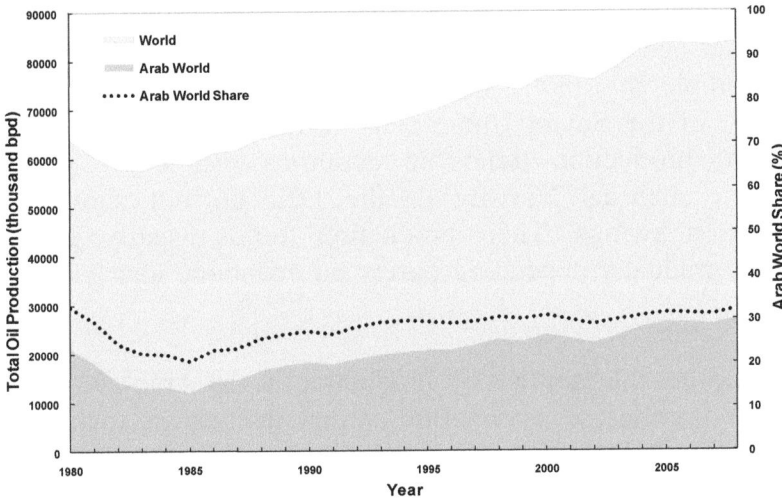

Figure 9.6: Arab world total oil production and its share to the world's total (1980-2008)
Source: EIA (http://www.eia.doe.gov/international).

The Arab world overall total oil production and its share to the world's total, since 1980, are shown in Figure 9.6. The data shows that, in real terms, the oil production in the Arab countries fluctuated, dropping to a low point in the mid 1980s before staging an almost uninterrupted recovery reaching 26.9 million bpd in 2008. If taken as one entity, the Arab world will be placed at the top of the top ten list in terms of total oil production, with three times more production than the second placed country, Russia.

You may wonder why the total oil production trend in the Arab world appears to be erratic. Despite continuous production increase in the world, the trend is not repeated in the Arab world where the additional production has failed to catch up. At first this seems odd; however once we consider the political turmoil in the area, the peaks and valleys can be all explained. For example the steep decline noticed in the early 1980s is due to the Iraq-Iran war, and the slight decline in early 1990s is due to the Iraqi invasion of Kuwait. In both instances a significant portion of oil production operations were halted. The decline though was not as steep as anticipated, due to Saudi Arabia opening its taps to compensate for the production shortage. This story is not unique, and it has occurred several times already but on a different scale, for example in Russia, whose short-lived dip was due to the collapse of the Soviet Union and the turmoil that followed. In contrast, production trends in countries that enjoy political stability, such as Norway or the UK, do not show these production swings. Their production trends usually grow or decline gradually depending purely on economic and geological factors.

Furthermore, the fact that oil production in the Arab countries is totally controlled by national oil companies that are managed as governmental departments, instead of commercial companies that are accountable for their share holders, resulted in considerably lower rates of investment in further production facilities. This is because the national oil companies are treated as cash cows by the governments, who use oil revenues for other politically motivated projects rather than investing in additional oil production. Consequently, some of the short-lived peaks observed so far (or maybe will be observed in the near future) were due to the lack of investment rather than depletion of oil reserves. This situation is not unique to the Arab countries, but applies to other countries where oil production is controlled by national oil companies. A prime example is Mexico as discussed earlier in Section 6.3.2.

The situation in the Arab world is similar to the global situation, where the crude oil constituent contributes the lion's share accounting for 89.5% of the total oil production; further 10.4% comes from NGL production and only a tiny 0.1% from refinery process gain.[35] It can be deduced that the NGL contribution, in percentage terms, to the total Arab oil production exceeds the global contribution; this reflects the relatively large size of the Arab natural gas production operations, which inevitably yields higher NGL production (refer to Section 10.4). On the other hand the 'refinery process gain' contribution in percentage terms to the total Arab oil production is far below the global contribution percentage; this reflects the relatively small capacity of the Arab oil refining sector (refer to Section 9.6 for more details).

The share of the Arab oil to the world total oil production mirrored the Arab total oil production trend qualitatively but not quantitatively. It can be seen that the share has stabilised since the early 1990s, where it has lingered at just over 30% and has recorded 32.1% in 2008, which emphasises the significance of the Arab world in terms of total oil production, since it produces almost a third of the world's total production.

The data discussed above though does not tell the whole story. It conceals the fact that the Arab world is far more important to oil markets than the numbers above suggest. This is because the Arab world accounts for a much larger share of the oil net exports (see Section 9.8), since many of the leading oil producers are oil thirsty and are therefore leading oil consumers. This can easily be seen by looking at the situation of the USA and China, who both occupy high spots in the top ten list of oil producing countries, but since both countries also rank high in the top ten oil consumers, they find themselves inevitably placed in the top ten list of net oil importers. Several previous and current net oil exporting countries headed, or are heading, in a similar direction, where the rapid increase in oil consumption is forcing them to cut oil exports gradually until they turn eventually into net importers. Examples include Indonesia and the UK, who has turned already into a net oil importer, and Mexico and Iran, who are heading that way! Refer to Sections 9.7 and 9.8 for more details.

The total oil production for all Arab countries in 2008, including

---

[35] EIA, (http://www.eia.doe.gov/international).

Table 9.14: Total oil production in the Arab countries (2008)

Country	Crude Oil Production	NGL Production	Refinery Processing Gain	Total Oil Production	Rank	Share
	thousand bpd					%
Bahrain	35.00	11.00	2.52	48.52	14	0.06
Iraq	2358.74	13.41	-3.24	2368.91	4	2.83
Jordan	0.02	0.00	-0.28	-0.26		
Kuwait	2585.75	142.83	12.80	2741.38	3	3.27
Lebanon	0.00	0.00	0.00	0.00		
Oman	757.32	5.42	-1.74	761.00	8	0.91
Palestine	0.00	0.00	0.00	0.00		
Qatar	924.03	255.53	11.04	1190.61	7	1.42
Saudi Arabia	9261.25	1433.70	6.17	10701.12	1	12.76
Syria	387.62	60.00	-3.24	444.38	11	0.53
UAE	2681.02	356.24	9.20	3046.47	2	3.63
Yemen	298.33	0.00	1.79	300.12	12	0.36
Algeria	1872.16	356.90	-1.73	2227.33	5	2.66
Comoros	0.00	0.00	0.00	0.00		
Djibouti	0.00	0.00	0.00	0.00		
Egypt	602.53	35.00	-6.95	630.58	9	0.75
Libya	1714.69	140.00	-0.63	1854.06	6	2.21
Mauritania	12.83	0.00	0.00	12.83	15	0.02
Morocco	0.50	0.00	-1.19	-0.69		
Somalia	0.00	0.00	0.00	0.00		
Sudan	520.85	0.00	2.41	523.25	10	0.62
Tunisia	84.18	3.00	-0.25	86.93	13	0.10
ARAB WORLD	24096.82	2813.05	26.69	26936.56		32.13

Source: EIA (http://www.eia.doe.gov/international).
Note 1: Totals may not add up due to rounding.

details of the constituents of the oil production for each country, is reported in Table 9.14. A complete set of data from 1980 can be obtained from "*http://www.2050consulting.com/books*". From the above table we can detect that oil production processes are active in 17 Arab countries. We can classify oil producing countries into five tiers in terms of total oil production: major, significant, medium, minor oil producing countries or countries with no oil production operations. The Arab countries fall into this classification as follows:

- Seven countries belong to the first tier, with a total production exceeding 1 million bpd each, and a combined share of approximately 28.8% of the world's total oil production. These countries in descending order in terms of total oil production are: Saudi Arabia, UAE, Kuwait, Iraq, Algeria, Libya and Qatar, with the first three also ranked in the top ten list globally including top spot (see Table 9.13). Oil production plays a significant role in these countries respective economies and it is the main contributor to their GDPs. All these countries are members of OPEC, and in theory – though not always in practice – their production is bound by OPEC oil production quotas, and as such they do not produce to their full capacity. Remembering it is excluded from the above quotas, the contribution of NGL to the total is significant in most of these countries, reaching a considerable 21% in Qatar, 16% in Algeria and above 10% Saudi Arabia and the UAE. Therefore when NGL is excluded, Libya will bypass Algeria in the ranking, while Qatar will be demoted to the second tier. Incidentally the ranking of these Arab countries is not similar in both oil reserves and production, which is the normal situation in all producing countries where, having huge reserves does not translate to being a top producer, and vice versa. For example other leading oil producing countries such as Mexico has relatively small reserves but is a leading oil producer! With improving political conditions in Iraq and Libya, the oil production from the two countries is expected to soar.

- Five countries belong to the second tier, with a total oil production exceeding 250 thousand but less than 1 million bpd each, and a combined share of approximately 3.2% of the world's total oil production. These countries in descending order in terms of total oil production are: Oman, Egypt, Sudan, Syria, and Yemen. Oil production plays a significant role in these countries respective economies contributing substantially to their GDPs. However, the combined oil production from these countries plays a minor role in the global oil market, and their combined share ranks them tenth, in terms of oil production. None of the above countries is an OPEC

member, though they coordinate closely with OPEC to control the oil market. Recently Egypt turned into a net oil importer, and Syria is also heading that way (refer to Section 9.8).
- One country, Tunisia, belongs to the third tier, with a total oil production exceeding 50 thousand but less than 250 thousand bpd, with a share of approximately 0.1%. Here also, oil production plays a significant role in Tunisia's economy contributing substantially to its GDP, despite it being a net oil importer (refer to Section 9.8).
- Four countries, Bahrain, Mauritania, Morocco and Jordan, belong to the fourth tier, with a total oil production less than 50 thousand bpd each, and a combined share of 0.1%. These countries in descending order in terms of total oil production are: Bahrain, Mauritania, Morocco and Jordan. Here also, oil production plays a significant role in the first two countries respective economies contributing substantially to their GDPs, even though Mauritania is a net oil importer (refer to Section 9.8). Morocco and Jordan operate very minor crude oil production, their ageing refineries cause a negative refinery process gain, which leads to a net negative oil production. Oil production contribution to these countries GDP is minimal.
- Finally five countries have no oil production operations, and thus are placed in the fifth tier. These countries are Lebanon, Palestine, Somalia, Djibouti, and Comoros.

A fundamental concept to be considered is to differentiate between production capacity and actual production. The latter is always reported, and what we discussed in this section so far. However, the former is what really matters to oil markets, since capacity, or to be precise spare capacity, plays a major role in the stability of the oil markets, being is vital in determining oil prices and subsequently the future of proposed production projects, hence providing additional future capacity!

Spare capacity numbers are often hard to verify, as they rely on political and economic factors which affect their disclosure. The data presented earlier does not reveal the exact production spare capacity. The majority of this capacity is controlled by one country, Saudi Arabia, which usually plays the role of market policeman opening and closing its taps to try and stabilise the oil markets when required. It is widely reported that an obvious relationship can easily be derived between spare capacity and oil price with low spare capacity causing high prices and vice versa.

The Middle East Perspective

The Middle East region is the source of approximately 30.7% of the world's total oil production, 83.9% of which are in its Arab countries. The region's countries occupy four places in the top ten list of the leading oil producing countries including the top spot. This indicates the strategic significance of the region as the major oil producing area in the world.

Source: EIA, (http://www.eia.doe.gov/international), based on 2008 data.

OPEC Perspective

OPEC member states are the source of approximately 43.7% of the world's total oil production, 65.8% of which are in its Arab member states. The organisation's members occupy five places in the top ten list of the leading oil producing countries including the top spot. This illustrates the power the organisation has in the oil markets, and reaffirms its strength in oil price and thus global energy policies and economics.

Source: EIA, (http://www.eia.doe.gov/international), based on 2008 data.

9.4.1 Converted Oil Production

Processes to convert 'produced oil' to gaseous or solid fuels are technically possible, and have been discussed in Sections 3.9 and 4.5. These processes are secondary transformation processes, thus their production numbers are already included in oil primary production quantities. In order to minimise confusion and avoid double counting, if any gas produced from oil gasification is not accounted for as gas production, it is considered as part of oil production. Similarly solid fuels derived from oil, result from oil refining processes and are already included in oil production original data.

9.4.2 Other Liquid Fossil Fuels Production

As discussed in Section 2.8, liquid fossil fuels can be derived from natural gas or coal. However since these processes are secondary transformations of the natural gas or coal, where the original feedstocks of the natural gas or the coal have already been accounted for in the original natural gas or coal production numbers, the production of this 'secondary' liquid is not included in any production numbers. However it is often included in consumption numbers, as at that stage, it is harder to distinguish the source of the liquid fuel. Refer to Sections 10.4.1 and 11.4.1 for more details.

9.5 Oil Reserve to Production Ratio (R/P)

The reserve to production ratio (R/P) is often used as a preliminary indicator of how long current reserves will last at current production rates. It is calculated simply by dividing the reserves remaining at the end of any year by the production in that year. R/P ratio is a supposed simple magic number that tells us how many years of oil we have left. It is a completely theoretical number, which gives the number of years current reserves would last if the oil reserves estimates are as projected and if the oil production from them remained at a steady constant rate, assuming that constant rate was enough to satisfy our needs.

The R/P ratio is flawed as an indicator of how long a reserve will last, as it neglects both production and reserve growth, as well as reduction effects. In fact it tells us nothing useful about either how many years of oil we have left, or when we should start worrying about running out of oil. To illustrate, three simple facts explain why this number is completely meaningless and should not be used, unless one wants to misrepresent the facts or intentionally mislead: the first fact is that oil consumption growth is tied to economic growth and environmental restrictions, thus we cannot maintain the current level of oil consumption. This leads to the second fact, that oil production cannot remain the same, as it is tied to consumption. Also it physically cannot stay at the same current level, since it will drop with time due to geological and technical factors. Finally the third fact, which is that oil reserves cannot remain constant as they are exploited as production continues or can grow with advances in technology and changing economical conditions.

Furthermore, reported R/P ratios usually use only conventional oil reserves as their basis and thus ignore the fact that some of the oil produced in the world is coming from unconventional resources. In addition the ratio calculation often does not include data from NGL reserves or production or both, and completely ignores refinery process gain.

However, despite this, the R/P ratio has its defenders, they argue that it is useful, as it translates the complicated oil statistics into a single simple quantity (measured in years), which is understood by the general public.

In this book R/P ratios are not considered as a credible analysis tool. They are only mentioned here to present a complete picture, while also highlighting the deficiencies that render them useless.

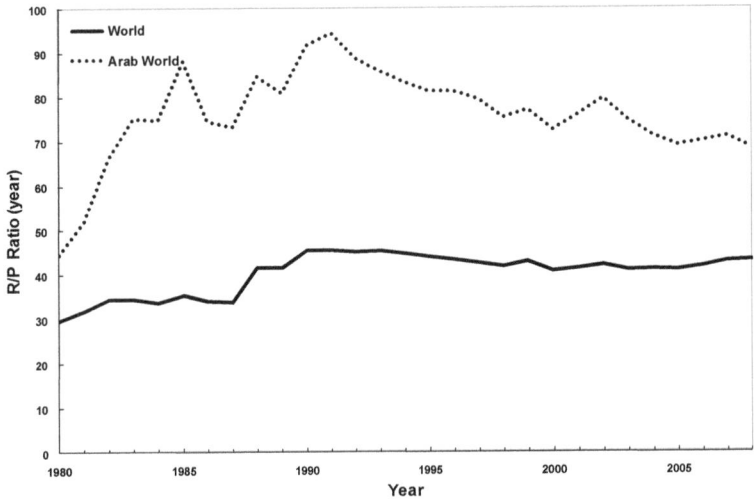

Figure 9.7: World oil reserves to production ratio (1980-2008)

Source: Calculated based on data in Sections 9.1 and 9.4.

Figure 9.7 shows the world's R/P ratios since 1980, calculated from data obtained from the EIA. R/P calculations are performed based on conventional oil reserves, including condensates and NGL reserves, but excluding unconventional oil reserves. The oil production data on the other hand are based on total crude oil production including NGL production but excluding other liquids and refinery process gain. However the oil production data are not adjusted to exclude unconventional production, since most oil production data presented does not report this production separately. The figure also shows the R/P ratio for the Arab world.

Figure 9.7 suggests that no imminent decline in oil production is envisaged as the global R/P ratio has not only remained more or less stable since early 1990s, but also increased by 10 years since 1980 despite continuing oil production. This immediately raises eyebrows as it seems very odd that despite changes in both oil reserves and production, which should result in varying R/P, the ratio hardly changed, one may think of reverse engineering? Enough said!

The R/P ratio for the Arab world did in fact increase. This was due to political instability that led to severe reduction in Iraq's oil production, as well as new oil discoveries in the Sudan, which resulted in an increase in the Arab world total oil reserves. In addition, due to OPEC maintaining (and reluctantly enforcing) its production quota system, no substantial increase in actual oil production in its Arab member states occurred. This led to less production capacity being planned, as when projects reached their end of life, there was no incentive to invest in extra production capabilities to install spare capacity that cannot be used for actual production. A closer inspection shows that the ratio fluctuated several times by 10 years up or down in a matter of few years, which does not reflect physical reality but rather political and economic instabilities.

To further illustrate the uselessness of the R/P ratio consider the case of a country like Jordan with hardly any oil reserves, but also hardly any oil production. Calculating R/P for Jordan gives almost 140 years, which is higher than the ratio of Saudi Arabia, which is just over 80! On paper this sounds impressive, but in reality it clearly demonstrates the flaw in the ratio and highlights the major weakness of using this indicator, since who cares if the tiny oil reserves of a country, such as Jordan, with hardly any oil production will last for over a century?

9.6 Refined Petroleum Production and Oil Refining Capacity

Most crude oil produced in the world is sent to refineries, where it is processed and converted into more useful petroleum products, such as gasoline, diesel, kerosene, jet fuel, naphtha, LPG, fuel oils, lubricating oils, paraffin wax, asphalt and petroleum coke. As stated in Section 2.7, oil refining technologies and details of refined petroleum production are outside the scope of this book. However, it is necessary to briefly visit the latter subject, at this instance, to remove any ambiguities and misunderstandings that are created by mass media outlets when they report issues concerning oil production.

A common mistake made in many reports is that refined petroleum production is added to the crude oil production and NGL production to reach a grand total. This should never be performed as it constitutes double counting and is fundamentally mathematically incorrect. By definition, refined petroleum products are secondary products and as such are transformations of the 'same' produced crude oil. However, since oil production is reported in volumetric units, any differences in volume are accounted for by the refining process gain or loss. This simple - but costly - mistake has been committed numerous times and has often resulted in gross overestimations in total oil production quantities.

However, even though refined petroleum production is not needed to estimate total oil production, the quantities of refined petroleum production are needed to establish meaningful estimations of both oil consumption and oil trade. The exact refined petroleum production values are not of interest in this book, but they are implicitly taken into account in determining the apparent consumption and the net oil trade data. Thus to get a full understanding of oil refining data, one has to keep referring to Sections 9.7 and 9.8 to consult oil consumption and oil trade data.

Obviously the refined petroleum production is limited by the available oil refining capacity. Providing this capacity has always been expensive and technically difficult. Furthermore, oil refining projects often take a long time and require extensive planning. Oil refining capacity is fundamental in determining the availability of usable refined petroleum products to fulfil demand and plays a crucial role in setting their prices. So, with the long term demand for oil showing no signs of decreasing, providing oil refining capacity is not only essential but critical to the global economy.

Oil refining capacity is not the only important factor. Crude oil characteristics are also important factors for technical reasons. The explanation for the latter point is, because oil refineries are designed to handle specific types of oil crude. Thus, the change in the characteristics of the oil stocks (e.g. increasing supply of heavier oil stocks) and the tendency and desire of oil consuming countries to diversify and source crude oil from different suppliers have made it necessary to provide flexibility in designing refineries. This has forced the drive to provide constant modifications to existing ones, so that refineries will to be able to handle the changing nature of crude oil supplies and characteristics. Consequently this not only called for the construction of new refineries, but also for non-stop upgrading, modification and expansion of existing refineries as well.

Historically, to fulfil the increasing demand, the number and capacity of oil refineries in the world increased steadily. However in the 1980s, as a result of the slump in oil price, mounting environmental concerns, political instabilities, and the fact that many oil refineries were aging or reaching their decommissioning date, numerous refineries - especially smaller ones - had to cease operation, thus reducing the total number of refineries as well as total capacity. As a result, the number of total refineries dropped significantly, by over 100 since 2001, to around 700 in 2008.[36] The recovery in oil demand in the 1990s only reversed part of the trend, so the number of refineries continued to decline due to closure and consolidation. However the refining capacity trend turned around, with additional capacity being added due to expanding, modernising and upgrading of existing refineries or the occasional establishing of new ones.

For example, the number of oil refineries in the USA declined from 301, in 1982, to only 150 in 2009. However the oil refining capacity climbed back to approximately 17.7 million bpd in 2009, after declining from 17.9 million bpd in 1982 to just over 15 million bpd in 1993.[37] This increase in capacity occurred despite the fact that very few small and simple refineries have been constructed in the USA since 1977, due to stringent environmental regulations as well as the growing 'not-in-my-backyard' attitude.[38]

Oil refinery capacity is usually measured as the capacity of the atmospheric distillation units. Secondary units such as catalytic cracking, thermal cracking or reforming, enhance the quality and value of the refined production; however the production from these secondary units should not be added to distillation units output to reach a grand total, as this would constitute counting the same stock input into the refineries twice.

Almost all oil refineries do not operate to their full capacities. This is due to both technical and economic conditions such as trade climate, weather conditions or need for maintenance. Thus, exact refined petroleum production depends on demand and is firmly linked to the health of the economy. It is estimated that in 2008 the world oil refineries operated to 84% of their name plate (i.e. stated) capacity.[39]

It is stressed again that details of specific refined products are not

---

[36] I Billege, 700 Oil Refineries Supply Oil Products to World, NAFTA, 60 (7-8) 401-403, 2009.
[37] EIA (http://tonto.eia.doe.gov/dnav/pet/pet_pnp_cap1_dcu_nus_a.htm).
[38] EIA (http://tonto.eia.doe.gov/ask/crudeoil_faqs.asp#last_refinergy_built).
[39] I. Billege, 700 Oil Refineries Supply Oil Products to World, NAFTA, 60 (7-8) 401-403, 2009.

covered in this book. These data can be found elsewhere in reports, bulletins, and online databases.

The following discussion covers refining capacity in terms of overall refined petroleum.

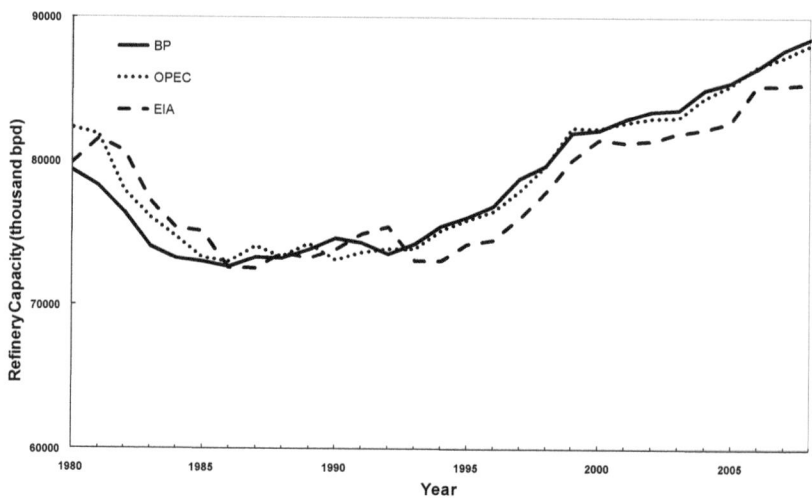

Figure 9.8: World nominal oil refining capacity (1980-2008)

Source: EIA (http://www.eia.doe.gov/international); OPEC (Annual Statistical Bulletins 1999-2008); BP (Statistical Review of World Energy 2001-2009).

World refining capacity (measured as described above) is shown in Figure 9.8, which reports the nominal oil refining capacity since 1980 as published by the EIA, BP, and OPEC. It is evident that all the sources' reported figures are close, where the increase in capacity is seen to range approximately between 7% and 12% for the period between 1980 and 2009, though the increase is higher if compared to the low point of mid 1980s and stands at approximately between 18% and 22%. As with oil reserves or production data, the slight differences are insignificant and are due to different practices in data reporting. From the EIA's latest data, the total world oil refining capacity is estimated to total 85.8 million bpd in 2009. Note that this is just over 2% more than total oil production for 2008. This number illustrates that despite low economic margins the world will require additional refining capacity to process the projected additional oil production.

The world's top ten countries in terms of oil refining capacity are listed in Table 9.15. These ten countries account for a massive 55.8% of the world's total oil refining capacity. It can be deduced that the countries in this list are not the world's top leading countries in terms of oil production or reserves, but the list is more similar to the world's top ten in terms of oil consumption with nine of the countries appear in both lists (see Table 9.17). This is not surprising, as countries that consume more oil need large oil refining capacities. However, this also demonstrates a missed opportunity by oil producing countries. If they had taken strategic decisions to invest in more oil refining capacity, exporting refined petroleum instead of crude oil, they would have reaped significant benefits as there are more revenues and the definite added value of refined petroleum products. So if the opportunities were even now to be taken, then leading oil producing countries would generate more income, enjoy more power and control the refined petroleum products prices. We can see that, to date, this strategic decision has not been taken and that only one of the top ten countries in Table 9.15 is a member of OPEC.

Table 9.15: World nominal oil refining capacity – top ten countries (2009)

Rank	Country	Refinery Capacity	Share
		thousand bpd	%
1	USA	17610.00	20.52
2	China	6446.00	7.51
3	Russia	5428.50	6.32
4	Japan	4690.70	5.46
5	South Korea	2606.50	3.04
6	Germany	2417.52	2.82
7	Italy	2337.23	2.72
8	India	2255.54	2.63
9	Saudi Arabia	2080.00	2.42
10	Canada	2029.45	2.36
	TOTAL	47901.44	55.81
	WORLD	85833.95	

Source: EIA (http://www.eia.doe.gov/international).
Note 1: Totals may not add up due to rounding.

Figure 9.9 illustrates the Arab world's total refining capacity and its share of the world's total capacity. The refining capacity data reveals that the share of the Arab world, in terms of oil refining capacity, is relatively small when compared to the Arab world's share, in terms of oil reserves or production, which stands at 8.4% of the global capacity, i.e. 7.2 million bpd in 2009. If treated as a single entity, the Arab world will be placed a distant second in the top ten list, in terms of oil refining capacity, trailing the USA by a massive 59%.

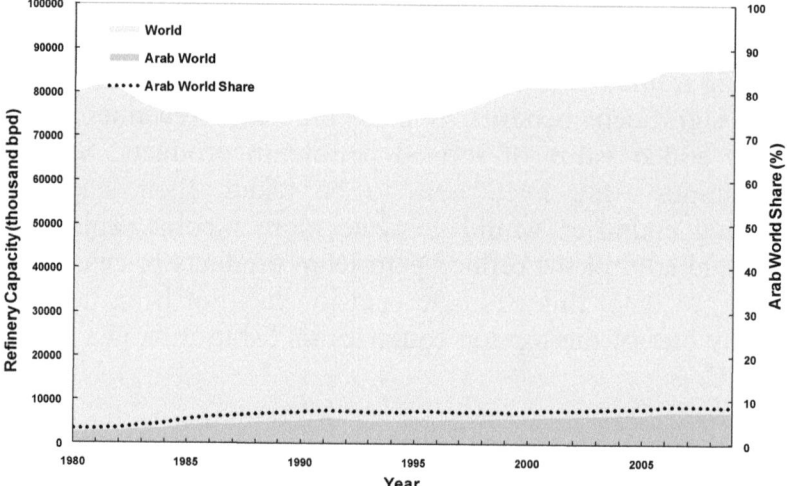

Figure 9.9: Arab world nominal oil refining capacity and its share to the world's total (1980-2009)

Source: EIA (http://www.eia.doe.gov/international).

Interestingly, the data shows that the Arab world's share has remained stable since 1980, hovering at around 8%. This indicates that the Arab countries did not take any advantage of their increased oil revenues to boost their global oil refining share, but seem to be content to maintain the status quo.

Please remember that the oil refining capacity discussed so far is the nominal capacity, and that the actual capacity is in fact considerably lower, since many countries report the refineries name plate capacity and fail to update the data if their refining capacities are reduced. For example, much of the oil refining capacity in Iraq was destroyed in the war but it is still reported as operational.

Table 9.16: Nominal oil refining capacity in the Arab countries (2009)

Country	Refinery Capacity thousand bpd	Rank	Share %
Bahrain	262.00	8	0.31
Iraq	597.50	5	0.70
Jordan	90.40	14	0.11
Kuwait	889.20	2	1.04
Lebanon	0.00		
Oman	85.00	15	0.10
Palestine	0.00		
Qatar	200.00	10	0.23
Saudi Arabia	2080.00	1	2.42
Syria	239.87	9	0.28
UAE	781.25	3	0.91
Yemen	140.00	12	0.16
Algeria	450.00	6	0.52
Comoros	0.00		
Djibouti	0.00		
Egypt	726.25	4	0.85
Libya	378.00	7	0.44
Mauritania	0.00		
Morocco	154.90	11	0.18
Somalia	0.00		
Sudan	121.70	13	0.14
Tunisia	34.00	16	0.04
ARAB WORLD	7230.07		8.42

Source: EIA (http://www.eia.doe.gov/international).
Note 1: Totals may not add up due to rounding.

Reflecting on the previous figure is puzzling as it indicates that no efforts have been made to increase the capacity of this strategic and profitable sector. The situation can be considered poor judgement by the Arab governments when one considers their investment in often less profitable sectors (Citi Bank or Manchester City FC to name two examples!) Equally when comparing the Arab oil refining capacity to the capacity of some oil importing countries that made strategic decisions to invest in oil refining and are now making substantial profits, exerting influence and creating many jobs as a result. Such countries include Singapore, with oil refining capacity of 1344 bpd in 2009, and total oil consumption of 896 bpd in 2008, and the Netherlands with oil refining capacity of 1208 bpd in 2009, and total oil consumption of 963 bpd in 2008.[40]

The total oil refining capacity for all Arab countries in 2009, including detailed data per country, is reported in Table 9.16. A complete set of data from 1980 can be obtained from "*http://www.2050consulting.com/books*". The table above lists oil refining capacity in all 22 Arab countries. In terms of oil refining capacity, countries can be classified into five tiers: countries with major, significant, medium, minor and no oil refining capacity. The Arab countries fall into this classification as follows:

- One country, Saudi Arabia, belongs to the first tier, with a total oil refining capacity exceeding 1 million bpd with a share of approximately 2.4% of the world's total capacity. It is the only Arab and OPEC country that is also ranked in the top ten list, globally occupying the ninth place (see Table 9.15). The Saudi oil refining sector is therefore important to the country's GDP. However its actual contribution is far below its potential as the country so far has made little efforts and shown little interest, in developing a large oil refining export sector to strengthen its crude oil production potential or to compete with established oil refining countries to guarantee favourable market share of Saudi oil. To date, most of the Saudi oil refining capacity planning has been performed to meet the large domestic oil consumption. This has been encouraged by the subsidies given by Saudi authorities to maintain cheap oil pricing, and only recently have some shifts in this position been detected.
- Seven countries belong to the second tier, with total oil refining capacity exceeding 250 thousand but less than 1 million bpd each, and a combined share of approximately 4.8% of the world's total capacity. These countries in descending order in terms of total oil refining capacity are: Kuwait, UAE, Egypt, Iraq, Algeria, Libya, and

---

[40] EIA, (http://www.eia.doe.gov/international).

Bahrain. Six of these countries are net oil exporters, and five of them are members of OPEC. However these five members rely on exporting crude oil to earn the majority of their export revenues, and in comparison, their exports of refined petroleum products are minimal if any. The two non-OPEC members are Egypt and Bahrain. Note that Bahrain imports most of its crude oil, but after its production peaked, it took the decision to invest in oil refining to earn the added value profits, supplementing its falling crude oil revenues. Thus due to its relatively large oil refining capacity, it now exports a substantial amount of its refined production, maintaining its position as a significant net oil exporter. Egypt is an exception since it has turned into a net importer since 2007 (refer to Section 9.8 for more details). Due to its large population, its relatively large oil refining capacity is needed to satisfy domestic demand leaving little extra for exporting.

- Seven countries belong to the third tier, with a total oil refining capacity exceeding 50 thousand but less than 250 thousand bpd and a combined share of approximately 1.2% of the world's total capacity. These countries in descending order in terms of total oil refining capacity are: Syria, Qatar, Morocco, Yemen, Sudan, Jordan, and Oman. With the exception of Morocco and Jordan, all the countries in this tier are net oil exporters, with Qatar being the only OPEC member. Here again it is seen that the refining capacity is designed to fulfil domestic demand and that none has a significant excess oil refining capacity for exporting. On the contrary, some such as Syria and Jordan need to import some of their refined petroleum consumption.[41]
- Two countries, Tunisia and Mauritania, belong to the fourth tier with a total capacity less than 50 thousand bpd with a share of less than 0.1% of the world's total oil refining capacity. Note that the data for Mauritania is not always reported and is not included by the EIA for example. This is due to the frequent disuse of the only

---

[41] OPEC, Annual Statistical Bulletin 2009.

refinery in the country throughout its troubled history.
- Finally five countries belong to the fifth tier with no oil refining capacity. These countries are Lebanon, Palestine, Somalia, Djibouti, and Comoros. Previously Lebanon had a moderate oil refining capacity, however the civil war of 1975-1990 forced its refineries to become inoperative. Since the end of the civil war, several announcements were made to re-establish oil refining capacity, but nothing has come to fruition so far. Similarly, Somalia had a small oil refinery prior to its civil war, but since the outbreak of its endless war in 1991, the refinery has effectively been out of operation. Significant investment and major rehabilitation work will be needed to get it operational again. Djibouti, on the other hand, is planning a big oil refinery that, if constructed, will place the country in the third tier with a significant excess capacity for exporting, but to date, this refinery is still at the planning stage despite numerous promising high profile announcements.

The Middle East Perspective

The Middle East region is home to approximately 8.2% of the world's total oil refining capacity, 76.3% of which are in its Arab countries. Only one country in the region occupies a place in the top ten list of leading oil refining countries – the ninth spot. The picture emerging is thus very similar to that of the Arab world, and illustrates that the region's refining capacity is small, and that the region has a huge potential to expand it.

Source: EIA, (http://www.eia.doe.gov/international), based on 2009 data.

OPEC Perspective

OPEC member states are home to approximately 11.4% of the world's total oil refining capacity, 54.7% of which are in its Arab member states. The organisation's members occupy only one place in the top ten list of leading oil refining countries – the ninth spot. This highlights the weakness of the organisation, where, as already discussed in the cases of the Arab world or the Middle East, OPEC's oil refining capacity is rather low and needs expansion. On the other hand, one has to note though that a more balanced picture appears when the data in relation to Venezuela is examined, which shows that the country has a large refining capacity and has made a strategic decision to expand it, both on its territory and abroad.

Source: EIA, (http://www.eia.doe.gov/international), based on 2009 data.

9.7 Oil Consumption

Despite calls for diversity in the energy markets, oil still accounts for the largest proportion of the total world energy consumption, particularly in the transport sector where its share is an overwhelming 94% (see Section 2.9). Whilst there continue to be complaints from all sectors of society calling for reduction in oil consumption, no one seems to agree on a specific procedure of how to exactly measure the total consumption or even how to define the oil consumption rate.

The procedure widely used in estimating oil consumption refers to the 'apparent' consumption, which is a calculated amount that includes both crude oil and refined products consumption. It encompasses domestic crude oil and refined products consumption utilised, both as fuel or chemical raw material feedstock, it thus includes refinery fuel and loss, as well as bunker fuels. At present, consumption data includes oil produced from conventional and unconventional deposits, as at this stage it is impossible to define or distinguish the source of oil.

Crucially, as emphasised earlier (see Section 9.6) estimating oil consumption should not sum the crude oil and refined products simply together, as this will lead to gross overestimation in consumption. This is because the consumption will be more or less accounted for twice, since the majority of crude oil will be *consumed* at the oil refineries, where it will be transformed into more useful products, which will be consumed as fuel or raw materials. To illustrate this point, consider a country with no oil refining capabilities, its consumption of crude oil will be none, as all the oil it consumes is in the form of refined products. So if we only consider crude oil consumption, we will reach a totally wrong conclusion regarding that country's oil consumption. We will need to include the refined products consumption in order to come to a correct estimation. Therefore, the methodology of oil consumption estimation needs to take refined products consumption as its starting point, adding consumption of crude oil used as refinery fuel to it as well as crude oil that has not been refined but consumed in an alternative way, for example as industrial raw material, or wasted.

Note that oil consumption is a calculated quantity. It excludes oil transformed into other fossil fuels (see Section 3.9 and 4.5), as this oil consumption is accounted for via the consumption of gaseous or solid fuels, and there are no numbers released that distinguish this consumption. On the other hand consumption includes other fossil fuels transformed into oil and also biofuels, as again there is no way to distinguish the source of liquid fuels once it is consumed. As already explained, this is in contrast to production, where oil transformed into gas or solid is accounted for in oil production, whereas gaseous or solid fuels transformed into liquid are accounted for by their respective fossil fuels production.

Crucially, oil consumption numbers include liquids transformed from other fossil fuels as it is hard to identify the source of the oil at consumption stage. However consumption numbers exclude oil transformed into gaseous or solid fossil fuels, where that consumption is accounted for as part of the relevant fossil fuel consumption.

Recent data has shown that the growth in the world's oil consumption is slowing down, albeit it is still inching upwards on a long-term increasing trend. Figure 9.10 shows the reported apparent oil consumption from different sources since 1980. As with other oil data, it can be seen that all major data sources agree in their reported trends of consumption, and the data they present differ quantitatively only slightly. It can be noted that unlike trends observed in oil reserves or production data, the world's oil consumption does not show a continuous increase since 1980, though compared to 1980, the consumption increased by over 30% in real terms. It is obvious from the figure that several blips are present, with hints of reaching a plateau sometimes being detected or even foreseen to be occurring in the future, with consumption in 2008 reaching approximately 85.5 million bpd according to the EIA – down from a peak of 85.9 million bpd in 2007. Reasons for this include energy diversification by using other energy resources (for example natural gas instead); energy conservation; bad economic conditions, such as the credit crunch of 2008-2009; and more recently the relatively high oil prices, which even having climbed down from their 2008 peak, values are still pretty high in actual terms.

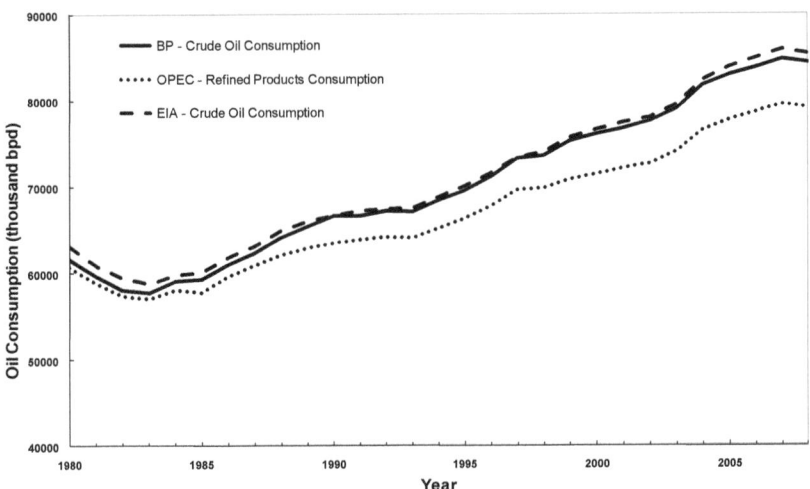

Figure 9.10: World oil consumption (1980-2008)

Source: EIA (http://www.eia.doe.gov/international); OPEC (Annual Statistical Bulletins 1999-2008); BP (Statistical Review of World Energy 2001-2009).

In the last 20 years China and India have been portrayed in the media as the main reasons for the increase in the world's oil consumption. While both countries almost tripled their consumption during that period, the combined consumption of two countries accounts for only 12.6% of the world's total oil consumption in 2008. In terms of oil consumption, the chief culprit was, and undoubtedly still is, the USA, which consumed 22.8% of the world's total in 2008. The oil consumption data for the three countries are presented in Figure 9.11. The facts presented speak for themselves.

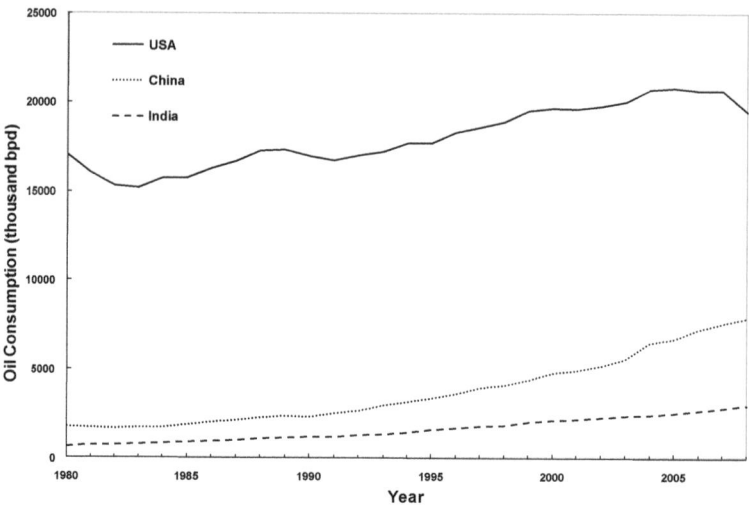

Figure 9.11: Oil consumption in the USA, China and India (1980-2008)
Source: EIA (http://www.eia.doe.gov/international).

With the increasing public awareness regarding climate change and the need to reduce oil consumption, countries are coming up with different, creative, reporting methods to mask their true oil consumption, attempting to look 'green' while trying to reduce their guilt and responsibility. As a result, measuring the oil consumption rate has become an increasingly contentious issue on the global stage. Due to differences in economic and social circumstances, countries tend to prefer using different measures, which portray them in a positive way, where they appear to be reducing their oil consumption rate, and thus are not accused of being the main offenders when policies to combat global climate change are discussed. Therefore, rich countries, such as the USA, prefer measuring the oil consumption per GDP, while China and India favour measuring consumption per capita. Neither method gives a true indication of actual oil consumption, as the numbers can be skewed to punish countries with small populations and large refining capacities, such as Singapore, and the Caribbean islands of Curacao and Aruba. The figures appear to indicate they consume a lot of oil even though, in reality most what they do is intermediate transformation, i.e. processing crude oil into refined products, which are then exported and are consumed by others.

Table 9.17 lists the top ten countries in the world, in terms of apparent oil consumption. These ten countries combined, account for approximately 58.4% of the world's total oil consumption. As expected these countries are amongst the richest in the world in terms of total GDP and have often large populations, which inevitably lead to higher oil consumption in real terms. As already discussed in Section 9.6, the list is very similar to top ten oil refining capacity, with nine countries placed in the top ten in both lists (see Table 9.15). Surprisingly though, one OPEC country is on the list. This country is Saudi Arabia, whose oil consumption is disproportionally large due to its policy of offering cheap oil to its residents, the reason for which will become clear as we go on.

Table 9.17: World oil consumption – top ten countries (2008)

Rank	Country	Oil Consumption	Share
		thousand bpd	%
1	USA	19497.96	22.81
2	China	7850.00	9.18
3	Japan	4784.85	5.60
4	India	2940.00	3.44
5	Russia	2900.00	3.39
6	Germany	2569.28	3.01
7	Brazil	2520.00	2.95
8	Saudi Arabia	2380.00	2.78
9	Canada	2259.20	2.64
10	South Korea	2174.91	2.54
	TOTAL	49876.21	58.36
	WORLD	85466.33	

Source: EIA (http://www.eia.doe.gov/international).
Note 1: Totals may not add up due to rounding.

The Arab world's apparent oil consumption and its share to the world's total since 1980 are shown in Figure 9.12. The figure reveals that the Arab world's share of the world's total consumption is relatively small, barely exceeding 7% and reaching 7.4% in 2008, with a consumption of just under 6.4 million bpd. Even if treated as on entity the Arab world will be placed third in the top ten list, with less than a third of the USA consumption. This reflects the relatively low state of development in the Arab countries, especially industrial development, which is growing at a slow pace in comparison to India and China who are both driving the growth of consumption. This also reflects the relatively small size of the oil refining sector and the stalled state in expanding the oil and gas production capacity generally. So if we are to assume that when the Arab world becomes industrialised, it will also become a large oil consumer, as is the case in highly industrialised developed countries, where the oil consumption is considerably higher due to more demand by industries. Do not be tricked though, this relatively low oil consumption does not reflect any green credentials, as many Arab countries have very weak environmental protection legislation, indeed many offer huge subsidies to maintain oil at low prices and thus encourage, rather than discourage oil consumption.

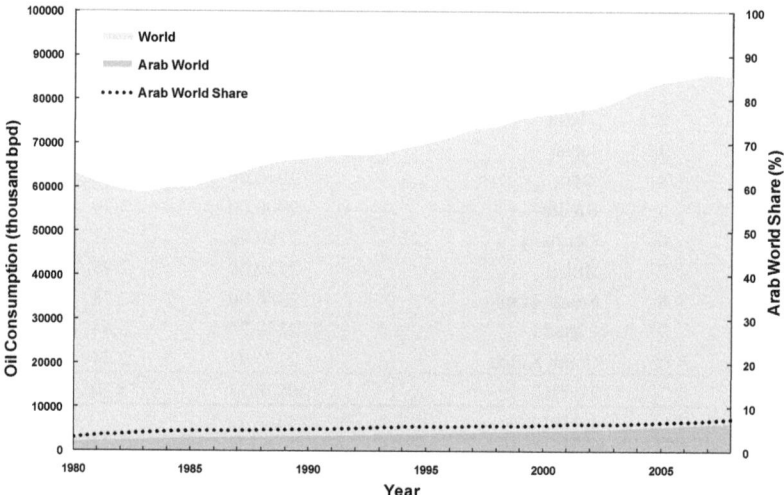

Figure 9.12: Arab world oil consumption and its share to the world's total (1980-2008)

Source: EIA (http://www.eia.doe.gov/international).

The apparent oil consumption for all Arab countries in 2008, including details of the latest consumption estimates per country, is reported in Table 9.18. A complete set of data from 1980 can be obtained from *"http://www.2050consulting.com/books"*. The table lists oil consumption in 22 Arab countries. In terms of oil consumption, we can classify the countries into four tiers: major, significant, medium and minor oil consuming countries. The Arab countries fall into this classification as follows:

- One country, Saudi Arabia, belongs to the first tier, with a total consumption exceeding 1 million bpd with a share of approximately 2.8% of the world's total oil consumption. It is the only Arab and OPEC country that is also ranked in the top ten list globally occupying the eighth place (see Table 9.18). The large Saudi oil consumption is due to the subsidies given by the Saudi authorities to their citizens in terms of cheap oil. This discourages them from reducing their fuel consumption and due to the disinterest in pursuing energy efficiency measures and the high standard of living, renders oil very affordable. Furthermore, the massive size of the country's oil production operations means that a substantial amount of oil is needed to provide the fuel for both oil production and refining.

- Seven countries belong to the second tier, with a total consumption exceeding 250 thousand but less than 1 million bpd each and a combined share of approximately 3.5% of the world's total oil consumption. These countries are in descending order in terms of total oil consumption: Egypt, Iraq, UAE, Kuwait, Algeria, Libya, and Syria. Five of these countries are OPEC members. These five countries, alongside Syria, are net oil exporters, who similar to Saudi Arabia, subsidise cheap oil prices locally thus encouraging high consumption. Egypt is the exception as already explained, since it has turned into net oil importer, however it still adopts similar policies to net oil exporting countries and subsidises cheap oil price. Therefore the consumption of these seven countries is higher, compared to net oil importing countries with similar population or development level. However, oil consumption correlation with the standard of living is evident, with Egypt, which has lower living standards, consuming lower quantities per capita than the other relatively richer countries.

Table 9.18: Oil consumption in the Arab countries (2008)

Country	Oil Consumption thousand bpd	Rank	Share %
Bahrain	38.00	17	0.04
Iraq	637.66	3	0.75
Jordan	108.27	12	0.13
Kuwait	324.98	5	0.38
Lebanon	92.25	13	0.11
Oman	80.95	16	0.09
Palestine	23.00	18	0.03
Qatar	129.15	11	0.15
Saudi Arabia	2380.00	1	2.78
Syria	256.19	8	0.30
UAE	463.01	4	0.54
Yemen	149.19	10	0.17
Algeria	298.51	6	0.35
Comoros	0.80	22	0.00
Djibouti	12.51	20	0.01
Egypt	696.89	2	0.82
Libya	273.32	7	0.32
Mauritania	20.95	19	0.02
Morocco	189.54	9	0.22
Somalia	5.11	21	0.01
Sudan	85.69	15	0.10
Tunisia	89.59	14	0.10
ARAB WORLD	6355.54		7.44

Source: EIA (http://www.eia.doe.gov/international).
Note 1: Totals may not add up due to rounding.

- Eight countries belong to the third tier, with a total consumption exceeding 50 thousand but less than 250 thousand bpd each, and a combined share of approximately 1.1% of the world's total oil consumption. These countries are in descending order in terms of total oil consumption: Morocco, Yemen, Qatar, Jordan, Lebanon, Tunisia, Sudan, and Oman. Four of these countries are net oil exporting countries, with Qatar being the only OPEC member. The consumption patterns of Qatar and Oman resemble other Arab oil exporting countries with high living standards, and thus are relatively high per capita. Total consumption in Yemen and Sudan resembles slow developing countries with low living standards elsewhere. Although they are currently net oil exporters, if their consumption levels were to increase to those of richer Arab countries, they would immediately turn into oil importers. The other four countries are net oil importers, whose consumption can easily be correlated to their level of industrial development and standard of living.

- Finally, six countries belong to the fourth tier, with a total consumption less than 50 thousand bpd each, and a combined share of just over 0.1% of the world's total oil consumption. These countries, in descending order in terms of total oil consumption are: Bahrain, Palestine, Mauritania, Somalia, Djibouti and Comoros. Only Bahrain is a net oil exporter and its consumption pattern resembles other Arab oil exporting countries with high living standards in the other tiers. The other five countries have very small economies and are relatively poor, which is reflected in their low oil consumption.

The Middle East Perspective

The Middle East region consumes approximately 7.8% of the world's total apparent oil consumption, 70.1% of which are in the Arab countries of the Middle East. Only one country in the region occupies a place in the top ten list of leading oil consuming countries – the eighth spot. This demonstrates that the Middle East as a region shows an oil consumption pattern similar to that of the Arab world.

Source: EIA, (http://www.eia.doe.gov/international), based on 2008 data.

OPEC Perspective

OPEC member states consume approximately 10.2% of the world's total apparent oil consumption, 51.8% of which are in the Arab member states. Only one member state occupies a place in the top ten list of leading oil consuming countries – the eighth spot. This illustrates that its small share in oil consumption contributes to its continuous dominant position in the oil markets as the major exporter, and reaffirms its strength in shaping global energy policies and economics at present and in the near future.

Source: EIA, (http://www.eia.doe.gov/international), based on 2008 data.

Table 9.19: Comparison of oil consumption data presentation in the Arab countries (2008)

Country	Oil Consumption barrel/thousand US$ GDP	Rank	Oil Consumption barrel/capita	Rank
Bahrain	0.65	17	19.1	5
Iraq	2.55	2	8.0	10
Jordan	1.86	5	6.2	11
Kuwait	0.75	14	44.1	2
Lebanon	1.15	10	8.4	9
Oman	0.49	21	8.6	8
Palestine	1.26	9	2.1	18
Qatar	0.46	22	56.6	1
Saudi Arabia	1.85	6	30.3	4
Syria	1.70	7	4.6	12
UAE	0.64	18	35.2	3
Yemen	2.02	4	2.3	17
Algeria	0.68	16	3.2	13
Comoros	0.55	19	0.4	21
Djibouti	4.65	1	8.8	7
Egypt	1.56	8	3.1	15
Libya	1.11	11	15.8	6
Mauritania	2.42	3	2.4	16
Morocco	0.78	13	2.0	19
Somalia	0.72	15	0.2	22
Sudan	0.54	20	0.8	20
Tunisia	0.80	12	3.1	14
ARAB WORLD	1.21		6.6	

Source: EIA (http://www.eia.doe.gov/international); CIA – The World Factbook
Note 1: Totals may not add up due to rounding.

As discussed above, inspecting the rate of oil consumption per GDP or per capita, leads to unreliable results, e.g. the increase in the oil price, since 2006, results in an apparent disproportional reduction in oil consumption per GDP. Similarly the explosion in the population of the Arab countries results in a disproportional reduction in oil consumption per capita.

This is shown in Figure 9.13 and Table 9.19, which clearly demonstrates how consumption data can be manipulated and presented as required! It is apparent how easily the rankings can become muddled.

Absolute consumption

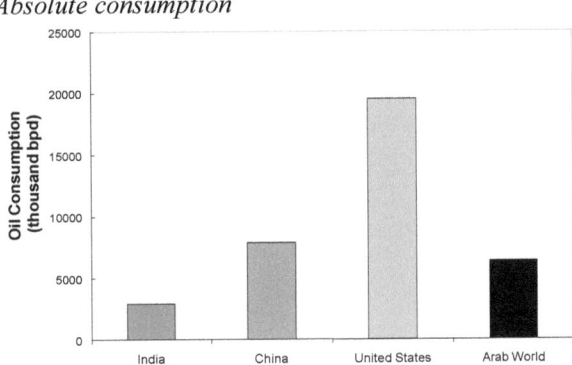

Consumption per GDP

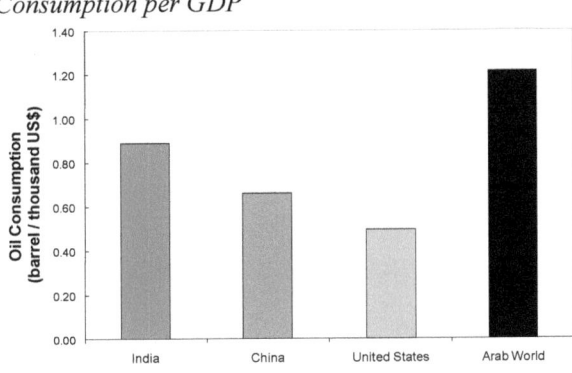

Consumption per Capita

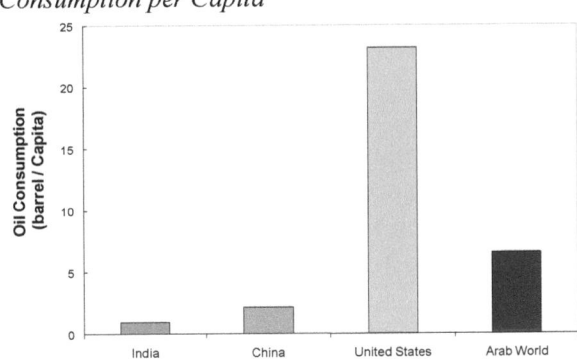

Figure 9.13: Comparison of oil consumption data presentation in the USA, China, India and the Arab world (2008)

Source: EIA (http://www.eia.doe.gov/international), CIA – The World Factbook

9.8 Oil Trade

As is often the norm with most commodities, we have seen

already that the main oil producing countries are not main oil consuming countries. Therefore following the basic rules of supply and demand, oil is traded on the open market and transported from producing regions to consuming regions both internationally or within national borders.

There is more trade in oil internationally, than in any other commodity. This is true whether oil trade is measured by volume, by value, or by the capacity needed to move it. To illustrate the enormity of this market we can calculate, using EIA data, that in the year 2008 almost 54% of total oil production was traded internationally.

Unlike most commodities, the price of oil is not simply determined by forces of supply and demand, but as mentioned again and again, several other circumstances including political situations, weather conditions, and environmental laws play significant roles. Therefore predicting oil price is an extremely complicated issue, and one which is outside the scope of this book.

Besides price, other factors affect oil trade. These include logistics, transportation and storage capacity. The cost of the latter two plays an important role in defining trade patterns and trade partners.

Although crude oil dominates the oil trade, accounting for almost two thirds of the total, according to latest data from both OPEC and the EIA, one must include both trade in crude oil and refined products, which account for the final third, to ascertain the total oil trade situation. The reason for this was explained earlier when we discussed refined petroleum production (Section 9.6) and oil consumption (Section 9.7), and stems from the fact that numerous countries in the world do not possess oil refining capabilities and thus do not import any crude oil. As oil consumers, they import all their oil requirements in the form of refined petroleum products. However other countries possess excess oil refining capacity and are large refined products exporters, even though they do not posses adequate oil resources or have much production capacity, but are actually crude oil importers.

It follows that, refined products trade is even more complicated as, besides price and capacity, each country has local specifications that refined products must adhere to which choices when these products are traded. In addition to this, local taxes play a primary role in determining the price of refined products, thus the overall cost of the product does not depend heavily on its imported cost. In fact, in Western countries, a significant proportion of the gasoline price, for example, is comprised of taxes levied on it. The tax can contribute to as much as 63% of the total price a consumer pays in Germany and the UK. Even in low taxing USA, tax constitutes 16% of the total price.[42]

As with reporting other oil data, reporting oil trade numbers is confusing as there are no agreed standards, also due to political and commercial factors, not all the data is disclosed. The data can be reported in many different ways. The following are just few examples: pure crude oil exports or imports; pure refined petroleum products exports or imports; net exports or imports for a country for either the crude oil or the refined petroleum products; an overall net value of exports or imports, combining the crude oil and refined petroleum products.

Since any oil exported by a country is imported by another, theoretically speaking, the data for world's exports and imports should be identical. However, this is not the case, with data discrepancies often occurring, and each data source using different methods to report trade data. To illustrate this, note that BP in its Statistical Review reconciles the values of exports and imports, by adding a term called 'unidentified' quantity, thus reports identical exports and imports data. However, OPEC only reports exports data. The EIA reports slight differences between exports and imports data and attributes this to various factors such as oil in bunkers and strategic oil reserves that are stored,

---

[42] Romain Davoust, Gasoline and Diesel Prices and Taxes in Industrialized Countries, European Governance and the Geopolitics of Energy, December 2008.

being released only if needed.

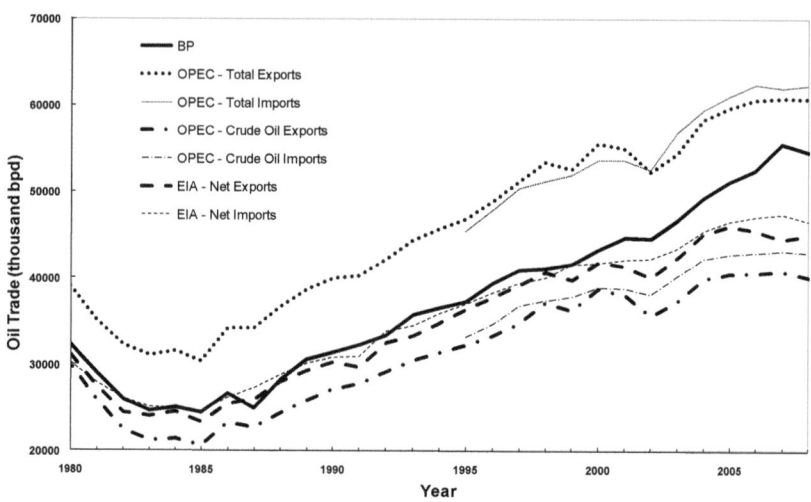

Figure 9.14: World oil trade (1980-2008)

Source: EIA (http://www.eia.doe.gov/international); OPEC (Annual Statistical Bulletins 1999-2008); BP (Statistical Review of World Energy 2001-2009).

As the world's oil consumption is continuing to increase and the production is not keeping up, there is significant pressure on the world's oil exports. Figure 9.14 shows the reported oil exports and imports from different sources since 1980. All sources agree that the net exports and imports have increased steadily by approximately 56% to 69% between 1980 and 2008. As with other data reported, the major sources estimates do not differ significantly, with the differences often being due to variations in compiling and reporting data, as explained earlier. According to the EIA's 2008 data, the net exports and imports of crude oil and refined products stood at 44.9 and 46.6 million bpd respectively.

Therefore, to simplify the presentation in the remainder of this section, the oil trade data are reported in terms of an overall net value, of either imports or exports, depending on the country. This will provide the most relevant data that gives a full picture of a country's trade. This net value is simply calculated as the difference between total oil production (as defined in Section 9.4) and the apparent oil consumption (as defined in Section 9.7), as such, it accounts for all traded crude oil and refined petroleum products. Comprehensive data that details global imports and exports of crude oil and refined products are not readily available and are often out of date. According EIA's latest comprehensive set of data from 2006, crude oil and refined petroleum products exports accounted for 67.8% and 32.2% of overall net oil exports respectively, while crude oil and refined petroleum products imports accounted for 68.9% and 31.1% of overall net oil imports respectively.

It has to be noted that in abstract terms net oil exports can on rare occasions exceed oil production quantities reported. Obviously this is physically impossible, but can mathematically happen due to refinery process gain. The fact that oil production, consumption and trade data are often reported in volumetric quantities rather than mass or energy quantities.

Table 9.20: World net oil exports – top ten countries (2008)

Rank	Country	Net Oil Trade (Export) thousand bpd	Share %
1	Saudi Arabia	8321.12	18.52
2	Russia	6889.76	15.33
3	UAE	2583.46	5.75
4	Kuwait	2416.41	5.38
5	Iran	2393.90	5.33
6	Norway	2245.78	5.00
7	Angola	1950.09	4.34
8	Algeria	1928.82	4.29
9	Nigeria	1882.95	4.19
10	Venezuela	1882.90	4.19
	TOTAL	32495.20	72.32
	WORLD	44932.35	

Source: Calculated based on data in Sections 9.4 and 9.7.
Note 1: Totals may not add up due to rounding.

Table 9.20 lists the top ten countries in the world, in terms of overall net oil exports. These ten countries combined, account for a massive 72.3% of the world's total net oil exports. This list is remarkably similar to the world leading countries list for oil reserves (see Table 9.1), with seven countries belonging to both lists and also shares six countries with the top oil producing countries list (see Table 9.13). This situation is not surprising as countries with large oil reserves are expected to be large producers, exporting their excess production. OPEC members dominate the list, with eight of them placed in it including the top net oil exporter, Saudi Arabia.

Table 9.21: World net oil imports – top ten countries (2008)

Rank	Country	Net Oil Trade (Import)	Share
		thousand bpd	%
1	USA	11790.72	25.31
2	Japan	4652.08	9.99
3	China	3876.87	8.32
4	Germany	2448.08	5.26
5	South Korea	2158.79	4.63
6	India	2056.49	4.42
7	France	1915.54	4.11
8	Spain	1533.92	3.29
9	Italy	1495.37	3.21
10	Taiwan	946.60	2.03
	TOTAL	32874.46	70.58
	WORLD	46578.01	

Source: Calculated based on data in Sections 9.4 and 9.7.
Note 1: Totals may not add up due to rounding.

Table 9.21 lists the top ten countries in the world, in terms of overall net oil imports. These ten countries combined, account for a massive 70.6% of the world's total net oil imports. This number is significant as it is close to the share of the trade of the top ten net oil exporters, demonstrating that over two thirds of the world oil trade is dominated by ten partners. The net oil importing countries list shares six countries with the world's leading oil consuming countries list (see Table 9.17). This is not surprising as countries with large oil consumption are expected to be large oil importers. This is further supported by the fact that none of the top net importers is on the list of leading countries, in terms of oil reserves, though two of them, the USA and China, are amongst the top ten oil producers. However since their consumption exceeds their production markedly, they are net oil importers.

Note that the numbers in the two tables above are for net oil exports and net oil imports. The net values are defined effectively as the difference between total exports and total imports, as some countries can be exporting and importing simultaneously either from different regions, or different products, which means that some countries can export from one region and import to another one within the same country, as is the case of the USA. Other countries, e.g. Iran, can export crude oil, but have no sufficient refining capacity so they import refined products. While some countries can do both! In this section the matter of concern is whether a country is a net exporter or importer; the exact details of the trade are not of interest in this book and specialised trade journals are better sources to obtain detailed data.

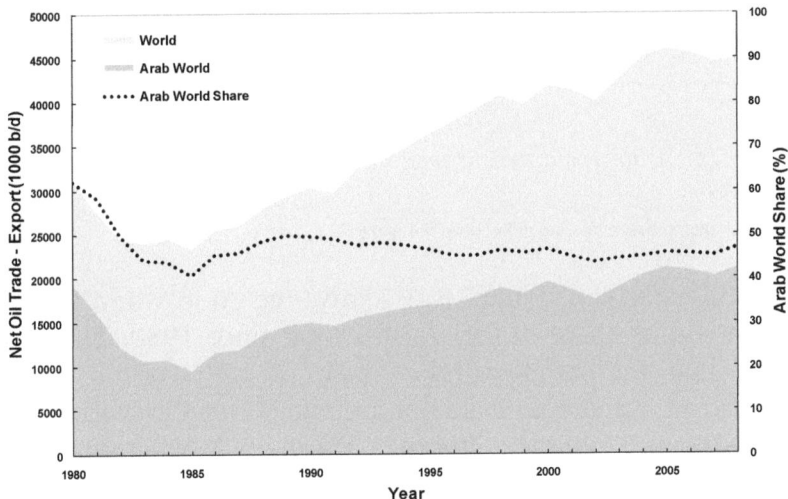

Figure 9.15: Arab world net oil exports and its share to the world's total (1980-2008)
Source: Calculated based on data in Sections 9.4 and 9.7.

The Arab world's oil exports and its share to the world's total, since 1980, are shown in Figure 9.15. The figure reveals that the Arab world's share of the world's total has declined from over 60% in 1980, to less than 40% in the mid 1980s, before recovering slowly and stabilising at around 45%, recording 46.9% in 2008, standing at just over 21 million bpd. If treated as one entity, the Arab world would be placed top in the leading net oil exporters list, with its exports amounting to over thrice more than the second placed country, Russia, thus confirming its position as the undisputed leader of oil the exporting world.

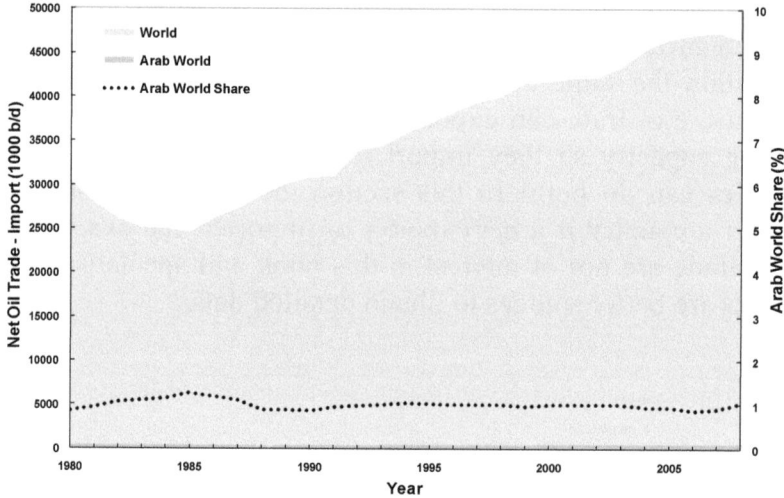

Figure 9.16: Arab world net oil imports and its share to the world's total (1980-2008)

Source: Calculated based on data in Sections 9.4 and 9.7.

In a similar fashion, Figure 9.16 shows the Arab world's net oil imports and its share of the world's total since 1980. The figure reveals that the Arab world's share of the world's total is insignificant, remaining stable at around 1% of the world's total net imports, recording slightly over 1% in 2008, where it stood at 490 thousand bpd in 2008. If treated as one entity, the Arab world is not a net importer and thus does not appear on any net oil importers lists.

Both figures above treat intra-Arab trade as international trade, since each Arab country reports its trade numbers separately. If the Arab world is taken as one entity then the intra trade has to be excluded. Then a recalculation of the numbers shows that the Arab world would be a net exporter, with a total global share of

46.4% in 2008, and obviously its net oil imports would become zero.

Table 9.22: Net oil exports in the Arab countries (2008)

Country	Net Oil Trade (Export) thousand bpd	Rank	Share %
Bahrain	10.53	12	0.02
Iraq	1731.25	5	3.85
Jordan	0.00		
Kuwait	2416.41	3	5.38
Lebanon	0.00		
Oman	680.05	8	1.51
Palestine	0.00		
Qatar	1061.47	7	2.36
Saudi Arabia	8321.12	1	18.52
Syria	188.19	10	0.42
UAE	2583.46	2	5.75
Yemen	150.93	11	0.34
Algeria	1928.82	4	4.29
Comoros	0.00		
Djibouti	0.00		
Egypt	0.00		
Libya	1580.74	6	3.52
Mauritania	0.00		
Morocco	0.00		
Somalia	0.00		
Sudan	437.56	9	0.97
Tunisia	0.00		
ARAB WORLD	21090.53		46.94

Source: Calculated based on data in Sections 9.4 and 9.7.
Note 1: Totals may not add up due to rounding.

The net oil exports for all Arab countries in 2008, including details of the latest estimates per country, are reported in Table 9.22. A complete set of data from 1980 can be obtained from "http://www.2050consulting.com/books". The above table shows that 12 Arab countries are net exporters. In terms of net oil exports, we can classify the countries into five tiers: major, significant, medium and minor oil exporting countries, with the fifth tier being net importing countries, whose net oil exporting equals zero. The Arab countries fall into this classification as follows:

Table 9.23: Net oil imports in the Arab countries (2008)

Country	Net Oil Trade (Import) thousand bpd	Rank	Share %
Bahrain	0.00		
Iraq	0.00		
Jordan	108.53	2	0.23
Kuwait	0.00		
Lebanon	92.25	3	0.20
Oman	0.00		
Palestine	0.00		
Qatar	0.00		
Saudi Arabia	0.00		
Syria	0.00		
UAE	0.00		
Yemen	0.00		
Algeria	0.00		
Comoros	0.80	9	0.00
Djibouti	12.51	5	0.03
Egypt	66.31	4	0.14
Libya	0.00		
Mauritania	8.12	6	0.02
Morocco	190.23	1	0.41
Somalia	5.11	7	0.01
Sudan	0.00		
Tunisia	2.66	8	0.01
ARAB WORLD	486.51		1.04

Source: Calculated based on data in Sections 9.4 and 9.7.
Note 1: Totals may not add up due to rounding.
Note 2: Palestine is net importer, numbers are not available.

- Seven countries belong to the first tier, with net exports exceeding 1 million bpd each, with a combined share of approximately 43.7% of the world's total net oil exports. These countries in descending order, in terms of total net oil exports are: Saudi Arabia, UAE, Kuwait, Algeria, Iraq, Libya, and Qatar. All seven countries are OPEC members and four of them are ranked in the top ten countries globally, with Saudi Arabia occupying the first place in the list (see Table 9.20). The large Saudi net oil exporting capacity is due to its massive production and, despite its large consumption, it still possesses huge production excess. It is worth noting that Saudi Arabia has even more exporting capacity, but it acts as the market stabiliser by maintaining significant spare production capacity, which if utilised will all be translated into additional exports. The vast majority of the exports of all the seven countries are crude oil, with refined petroleum products contributing only a fraction that varies from country to country. Refined products contribution is highest in Kuwait, where it accounts for almost a third of net exports and lowest in Iraq where it is virtually nil. As with oil production predictions, the improving political conditions in Iraq and Libya suggest that the oil exports from the two countries will be expected to soar.

- Two countries belong to the second tier, with net exports exceeding 250 thousand but less than 1 million bpd each, with a combined share of approximately 2.5% of the world's total net oil exports. These countries in descending order are Oman and Sudan. The two countries are at different stages of their oil industries and are resetting their exporting policies, with the former in decline, while the later still on the rise. However, with the potential secession of southern Sudan, Sudanese exports may decline until the new South Sudan establishes new export routes.

- Two countries belong to the third tier, with net exports exceeding 50 thousand but less than 250 thousand bpd each, with a combined share of approximately 0.8% of the world's total net oil exports. These countries in descending order are Syria and Yemen. Both countries are suffering from declines in net oil exports and are threatened to become net oil importers in the near future, following in the footsteps of Egypt and Tunisia.

- One country, Bahrain, belongs to the fourth tier, with total net exports less than 50 thousand bpd, and a share of less than 0.1% of the world's total net oil exports. Though the country still carries huge symbolism, being the first Arab exporter in the Gulf region, it is the only Arab country where refined petroleum products constitute the majority of its exports.

- Finally the remaining ten countries belong to the fifth tier with no net oil exports. These countries are Jordan, Lebanon, Palestine, Egypt, Mauritania, Morocco, Tunisia, Djibouti, Somalia and Comoros.

The net oil imports for all Arab countries in 2008, including details of the latest net oil imports estimates for each country, are reported in Table 9.23. A complete set of data from 1980 can be obtained from *"http://www.2050consulting.com/books"*. The table lists net oil imports in ten Arab countries. In terms of net oil imports, we can classify the countries into five tiers: major, significant, medium and minor oil importing countries; with the fifth tier being countries that are net exporters, whose net oil importing equals zero. The Arab countries fall into this classification as follows:

- None of the Arab countries belong to the first tier, with a total net oil imports exceeding 1 million bpd.
- None of the Arab countries belong to the second tier, with a total net oil imports exceeding 250 thousand but less than 1 million bpd.
- Four countries belong to the third tier, with total net oil imports exceeding 50 thousand but less than 250 thousand bpd each and a combined share of approximately 1.0% of the world's total net oil imports. These countries in descending order are Morocco, Jordan, Lebanon, and Egypt. The first three countries have minimal or no oil production, while Egypt has recently turned into oil importer (in 2007), despite significant oil production.
- Six countries belong to the fourth tier, with total net oil imports less than 50 thousand bpd each and a combined share of less than 0.1% of the world's total net oil imports. These countries in descending order are Palestine, Djibouti, Mauritania, Somalia, Tunisia, and Comoros. Note that although Tunisia and Mauritania are oil producers, their production does not satisfy all their consumption needs.
- Finally the remaining twelve countries belong to the fifth tier with no net oil imports. These countries are Bahrain,

Iraq, Kuwait, Oman, Qatar, Saudi Arabia, Syria, UAE, Yemen, Algeria, Libya and Sudan.

The Middle East Perspective

The Middle East's share of the world's net oil exports reaches 43.5% of the total, 87.8% of which are from the Arab countries of the region. The region's countries occupy only four positions in the top ten list of leading net oil exporting countries, including the top spot.

Predictably none of the Middle Eastern countries is in the list of leading net oil importers, and the region accounts for only 0.9% of the world's net oil imports, 46.6% of which are to Arab Middle Eastern countries.

The numbers illustrate the importance of the Middle East as the leading net exporting region in the world, with the situation very similar to the Arab world situation.

Source: Calculated based on data in Sections 9.4 and 9.7, *based on 2008 data.*

OPEC Perspective

OPEC member states dominate the world's net oil exports with a share of approximately 62.5% of the total, 69.9% of which are in its Arab member states. The organisation's members occupy eight positions in the top ten list of leading net oil exporting countries, including the top spot.

Predictably none of OPEC members is in the list of leading net oil importers, and the organisation accounts for only 0.2% of the world's oil imports, none of which is to its Arab member states. This surprising net overall oil import is due to Indonesia turning into net importer, and the potential of Iran following suit unless it manages to significantly increase its production.

The numbers speak for themselves and illustrate without a shadow of a doubt the might that OPEC has on oil exporting markets, which further exceeds its production power. This might is due to the spare production capacity which can easily be used as a whip to control oil markets.

Source: Calculated based on data in Sections 9.4 and 9.7, *based on 2008 data.*

9.8.1 Oil Transport

As already explained in Section 2.10, oil is typically transported via sea tankers or pipelines, with only a minuscule proportion using rail and road transport. The overwhelming majority of the world's oil trade, both crude oil and refined petroleum products, is transported by way of oil tankers, which accounted for over 80% of the world's total trade in 2008 (BGR, 2009).[43] This mode of transport is the cheapest means of transporting oil and refined petroleum products, compared to using pipelines. It has a distinct cost advantage that increases with the transport distance required; where it increases from twice over 1000 km to five times over 6000 km for the same amount of oil (BGR, 2009).[44] However we must bear in mind that, despite their benefits, they are vulnerable and can be targeted by vandalism, sabotage, terrorists or as legitimate targets in military disputes.

Oil pipelines are mainly used domestically to transport oil and refined petroleum products from production areas to export terminals, or from import terminals to consumption areas, with an extensive network of pipelines criss-crossing many countries. In terms of international trade, oil pipelines contributed a mere 20% of the total oil trade, mostly transporting crude oil via transit countries, to export terminals. The majority of oil pipelines connect Russia with the former Soviet Union republics, or the European Union countries and connect Canada and the USA. In the last few years several strategic oil pipelines have been constructed mainly to export the Central Asian oil either westwards via Georgia to Turkey or eastwards to China. Besides this, several pipelines are on the drawing board mainly to connect Russia to East Asian countries.

The situation in the Arab world mirrors the global image with almost all oil trade transported via oil tankers, with the Strait of Hormuz constituting a massive trade route bottleneck.

Nowadays, oil trade via pipelines is minimal in the Arab world. This is somehow sad when one takes into account that oil pipelines have a long history in the Arab world, with the currently disused Kirkuk-Haifa pipeline playing a significant role

---

[43] BGR, Energierohstoffe 2009 – Reserven, Ressourcen, Verfügbarkeit.
[44] Ibid.

in drawing the map of the Middle East and connecting Jordan to Iraq, to maintain the whole of the line route while under British control. However, in recent years, many international pipelines that connected the Arab countries ceased to operate, mainly due to the unrest in Iraq. This includes oil pipelines that connect Iraq to Syria, Lebanon, Saudi Arabia, and Kuwait. This also includes the historically famous but now defunct pipeline 'Tapline' that connected Saudi Arabia to Lebanon, via Jordan and Syria. At the moment, Iraq is the only Arab country that has oil exporting pipelines that flow to outside the Arab world, namely to Turkey, with plans for connections to Iran often mentioned. Between Arab countries, operational pipelines exist between Algeria and Tunisia, with planned pipelines being considered between Iraq and Jordan, and between Algeria and Morocco. Furthermore, the Suez-Mediterranean (Sumed) pipeline is currently operated by Egypt to facilitate oil trade between the Red Sea and the Mediterranean Sea, parallel to the Suez Canal, with proposed future plans to connect it to Saudi Arabia.

Chapter 10
NATURAL GAS – GLOBAL AND ARAB PERSPECTIVE

Similar to oil, natural gas is a finite commodity which, like other fossil fuels, will run out eventually. However, despite the loud 'peak oil' debate that is currently raging, as already discussed, very little similar debate is being heard regarding natural gas, with the talk of a 'peak natural gas' being quite muted. This is unlikely to last and the peak natural gas debate will be gaining prominence soon, where arguments analogous to the peak oil debate are expected to be vigorously contested by both sides.

This can be understood as, until the 1970s, natural gas was not considered to be a fuel source of value but was perceived as a nuisance and waste, and therefore was often flared. With the increasing usage of natural gas as an important and valuable fuel source this will inevitably change that old perception and will provoke the 'natural gas peak' debate sooner or later. It is unlikely that the debate will ever be as heated as peak oil debate since many energy analysts do not consider natural gas as a prominent fuel source but regard it instead as a stop gap measure that will bridge the energy usage between the decline in oil and the future prominence of nuclear energy or renewable energy sources.

Let us put that potential debate on hold for now. The aim of this chapter is to present a brief quantitative analysis of natural gas globally, then concentrate specifically on the Arab world. The analysis is presented in terms of four main aspects: reserves and resources; production; consumption and trade.

As in Chapter 9, this chapter is structured systematically with each section starting by presenting the global standpoint of a certain natural gas aspect; it is then followed by assessing and quantifying the contribution of the Arab world, finally it then discusses the importance of the Arab world in relation to that specific aspect. An overall analysis of the significance of the Arab world's natural gas, along with its positioning into the overall energy picture globally, as well as any implications of this positioning are discussed in Chapter 12.

In a nutshell, this chapter endeavours to answer three main questions:
- How much natural gas reserves and resources there are in the Arab world?
- How long will these natural gas reserves last?
- Will the Arab world establish itself as a major natural gas supplier in the world, attaining a similar position to its oil position?

After reading, I once again encourage the readers to draw their own conclusions –with the help of the hints thrown in along the way!

10.1 Conventional Natural Gas Reserves and Resources

10.1.1 Conventional Natural Gas Reserves

Unlike oil, there is an agreement between the majority of media outlets and major data sources that the world's natural gas conventional reserves are continuing to increase. This upward trend is evident in Figure 10.1, which shows the conventional natural gas reserves since 1980 reported by the EIA, BP, and OPEC, with all sources reporting a substantial increase in proved reserves in excess of 119-139% between 1980 and 2009. The data reported are for conventional natural gas reserves (as defined in Section 3.1). The data however is inconsistent regarding natural gas liquids (NGLs). They are excluded by the majority of data sources including the EIA, BP and OPEC, who treat them as part of the conventional oil reserves (as discussed in Section 9.1). NGLs are sometimes included as part of natural gas reserves by some data sources, especially in the mass media, who have the tendency of confusing data that often leads to inaccurate reporting. In this book NGLs are treated as part of conventional oil reserves as already discussed in Chapter 9.

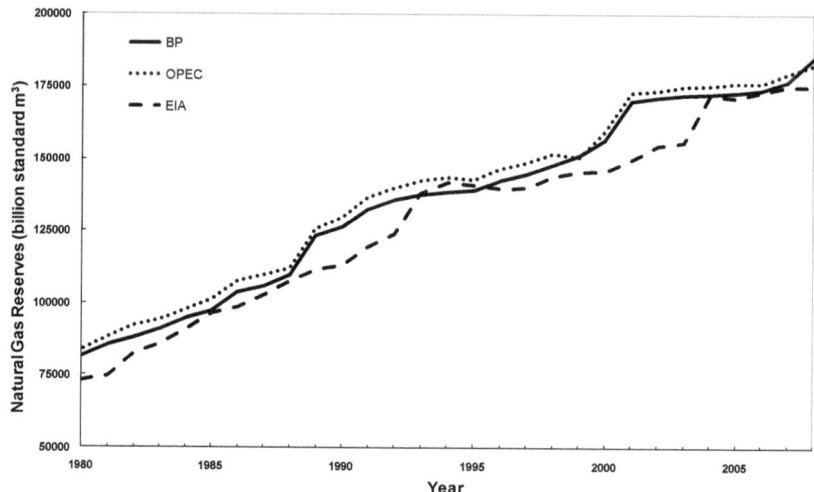

Figure 10.1: World conventional proved natural gas reserves (1980-2008)

Source: EIA (http://www.eia.doe.gov/international); OPEC (Annual Statistical Bulletins 1999-2008); BP (Statistical Review of World Energy 2001-2009).

The data in Figure 10.1 generally excludes unconventional gas, though this has started changing, with the USA including substantial amounts of its unconventional gas reserves into its reported conventional reserves, as published by the EIA for the USA for 2009. Thus these 2009 numbers include a proportion of unconventional gas, since the USA has re-evaluated substantially its reserves at the end of 2008 to include some unconventional plays of tight gas, shale gas, and coalbed methane, reclassifying them as conventional reserves. The same exercise is underway in Canada and Australia, and it is expected that few other countries may follow suit promptly. All the major data sources estimates are in broad agreement and differ insignificantly, with small unavoidable differences being due to different practices in data reporting. According to the EIA, the world's total proved natural gas reserves in 2009 stand at 177.1 trillion standard m^3.

Table 10.1 lists the conventional natural gas reserves in the leading ten countries in the world, which shows that they account for almost 78.4% of the global proved reserves. It is interesting to note that eight of these countries are OPEC members, which again highlights the fact that the organisation can be as dominant in the natural gas markets as in oil markets if a choice to implement a common policy on natural gas is to be put in place. However, to date this strategic choice has not been considered by the organisation. Interestingly, six countries in this top ten list also appear in the top ten list of leading countries in terms of conventional oil reserves. This is unsurprising since oil and natural gas deposits are often located in the same areas. This leads to both industries being very much integrated and inseparable with many fields containing simultaneous deposits of oil, NGLs and natural gas, but in different proportions.

Table 10.1: World conventional proved natural gas reserves – top ten countries (2009)

Rank	Country	Natural Gas Reserves billion standard m^3	Share %
1	Russia	47571.85	26.86
2	Iran	28078.72	15.85
3	Qatar	25256.83	14.26
4	Saudi Arabia	7318.99	4.13
5	USA	6731.59	3.80
6	UAE	6071.07	3.43
7	Nigeria	5214.78	2.94
8	Venezuela	4839.87	2.73
9	Algeria	4502.34	2.54
10	Iraq	3169.76	1.79
	TOTAL	138755.80	78.35
	WORLD	177102.19	

Source: EIA (http://www.eia.doe.gov/international).
Note 1: Totals may not add up due rounding.

Surprisingly, if the proposed GECF organisation becomes a gas OPEC, its members will only occupy seven places in the top ten list, i.e. one less than OPEC, with American influenced Saudi Arabia and Iraq and the USA, not being members. However, quantitatively, the organisation will be as powerful as OPEC, with over 74.4% of the conventional natural gas reserves, as the top three countries are the main members.

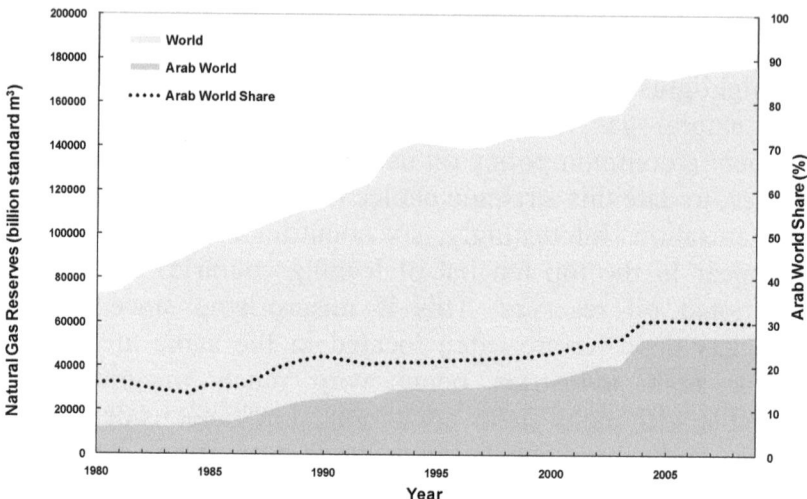

Figure 10.2: Arab world conventional proved Natural gas and its share to the world's total (1980-2009)

Source: EIA (http://www.eia.doe.gov/international).

Figure 10.2 shows the Arab world's overall conventional natural gas reserves and their share to the world's total since 1980. The data shows a continuous increase in the Arab world's proved natural gas reserves, which in 2009 were estimated at approximately 53.2 trillion standard m^3. So if taken as one entity, the Arab world will be placed at the top of the top ten list in terms of natural gas reserves, with slightly more reserves than the current top placed country, Russia. This data reveals the significance of the Arab world in terms of conventional natural gas reserves, as its share of the world's total stands at approximately 30% in 2009. This share is far below that of the Arab share in terms of conventional oil reserves showing that although significant, the Arab world is not dominant in terms of natural gas reserves, as the reserves are more evenly distributed globally.

Similarly to oil, natural gas reserves are only one part of the jigsaw that forms the natural gas market, with other parts including natural gas (actual and spare) production capacity, natural gas consumption, export and import facilities (availability and location), global economic conditions and political conditions. In the natural gas market, the Arab world's weight is larger than its reserves share, mainly because Iran is effectively cut off from the lucrative natural gas markets (especially LNG) for political reasons, endless economic sanctions and technology usage restrictions. This allows the Arab world a bigger influence than first anticipated. Read on, as the complete picture will become clearer in the next few sections.

Table 10.2: Conventional proved natural gas reserves in the Arab countries (2009)

Country	Natural Gas Reserves billion standard m^3	Rank	Share %
Bahrain	92.03	12	0.05
Iraq	3169.76	5	1.79
Jordan	6.03	17	0.00
Kuwait	1794.14	6	1.01
Lebanon	0.00		
Oman	849.50	9	0.48
Palestine	40.00	15	0.02
Qatar	25256.83	1	14.26
Saudi Arabia	7318.99	2	4.13
Syria	240.69	11	0.14
UAE	6071.07	3	3.43
Yemen	478.55	10	0.27
Algeria	4502.34	4	2.54
Comoros	0.00		
Djibouti	0.00		
Egypt	1656.52	7	0.94
Libya	1539.86	8	0.87
Mauritania	28.32	16	0.02
Morocco	1.50	19	0.00
Somalia	5.66	18	0.00
Sudan	84.95	13	0.05
Tunisia	65.13	14	0.04
ARAB WORLD	53201.86		30.04

Source: EIA (http://www.eia.doe.gov/international).
Note 1: Totals may not add up due rounding.

The Arab world's overall conventional natural gas reserves in 2009, including details of the latest reserves estimates for all Arab countries, is reported in Table 10.2. A complete set of data from 1980 can be found at *"http://www.2050consulting.com/books"*. The table above shows that proved natural gas reserves exist in

19 Arab countries. We can classify countries into five tiers in terms of total proved conventional natural gas reserves: countries with major, significant, medium, minor or no reserves. The Arab countries are classified as follows:

- One country, Qatar, belongs to the first tier, with proved natural gas reserves exceeding 10 trillion standard m^3, leading to a massive share of approximately 14.3% of the world's proved natural gas reserves, and placing it third in the top ten list (see Table 10.1). Qatar is a member of OPEC, and is home to the world's largest natural gas field, the North field, which is effectively the southern part of the combined larger 'North – South Pars' field, with the latter section reported often as the Iranian South Pars field. The reported Qatari proved natural gas reserves continue to increase despite continuous natural gas production. Unlike oil, no one seems to question this apparent reserve growth, and the reserves numbers are accepted universally.
- Seven countries belong to the second tier, with proved natural gas reserves exceeding 1 trillion standard m^3 but less than 10 trillion standard m^3 each, and a combined share of approximately 14.7% of the world's proved natural gas reserves. These countries in descending order in terms of proved oil reserves are: Saudi Arabia, UAE, Algeria, Iraq, Kuwait, Egypt, and Libya, with the first four also ranked in the top ten list globally (see Table 10.1). All these countries, with the exception of Egypt, are also members of OPEC. However, due to its relatively small oil reserves, natural gas is playing an increasingly vital role in shaping the Egyptian economy, positioning the country as a major player in energy markets. Here again, the reported proved natural gas reserves numbers of all seven countries continue to increase, despite continuous natural gas production. This situation is similar to that of Qatar, and here, still no doubts are being raised regarding the continuous reserve increase. It is vital to note that the share of all the seven countries combined just exceeds slightly that of Qatar, which explains why Qatar is punching above its weight and acts as a mini-

super power in the world, relying on the fact that it is a natural gas super power!
- Two countries belong to the third tier, with proved natural gas reserves exceeding 250 billion standard m^3 but less than 1 trillion standard m^3 each, and a combined share of approximately 0.8% of the world's proved natural gas reserves. These countries in descending order in terms of their reserves are Oman and Yemen, neither of which are members of OPEC. Natural gas contributes substantially to the two countries GDP and, with relatively small oil reserves, natural gas has more eminence as a major income source than in other Arab countries with larger reserves.
- Nine countries belong to the fourth tier, with minor proved natural gas reserves less than 250 billion standard m^3 each, and a combined share of just over 0.3% of the world's proved natural gas reserves. These countries in descending order in terms of their reserves are: Syria, Bahrain, Sudan, Tunisia, Palestine, Mauritania, Jordan, Somalia, and Morocco (Palestine data from BGR). As with the third tier countries, natural gas has more importance in these nine countries, especially for the latter four, where it is almost the sole fossil fuel source that is commercially feasible. For all these countries, since oil reserves are relatively on the small size, natural gas deposits are becoming valuable assets. Therefore active exploration of natural gas is underway even in Palestine, despite the ongoing Palestinian-Israeli conflict and in Somalia, though only offshore, despite its ongoing war and the increasing pirate activity off its coasts.
- Finally three countries have no proved natural gas reserves. These countries are Lebanon, Djibouti, and Comoros. Currently no exploration activities are being pursued in these three countries.

The Middle East Perspective

The Middle East region is home to approximately 41.4% of the world's total conventional proved natural gas reserves, 61.7% of which are in its Arab countries. The region's countries occupy six places in the top ten list of leading conventional natural gas reserves countries despite missing on top spot. This illustrates the significance of the region as a major natural gas source that can rival the dominance of Russia.

Source: EIA, (http://www.eia.doe.gov/international), based on 2009 data.

OPEC Perspective

OPEC member states are home to approximately 51.4% of the world's total conventional proved natural gas reserves, 54.5% of which are in its Arab member states. The organisation's members occupy eight places in the top ten list of conventional natural gas reserves countries despite missing on the top spot. This illustrates the potential dominant position the organisation can enjoy in the natural gas markets if it decides to coordinate policies, and reaffirms the power it can extract in shaping global energy policies and economics.

Source: EIA, (http://www.eia.doe.gov/international), based on 2009 data.

10.1.2 Conventional Natural Gas Resources

Unlike natural gas reserves data, which are published by many sources, natural gas resources data are not readily available. This is similar to the oil situation as discussed in Section 9.1.2. Therefore these data need to be estimated using the latest available set of data. To achieve this, an analogous method, resembling the method implemented to attain an estimate of conventional oil resources, is carried out here to estimate conventional natural gas resources on a global level. Obviously, this is a conservative approximation that gives an estimate of the resources.

The methodology used to estimate natural gas resources is intuitive, and uses the available data from different sources to calculate resources quantities. The original natural gas-in-place (OGIP) needs first to be evaluated, as using its value, other quantities can be evaluated based on the recovery factors selected. This calculation assumes uniform recovery factors for all global conventional natural gas resources, which is not true as each field has unique recovery factor; however its usage is justified to achieve approximate values in the absence of detailed data for every single natural gas field in the world.

As already discussed, the BGR[1] publishes what it labels 'resources estimates' on a country-by-country basis. These resources are in fact remaining recoverable resources, excluding proved reserves and cumulative production. Thus the sum of these resources plus the proved reserves and the cumulative production leads to a total recoverable resource estimate, which is considered to be the ultimate natural gas recoverable resources (UGRR). Using an ultimate recovery factor of 90%[2,3,4] the OGIP based on the BGR data can be calculated by dividing the UGRR over the ultimate recovery factor.

Other data sources provide only proved reserves and production data, thus using the sum of the reserves and the cumulative production data, the OGIP can be calculated using recovery factors, ranging between 50%[5] and 75%[6], by dividing the sum of the reserves and the cumulative production over an average recovery factor, which in this instance is the harmonic average of the recovery factor range above calculated to be 60%. The proved reserves data are obtained from the latest EIA data, while the cumulative production data are obtained from the BGR till 2007, and corrected to 2008 using the EIA latest production data.

The two calculated OGIP quantities are compared, while the recovery factors for all countries based on BGR reserves and resources data are back calculated. So, if the calculated recovery factor for a country is found to be above the average value of 60%, it suggests that the BGR resources value, and thus the OGIP for that country is considered underestimated, and the

---

[1] BGR, Energierohstoffe 2009 – Reserven, Ressourcen, Verfügbarkeit.
[2] Gas Strategies (http://www.gasstrategies.com/industry-glossary).
[3] GEO-SEQ Project Team, GEO-SEQ Best Practices Manual, Geologic Carbon Dioxide Sequestration: Site Evaluation to Implementation, , Lawrence Berkeley National Laboratory, 2004.
[4] SPE, Low Cost Methods for Improved Oil and Gas Recovery, SPE Distinguished Lecturer Series, (http://queensland.spe.org/images/Queensland/articles/37/SPEDL(2008-09).pdf).
[5] Uwe Remme, Markus Blesl & Ulrich Fahl, Global Resources and energy trade: An overview for coal, natural gas, oil and uranium, IER, Universität Stuttgart, 2007.
[6] Jean Laherrère, Distribution and Evolution of Recovery Factor, presented at 'Oil Reserves Conference' in Paris, November 11, 1997.

OGIP based on the EIA reserves data is used instead for that country. After reconciling the data for all countries a global OGIP value is obtained as shown in Table 10.3, and is then used to back calculate all other quantities.

Table 10.3: World and Arab world conventional natural gas in place and natural gas resources (2009)

Quantity (billion Standard m^3)	World	Arab World	Share
	Estimated	Estimated	%
Original Natural Gas in Place (OGIP)	590057	111314	18.9
Remaining Natural Gas in Place (RGIP)	500144	105135	21.0
Ultimate Natural Gas Recoverable Resources (UGRR)	531051	100182	18.9
Remaining Natural Gas Recoverable Resources (RGRR)	441138	94004	21.3
Remaining Natural Gas Resources Excluding Proved Reserves	264036	40842	15.5
	Reported	Reported	
Remaining Natural Gas Proved Reserves	177102	53162	30.0
Cumulative Production	89912	6179	6.9

Source: EIA (http://www.eia.doe.gov/international); BGR (Energierohstoffe 2009 – Reserven, Ressourcen, Verfügbarkeit).

One point to stress though is that, the estimated OGIP numbers are calculated not only considering proved reserves numbers, which come from producing and future natural gas fields, but also take into account the natural gas remaining in place in natural gas fields that ceased production, but still contain a significant amount of natural gas that can be produced in the future if conditions change, such as higher price or advances in technology. The inclusion of the latter is important to accurate natural gas resources estimation, but is unfortunately often overlooked by many estimators. Furthermore, even though ideally, all the above values must include only conventional natural gas, implicitly they include few quantities of unconventional natural gas that were not reported separately either as reserves or as production. This is due to both lax reporting standards and that the distinction between conventional and unconventional gas is not standardised and well defined.

Thus, based on the reconciled OGIP of 590 trillion standard m^3, the UGRR is calculated by multiplying the OGIP estimates by 90% ultimate recovery factor. The above OGIP estimates are in line with the higher conventional natural gas estimates of 603 trillion standard m^3 provided by the USGS, but significantly higher than USGS lower estimate[7] of 304 trillion standard m^3 although it is noticeable that the lower estimate of the USGS is very conservative. However, they are considerably lower than the estimated numbers given by the NPC[8] of 1400 trillion standard m^3, though these numbers include unconventional gas as well.

By deducting the cumulative production, the remaining natural gas-in-place (RGIP) can be calculated as 500 trillion standard m^3, then performing a simple calculation gives global UGRR estimate of 531 trillion standard m^3. By subtracting the cumulative production, the remaining natural gas recoverable resources (RGRR) can be calculated, and is shown to be 441 trillion standard m^3. The above remaining resources estimations are in line with the conventional natural gas estimates of 468 trillion standard m^3 provided by the IEA.[9]

The same methodology described above is used to estimate the OGIP, RGIP, UGRR and RGRR in the Arab world. The results are shown in Tables 10.3 and 10.4. As can be seen, the Arab world's share of RGRR decreases to approximately 21.3% which, even though lower than proved reserves share of 30.0%, remains very high and cements the position of the Arab world as a significant force in terms of conventional natural gas resources. The reason for this decline is that more natural gas to date has been produced from non-Arab natural gas fields, which even though considered exhausted now have the potential to be reproducing if technical and economic conditions change. As expected the ranking of the Arab countries, in terms of conventional natural gas resources, is identical to their ranking in terms of conventional natural gas reserves, though obviously the

---

[7] USGS, Digital Data Series 60, 2000.
[8] The National Petroleum Council (NPC), Working Document of the NPC Global Oil & Gas Study, Topical Paper #29, Unconventional Gas, 2007.
[9] IEA, World Energy Outlook 2009.

natural gas quantities differ.

Table 10.4: Conventional natural gas in place and natural gas resources in the Arab countries (2009)

Country	Original Natural Gas in Place (OGIP)	Remaining Natural Gas in Place (RGIP)	Ultimate Natural Gas Recoverable Resources (UGRR)	Remaining Natural Gas Recoverable Resources (RGRR)	Remaining Natural Gas Resources Excluding Proved Reserves	Remaining Natural Gas Proved Reserves	Cumulative Production
	billion standard m^3						
Bahrain	524.33	324.79	471.90	272.36	180.33	92.03	199.54
Iraq	8074.89	7975.61	7267.40	7168.12	3998.36	3169.76	99.28
Jordan	122.11	117.96	109.90	105.75	99.72	6.03	4.15
Kuwait	3373.83	3120.83	3036.45	2783.45	989.31	1794.14	253.00
Lebanon	0.00	0.00	0.00	0.00	0.00	0.00	0.00
Oman	2178.56	1943.86	1960.70	1726.00	876.50	849.50	234.70
Palestine	77.78	77.78	70.00	70.00	70.00	0.00	0.00
Qatar	43537.83	42938.15	39184.05	38584.37	13327.54	25256.83	599.68
Saudi Arabia	21490.11	20256.57	19341.10	18107.56	10788.57	7318.99	1233.54
Syria	639.33	541.89	575.40	477.96	237.27	240.69	97.44
UAE	11514.50	10627.56	10363.05	9476.11	3405.04	6071.07	886.94
Yemen	1100.22	1098.02	990.20	988.00	509.45	478.55	2.20
Algeria	10419.00	8596.09	9377.10	7554.19	3051.86	4502.34	1822.91
Comoros	0.00	0.00	0.00	0.00	0.00	0.00	0.00
Djibouti	0.00	0.00	0.00	0.00	0.00	0.00	0.00
Egypt	3858.67	3385.57	3472.80	2999.70	1343.18	1656.52	473.10
Libya	2857.00	2621.90	2571.30	2336.20	796.34	1539.86	235.10
Mauritania	46.67	46.67	42.00	42.00	13.68	28.32	0.00
Morocco	12.44	10.18	11.20	8.94	7.44	1.50	2.26
Somalia	451.11	451.11	406.00	406.00	400.34	5.66	0.00
Sudan	594.44	594.44	535.00	535.00	450.05	84.95	0.00
Tunisia	440.89	406.12	396.80	362.03	296.90	65.13	34.77
ARAB WORLD	111313.72	105135.12	100182.35	94003.74	40841.89	53161.86	6178.61

Source: EIA (http://www.eia.doe.gov/international), BGR (Energierohstoffe 2009 – Reserven, Ressourcen, Verfügbarkeit).

10.1.3 Reliability of Conventional Natural Gas Reserves and Resources Data

As with oil, further detailed scrutiny is required concerning natural gas conventional reserves and resources estimated quantities, as using reliable data is fundamental in obtaining correct figures. This is essential as it determines current and future energy policies as well as the impact on other natural gas issues such as production and trade.

Unlike conventional oil reserves data, the reliability of conventional natural gas reserves data has been questioned considerably less by experts, including peak oil theory advocates. They alongside mass media and data sources, often accept the published data without much fuss. This is strange when we consider that many of the arguments put forward by peak oil theory advocates to doubt conventional oil data (such as that most data reported has no independent auditing procedures, and that the data seem to stay constant for years before having a sudden jump) are also true in natural gas case. Still, no one seems to suggest that the natural gas reserves data are inflated or exaggerated. There are a few plausible explanations offered for this:

1. Since the majority of natural gas reserves estimations are recent, they are based on more reliable and up-to-date exploration data, thus experts have more confidence in them and makes questioning their reliability more difficult.
2. There is no talk of an imminent natural gas peak and no expert is predicting any shortage in production soon, therefore alarmists are yet not focusing on the subject.
3. OPEC does not impose natural gas production quotas, therefore the arguments put forward, using the quota excuse to raise suspicions of oil reserves, cannot be used to undermine natural gas data.
4. More natural gas deposits are located in areas where multinational oil and gas companies already operate. It is widely agreed that their numbers are more reliable as they are governed by international financial standards and are responsible to shareholders rather than governments.

It is important to note that several of the factors applied to questioning the reliability of conventional oil reserves and resources data *are* also applicable in the case of natural gas. Amongst these factors are:

1. Prior to the nationalisation of the oil and gas industry by producing countries, the bulk of the conventional natural

gas reserves' rights were owned by multinational oil and gas companies. Until the 1970s, natural gas reserves were not considered important, thus minimal royalty fees or taxes were paid on these reserves to the local governments and the fees were proportional to the size of the declared reserves. Therefore it was in the interest of these multinational companies to underestimate the size of their reserves to minimise their fees and tax bills, therefore we notice a great jump in reserves after nationalisation.

2. The multinational oil and gas companies deliberately withheld data regarding part of the discovered natural gas reserves so that they can choose the time of announcing to the financial markets the 'good news' of additional reserves discoveries when needed. This behaviour became more important if these companies were facing difficulties or when market conditions were tough, where these added reserves could enhance significantly their financial positions. So, although the share value for these publicly listed oil and gas companies is always proportional to the amount of the proved reserves, a declaration of a massive new reserve find can lift the share price significantly. If the companies can then prove to the financial markets a trend of continuous reserve growth rather than a series of one-off jumps, the share price can be maintained at a higher level.

3. As explained in Chapter 6, it is easy to prove that the natural gas reserves estimates can vary higher or lower without the modification of any technical or physical parameters, but by adjusting economic factors, as this will mean that the definition of what is 'reasonable reserves' has changed.

4. Until 1970s, countries had no incentives to invest in natural gas exploration activities. Natural gas was considered a nuisance rather than a valuable asset and therefore was always flared (i.e. burnt off). Countries found it difficult to justify investing any capital in looking for additional natural gas; they did not need to produce as at the time there was no market for it. So the financial

strains, they were under, restricted expansion in the natural gas industry for years to come.

It is universally accepted that, unlike oil fields, major giant natural gas fields, such as 'North - South Pars' field, have been discovered and that these fields have added huge amounts of reserves that compensated for all consumption and still added to the reserves. Contrary to the oil situation, it is widely accepted that there is still a huge potential for production from the current giant gas fields. Thus unlike oil, there is no argument regarding the lack of discovery of new major fields, when discussing natural gas reserves. It is worth re-iterating here though (as already discussed in Chapter 6) that enhanced natural gas recovery methods (EGR), increased recovery factors, better feasibility and improved technology are all very important factors in increasing the natural reserves estimates.

An interesting point is that, unlike the case of oil, where many peak oil theory advocates (especially ASPO affiliates) have the tendency to discredit almost all data sources and use only oil reserves data that they had back-calculated themselves (which contradicts all major data sources), this practice is absent when natural gas data are discussed. One explanation may be that peak oil theory advocates have no access to the data on natural gas reserves from ancient Petroconsultants database (see Section 8.1.7), unlike oil data, which they had access to, while they were working in that company.

Nevertheless, it is imperative to consider some limitations of the conventional natural gas data, which are in line with criticisms of the conventional oil reserves data. The most important criticism is that many data sources, including the EIA, BP and OPEC use the data of the Oil & Gas Journal which is based on survey responses and updates released by the natural gas countries without any independent verification. In addition, some odd and unexplainable discrepancies in conventional natural gas reserves data persist, which are often attributed to inaccurate, or out-of-date, data reporting. For example, all the major data sources report that South Africa has no natural gas reserves, whereas it already produces natural gas, as published by the same data sources who credit it with zero reserves!

10.2 Unconventional Natural Gas Reserves and Resources

We have already encountered the difficulties associated with estimating unconventional oil reserves and resources in Section 9.2; unfortunately, estimating the reserves and resources of unconventional natural gas are even more difficult than oil. It shares various problems with oil including it is very contentious subject, far from straight forward with various definitions used to define what is unconventional gas, and open to misinterpretations by mass media and conflicting points of view by organisations and politicians. On top of all these problems a further complication exists, which is some of unconventional natural gas categories are not even universally accepted as part of the unconventional natural gas endowment. Even when data regarding a category are reported, they are often inconsistent and vary considerably.

The story of unconventional gas is a carbon copy to that of oil. So the definition of what is considered to be unconventional natural gas keeps changing where, with improving technologies and long term natural gas price increase, a considerable proportion of what has traditionally been deemed unconventional natural gas, is making a gradual shift into being reclassified as conventional. This reclassification is unavoidable and has already been proven, over time, based on historic reclassification of both deepwater and Arctic natural gas, it is commonly accepted now that both of the above natural gas categories are included within the conventional natural gas resources and reserves.

In this book, five categories of unconventional natural gas are discussed: tight gas, shale gas, coalbed methane, methane hydrates and 'geopressurised and hydropressurised' gas. These five categories have already been introduced in Sections 3.2 to 3.6 but it is worth reemphasising here, that what set these categories aside from conventional natural gas are not physical or chemical characteristics, but rather the added technical difficulties and cost involved in extracting the gas from them, which lead to a lower economic feasibility. The resolution of some of these difficulties is paving the way for these deposits to be partially reclassified as conventional. Once this is acknowledged universally by organisations and governments, some of these deposits will finally be recognised as conventional by all data sources.

Note that in addition to the above mentioned categories, other forms of unconventional natural gas categories exist. However the gas can be much dispersed, found in very low concentrations or found in very scattered deposits, which render its usage as a fuel source virtually impossible or very remote for years to come. As a result, these natural gas categories are excluded from fossil fuels resource estimations, amongst these categories are peat gas (also known as swamp or marsh gas) and ultra deep gas.

So, despite being located and identified for years, until recently, most major data sources did not report data on the above unconventional natural gas reserves. They did not consider them to constitute a part of the proved reserves, even though they accounted for over 40% of the USA production in 2008 and are projected to account for over 50% in the next decade.[10] As mentioned in Section 10.1, this practice started to change in the last few years, when the USA government started including part of its unconventional natural gas reserves in its official data by reclassifying part of its tight gas, shale gas and coalbed methane resources into conventional reserves. This practice was quietly endorsed by all major data sources including Oil and Gas Journal, BP and OPEC, who all accepted the new numbers without explicitly indicating that some of the estimates relate to unconventional gas, and then they updated their reserves data. This is in stark contrast to the practices implemented regarding unconventional oil reserves (see Section 9.2.1). To date, unconventional natural gas data remain scarce and notoriously inconsistent with estimated values varying sometimes by several orders of magnitude. Generally, with the exception of partial data from the USA, Canada and Australia, no other country publishes

---

[10] Don Stowers, Unconventional gas outlook to 2020, Oil and Gas Financial Journal, Aug 1, 2009.

its data in the public domain. So, unlike oil (both conventional and unconventional) and conventional natural gas, there are no detailed data available on a country-by-country basis, though data on a regional level have been published by few scientific papers or commercial consultancies or have been reported by the BGR.

In this book, we regard a proportion of unconventional natural gas resources as potential reserves, but rather than amalgamating them with conventional reserves, they are labelled as 'unconventional reserves'. These reserves are estimated based on conservative recovery factors values and the estimation methodology can be applied generally to all unconventional natural gas resources with a fraction of them reported as unconventional reserves; however here it is only applied to tight natural gas, shale gas and coalbed methane, as they are the categories that have been proved to be recoverable under current economic conditions using available technology. Note however that we have to acknowledge that a proportion of unconventional natural gas reserves are inevitably being implicitly reported as part of the conventional natural gas reserves since there is no way to distinguish them if they are not reported explicitly as a separate category.

10.2.1 Tight Natural Gas

Plays (i.e. reservoirs) of tight natural gas are plentiful and have been identified all over the world. However, even though significant production from these deposits has been underway in the USA since late 1970s, the subject has only recently made media headlines when the USA re-evaluated its reserves and added over 35% into its proved reserves estimates in 2009. This action caught the media by surprise, and instead of writing about future natural gas shortages in the USA, they hurried up to hail the unconventional natural gas 'revolution', when they suddenly realised that over 30% of the USA natural gas production was already coming from tight natural gas deposits in 2008.[11] So, the massive interest has nothing to do with the actual availability of the reserves. Despite this massive interest the quality and

---

[11] Advanced Resources International, 2009, (http://www.adv-res.com).

availability of tight natural gas data leaves a lot to be desired.

Table 10.5: World tight natural gas reserves and resources

Data Source	Publication year	Reserves	Resources	Gas in Place
			trillion standard m^3	
BGR	2007	*89*	235	456
IFP	2005	*82*	219	422
Total	2009	*60*	161	310
Swindell	2008	Note 3	Note 3	Note 3

Source: BGR (Energierohstoffe 2009 – Reserven, Ressourcen, Verfügbarkeit); IFP (Armelle Sanière, Gas Reserves, Discoveries and Production, Panorama 2006); Total (http://www.total.com/en/our-energies/natural-gas-/exploration-and-production/our-skills-and-expertise/tight-gas-940885.html&textsize=1); Swindell (http://gswindell.com).
Note 1: Numbers in italic are calculated by the author.
Note 2: IFP numbers are average values from published range.
Note 3: Swindell does not report tight gas as part of his total world fossil fuel resources.

With the exception of extensive data of tight natural gas in the USA and Canada, there is no comprehensive set of data available on a country-by-country basis. The only data available on a global and regional level in the public domain are provided by the BGR, which obtains its data from scientific research using as a basis the work of Rogner.[12] In addition, several bits and pieces of data are scattered all over scientific papers and commercial consultancies reports but are often hard to find or access. Furthermore, these few global estimates vary significantly. Table 10.5 summarises some of the latest world tight natural gas original gas-in-place (OGIP) as reported by different sources with the estimated resources and reserves calculated using conservative recovery factors. The data in the above table are calculated using the published data of original gas-in-place (OGIP) and accounting for cumulative tight natural gas production from its deposits, estimated only from the USA, Canada and China production numbers, which are the only production data available.[13,14,15,16] The reserves are estimated

---

[12] H-H Rogner, An assessment of world hydrocarbon resources, Annu. Rv. Energy Environ., vol 22, p 217-62, 1997.
[13] Advanced Resources International, 2009, (http://www.adv-res.com).
[14] Navigant Consulting, North American natural gas assessment, 2008, (http://www.navigantconsulting.com).
[15] EIA, 2009, (http://www.eia.gov/dnav/ng/ng_prod_top.asp).
[16] National Energy Board, Canada, An energy market assessment July 2009: Canadian energy

using a conservative recovery factor of 20% which is quoted universally as a reasonable recovery factor, while the resources were estimated using an ultimate recovery factor of 52%.[17]

Despite the lack of detail, the data reported by the BGR is used here as the basis for further analysis as it is the only set of complete data available in the public domain. It is widely accepted by experts that tight natural gas OGIP and resources numbers are conservative but will increase with time when more countries start further exploration and assessment of their potential tight natural gas deposits.

Note that in this section the data for tight natural gas and shale gas are reported separately and not amalgamated, as frequently practiced by the BGR in the majority of its reporting. This practice is due to the fact that even though the two gas categories have similar geological characteristics, the production processes needed to extract natural gas from them differ significantly both technically and economically, and the overall extraction feasibility depends also on political conditions, deposits location, legal and environmental restrictions, public opinion, and labour availability to name but a few.

Currently commercial production from tight natural gas deposits takes place in many countries besides the USA including Canada, Argentina, Egypt, Algeria, and Oman, with planned commercial exploitation already underway in China, Venezuela and Australia. However, only the USA and Canada publish data that identify the gas production by the type of deposit, thus allowing the differentiation of tight natural gas production from the rest of the production.

Since there is no set of data detailing tight natural gas data per country, in this book individual country data is estimated using the regional data as reported by BGR[18] and then apportioning it to individual countries in each region, in the same ratio of the

---

demand and supply to 2020.
[17] Advanced Resources International, 2009, (http://www.adv-res.com).
[18] BGR, Energierohstoffe 2009 – Reserven, Ressourcen, Verfügbarkeit.

Table 10.6: World natural tight gas reserves and resources – top ten countries (2007)

Rank	Country	Reserves billion standard m³	Share %	Resources billion standard m³	Share %	Gas in Place billion standard m³	Share %
1	USA	3859.7	9.74	12642.6	11.82	25816.9	12.42
2	Russia	4398.2	11.10	11435.4	10.70	21991.1	10.58
3	Australia	3617.1	9.13	9404.5	8.80	18085.5	8.70
4	Venezuela	3255.8	8.21	8465.1	7.92	16279.1	7.83
5	Nigeria	2755.0	6.95	7163.0	6.70	13775.1	6.63
6	China	1846.0	4.66	4851.1	4.54	9358.6	4.50
7	Canada	922.1	2.33	3604.8	3.37	7628.7	3.67
8	Iran	1405.7	3.55	3654.7	3.42	7028.3	3.38
9	Indonesia	1450.0	3.66	3770.0	3.53	7250.1	3.49
10	Qatar	1259.4	3.18	3274.6	3.06	6297.2	3.03
	TOTAL	24769.1	62.49	68265.7	63.85	133510.6	64.24
	WORLD	39635.8		106919.0		207843.8	

Source: EIA (http://www.eia.doe.gov/international), BGR (Energierohstoffe 2009 – Reserven, Ressourcen, Verfügbarkeit).
Note 1: Totals may not add up due rounding.

conventional natural gas OGIP of these individual countries. This can be used as an initial approximation. In the future the data will need to be updated and reconciled to include actual data as they become available. The derived list of the top ten countries, in terms of tight natural gas reserves and resources, is shown in Table 10.6. Based on total OGIP estimation of 208 trillion standard m³, these ten countries account for almost 62.5% and 63.9% of the global estimated reserves and resources respectively. The data in the above table is calculated using the same methodology described in deriving Table 10.5 above. The resources and reserves are estimated from original tight natural gas-in-place (OGIP) data using identical recovery factors for all countries. Note that since a significant proportion of USA's tight natural gas deposits are under active development, recovery factors from its deposits, which are based on actual audited production and reserves numbers, are used as a guideline and are applied for the rest of the world deposits.

It is interesting to note that no single country or region dominates tight natural gas resources and reserves as deposits found in significant amounts in all world regions. The two leading countries are the USA, in terms of resources, with approximately 11.8% and Russia, in terms of reserves, with approximately 11.1% of the world's total. Note that five of the top ten countries, in terms of tight natural gas resources and reserves, are members of OPEC, though they account for just under 23.3% and 24.6% of the total resources and reserves respectively. As with oil unconventional deposits, this cannot only be attributed to pure geology, but it suggests that OPEC countries are making no efforts and allocating no money to explore unconventional tight natural gas deposits, since the majority of them already have abundant better quality, easier to extract and more profitable, conventional natural gas.

Table 10.7 shows the tight natural gas estimated resources and reserves in the Arab world, highlighting the Arab world's share of the world total resources and reserves. Using the methodology outlined above, the detailed country-by-country data in the table is estimated. In each country the resources and reserves are estimated from the data published by the BGR for the Middle East and Africa and, depending on the geographical location of each country, the country is allocated its share relative to its region's overall deposits, assuming similar proportion as with conventional natural gas reserves between the countries in that region. In a stark contrast to the conventional natural gas reserves and resources, where the Arab world is very significant, the resources of tight natural gas are minor in the Arab world, where its share is a rather small 8.4% and 8.1% of the reserves and resources respectively. Thus it is obvious that further development of these tight natural gas resources and reserves globally will reduce the Arab world share and in time render it less significant to global natural gas markets.

The table above suggests that tight natural gas reserves exist in 19 Arab countries. Applying the same tier classification as that of the conventional natural gas (see Section 10.1) on the reserves data in Table 10.7, the Arab world countries fall into the classification as follows:

Table 10.7: Natural tight gas reserves and resources in the Arab countries (2007)

Country	Reserves billion standard m³	Rank	Share %	Resources billion standard m³	Share %	Gas in Place billion standard m³	Share %
Bahrain	15.2	14	0.04	39.4	0.04	75.8	0.04
Iraq	233.6	5	0.59	607.3	0.57	1167.9	0.56
Jordan	3.5	17	0.01	9.2	0.01	17.7	0.01
Kuwait	97.6	8	0.25	253.8	0.24	488.0	0.23
Lebanon	0.0			0.0		0.0	
Oman	63.0	10	0.16	163.9	0.15	315.1	0.15
Palestine	2.2	18	0.01	5.8	0.01	11.2	0.01
Qatar	1259.4	1	3.18	3274.6	3.06	6297.2	3.03
Saudi Arabia	621.7	2	1.57	1616.3	1.51	3108.3	1.50
Syria	18.5	12	0.05	48.1	0.04	92.5	0.04
UAE	333.1	3	0.84	866.0	0.81	1665.4	0.80
Yemen	31.8	11	0.08	82.7	0.08	159.1	0.08
Algeria	301.4	4	0.76	783.6	0.73	1507.0	0.73
Comoros	0.0			0.0		0.0	
Djibouti	0.0			0.0		0.0	
Egypt	111.6	7	0.28	290.2	0.27	558.1	0.27
Libya	82.6	9	0.21	214.9	0.20	413.2	0.20
Mauritania	12.8	15	0.03	33.3	0.03	64.0	0.03
Morocco	0.9	19	0.00	2.4	0.00	4.5	0.00
Somalia	123.7	6	0.31	321.5	0.30	618.3	0.30
Sudan	17.2	13	0.04	44.7	0.04	86.0	0.04
Tunisia	12.8	16	0.03	33.2	0.03	63.8	0.03
ARAB WORLD	3342.6		8.43	8690.9	8.13	16713.2	8.04

Source: EIA (http://www.eia.doe.gov/international); BGR (Energierohstoffe 2009 – Reserven, Ressourcen, Verfügbarkeit).
Note 1: Totals may not add up due rounding.

- None of the countries belong to the first tier, with estimated tight natural gas reserves exceeding 10 trillion standard m³.
- One country, Qatar, belongs to the second tier, with estimated tight natural gas reserves exceeding 1 trillion standard m³ but less than 10 trillion standard m³. Its share is approximately 3.2% of the world's estimated tight natural gas reserves. Obviously, this adds to Qatar's significance as a natural gas 'super power', with even more reserves to add to its already massive endowment.

- Three countries belong to the third tier, with estimated tight natural gas reserves exceeding 250 billion standard m^3 but less than 1 trillion standard m^3 each. Their combined share is approximately 3.2% of the world's estimated tight natural gas reserves. These countries, in descending order in terms of their reserves, are Saudi Arabia, UAE and Algeria. To date, there are no reports of any active tight natural gas production from the first two countries, with reports of active production only being pursued in Algeria.
- 15 countries belong to the fourth tier, with minor estimated tight natural gas reserves less than 250 billion standard m^3 each and a combined share of around 2.4% of the world's total reserves. These countries, in descending order in terms of their reserves, are Iraq, Sudan, Somalia, Kuwait, Libya, Egypt, Oman, Yemen, Mauritania, Syria, Bahrain, Tunisia, Jordan, Palestine and Morocco. With the exception of Egypt and Oman, where there are reports of active tight natural gas production, the deposits in all other countries in this tier appear to be undeveloped.
- Finally three countries have no estimated tight natural gas reserves. These countries are Lebanon, Djibouti and Comoros. Currently no tight natural gas exploration activities are being pursued in these countries.

10.2.2 Shale Gas

Since 2009 a flurry of acquisitions by major multinational oil companies of shale gas producers created frenzy, especially in the USA, with endless columns published in the media devoted to shale gas in particular and unconventional natural gas in general. Having ignored the fact for decades, suddenly, the media highlighted the fact that the USA produced gas from shale deposits since the late 1970s, with over 6% of the total USA gas production coming from these shale gas deposits in 2008.[19] . The growth in its production is the highest amongst all natural gas categories.

---

[19] Advanced Resources International, 2009, (http://www.adv-res.com).

However, similar to tight gas, despite this new interest and change in attitude, and despite the current acknowledgment of the significance of shale gas as a credible natural gas resource, obtaining estimates of resources and reserves of shale gas are very difficult, with the data often very limited and extremely unreliable. Furthermore, no annual data compilations, global, regional, or national, are made available in the public domain.

As described earlier, the situation regarding shale gas data is identical to that of tight natural gas, with extensive data on shale gas available for only a few countries such as the USA and Canada. Again, there is no comprehensive set of data available on a worldwide national level. The only data available on a global and regional level in the public domain is provided by the BGR, which, as already discussed, obtains its data from scientific research using as a basis the work of Rogner.[20] Besides this, few scientific journals or commercial consultancy reports provide additional data, so the quality and availability of this data remains poor. Unsurprisingly the few global estimates available vary significantly. Table 10.8 summarises some of the latest world shale gas original gas-in-place (OGIP) as reported by different sources, with the estimated resources and reserves calculated using conservative recovery factors. The data in the above table are calculated using the published data of the OGIP. The only cumulative production data available is estimated from the USA and Canada numbers, which account for the shale gas production from their deposits.[21,22,23,24] The reserves are estimated using a conservative recovery factor of 20%[25], which is quoted universally as a reasonable recovery factor, while the resources were estimated using an ultimate recovery factor of 40%.[26]

---

[20] H-H Rogner, An assessment of world hydrocarbon resources, Annu. Rv. Energy Environ., vol 22, p 217-62, 1997.
[21] Advanced Resources International, 2009, (http://www.adv-res.com).
[22] Navigant Consulting, North American natural gas assessment, 2008, (http://www.navigantconsulting.com).
[23] EIA, 2009, (http://www.eia.gov/dnav/ng/ng_prod_top.asp).
[24] National Energy Board, Canada, An energy market assessment July 2009: Canadian energy demand and supply to 2020.
[25] Brian Gault & Garth Scotts, Improve shale gas production forecasts, E&P, 2007, (http://www.epmag.com/archives/features/306.htm).
[26] Advanced Resources International, 2009, (http://www.adv-res.com).

Table 10.8: World shale gas reserves and resources

Data Source	Publication year	Reserves	Resources	Gas in Place
			trillion standard m^3	
BGR	2007	*42*	84	210
IFP	2005	8	17	44
Swindell	2008	90	*181*	452

Source: BGR (Energierohstoffe 2009 – Reserven, Ressourcen, Verfügbarkeit); IFP (Armelle Sanière, Gas Reserves, Discoveries and Production, Panorama 2006); Total (http://www.total.com/en/our-energies/natural-gas-/exploration-and-production/our-skills-and-expertise/tight-gas-940885.html&textsize=1); Swindell (http://gswindell.com).
Note 1: Numbers in italic are calculated by the author.
Note 2: IFP numbers are average values from published range.

The data reported hereby the BGR is adopted as the basis for further analysis, despite its limitations, since it is the only set of complete data available in the public domain. The BGR numbers are considered to be conservative and the resources numbers are expected to increase with time when more countries start exploring and assessing their deposits.

The majority of current commercial production from shale gas deposits takes place mainly in the USA, while other countries such as Canada have just started. Moreover, planned commercial exploitation is already underway in many countries, such as China and Australia. However, only the USA and Canada publish data that identifies the gas production by the type of deposit, thus allowing the differentiation of shale gas production, from the rest of the production.

Since there is no set of data detailing shale gas data per country, the exact method and data sources described in deriving Table 10.6 are applied here. The derived list of the top ten countries, in terms of shale gas reserves and resources, is shown in Table 10.9. Based on the total OGIP estimation of 456 trillion standard m^3, these ten countries combined account for almost 77.5% and 77.6% of both the global estimated reserves and resources respectively.

It is remarkable to note that no single country or region

dominates shale gas reserves or resources, deposits are found in large amounts in all world regions, although the leading country, China, is home to around 20.6% the world's total reserves and resources. Even though four of the top ten countries, in terms of shale gas resources or reserves, are members of OPEC, they account for just over 17.1% of the total reserves and resources. As with tight natural gas deposits, this cannot be attributed to pure geology, but also to the fact that OPEC countries are currently not making any efforts or allocating any resources, to explore shale deposits since they are already loaded with better quality and more profitable conventional natural gas deposits.

Table 10.9: World shale gas reserves and resources – top ten countries (2007)

Rank	Country	Reserves billion standard m^3	Share %	Resources billion standard m^3	Share %	Gas in Place billion standard m^3	Share %
1	China	18762.7	20.60	37525.3	20.57	93813.3	20.55
2	USA	15078.5	16.55	30392.2	16.66	76333.5	16.72
3	Australia	11846.0	13.00	23692.0	12.99	59230.1	12.98
4	Venezuela	5279.7	5.80	10559.4	5.79	26398.5	5.78
5	Iran	4412.5	4.84	8825.0	4.84	22062.6	4.83
6	Canada	4677.2	5.13	9354.6	5.13	23386.9	5.12
7	Qatar	3953.6	4.34	7907.1	4.33	19767.8	4.33
8	Russia	3011.1	3.31	6022.2	3.30	15055.4	3.30
9	Argentina	1646.2	1.81	3292.3	1.80	8230.8	1.80
10	Saudi Arabia	1951.5	2.14	3902.9	2.14	9757.3	2.14
	TOTAL	70618.9	77.52	141473.3	77.55	354036.4	77.56
	WORLD	91100.0		182435.4		456441.8	

Source: EIA (http://www.eia.doe.gov/international); BGR (Energierohstoffe 2009 – Reserven, Ressourcen, Verfügbarkeit).
Note 1: Totals may not add up due rounding.

Table 10.10 shows the shale gas estimated resources and reserves in the Arab world, highlighting the Arab world's share of the world total resources and reserves. The data in the table are derived in a similar manner described in Table 10.7, above. Note that the resources of shale gas are small, though not minor, in the Arab world, where its share is just over 11.1% of the reserves and resources. This is far less than the Arab world's share, in terms of conventional natural gas reserves and resources, where as we have already established that the Arab world share is very significant. Thus it is obvious that further development of these shale gas resources and reserves globally will diminish the current Arab world share.

Table 10.10: Shale gas reserves and resources in the Arab countries (2007)

Country	Reserves billion standard m^3	Rank	Share %	Resources billion standard m^3	Share %	Gas in Place billion standard m^3	Share %
Bahrain	47.6	13	0.05	95.2	0.05	238.1	0.05
Iraq	733.3	5	0.80	1466.5	0.80	3666.3	0.80
Jordan	11.1	16	0.01	22.2	0.01	55.4	0.01
Kuwait	306.4	7	0.34	612.7	0.34	1531.8	0.34
Lebanon	0.0			0.0		0.0	
Oman	197.8	9	0.22	395.7	0.22	989.1	0.22
Palestine	7.1	17	0.01	14.1	0.01	35.3	0.01
Qatar	3953.6	1	4.34	7907.1	4.33	19767.8	4.33
Saudi Arabia	1951.5	2	2.14	3902.9	2.14	9757.3	2.14
Syria	58.1	11	0.06	116.1	0.06	290.3	0.06
UAE	1045.6	3	1.15	2091.2	1.15	5228.0	1.15
Yemen	99.9	10	0.11	199.8	0.11	499.5	0.11
Algeria	946.1	4	1.04	1892.2	1.04	4730.6	1.04
Comoros	0.0			0.0		0.0	
Djibouti	0.0			0.0		0.0	
Egypt	350.4	6	0.38	700.8	0.38	1752.0	0.38
Libya	259.4	8	0.28	518.9	0.28	1297.2	0.28
Mauritania	4.5	18	0.00	9.1	0.00	22.7	0.00
Morocco	1.1	19	0.00	2.3	0.00	5.7	0.00
Somalia	43.8	14	0.05	87.7	0.05	219.2	0.05
Sudan	54.0	12	0.06	108.0	0.06	269.9	0.06
Tunisia	40.0	15	0.04	80.1	0.04	200.2	0.04
ARAB WORLD	10111.3		11.10	20222.7	11.08	50556.6	11.08

Source: EIA (http://www.eia.doe.gov/international); BGR (Energierohstoffe 2009 – Reserven, Ressourcen, Verfügbarkeit).
Note 1: Totals may not add up due rounding.

The table above shows that proved natural gas reserves exist in 19 Arab countries. Applying the same tier classification as that of the conventional natural gas (see Section 10.1) on the reserves data in Table 10.10, the Arab world countries fall into the classification as follows:
- None of the countries belong to the first tier, with estimated shale gas reserves exceeding 10 trillion

standard m^3.
- Three countries belong to the second tier, with estimated shale gas reserves exceeding 1 trillion standard m^3 but less than 10 trillion standard m^3 each. They have a combined share of approximately 7.6% of the world's estimated shale gas reserves. These countries, in descending order in terms of their reserves, are Qatar, Saudi Arabia and UAE.
- Five countries belong to the third tier, with estimated shale gas reserves exceeding 250 billion standard m^3 but less than 1 trillion standard m^3 each. Their combined share is approximately 2.9% of the world's estimated shale gas reserves. These countries, in descending order in terms of their reserves, are Algeria, Iraq, Kuwait, Libya and Egypt.
- 11 countries belong to the fourth tier, with minor estimated shale gas reserves less than 250 billion standard m^3 each and a combined share of just over 0.6% of the world's reserves. These countries, in descending order in terms of their reserves, are Oman, Yemen, Sudan, Syria, Somalia, Bahrain, Tunisia, Jordan, Mauritania, Palestine and Morocco.
- Finally, three countries have no estimated shale gas reserves. These countries are Lebanon, Djibouti, and Comoros. Currently no shale gas exploration activities are being pursued in these three countries.

10.2.3 Coalbed Methane
The occurrence of coalbed methane in coal mines is a well known phenomenon. For decades, mining companies had to deal with, what was termed then, as a risk to the mining industry. However, this attitude changed dramatically in 1980s with the transformation of coalbed methane from an annoying nuisance to a valuable asset in energy markets. This transformation is now complete, being well recognised and accepted universally. In addition the need to curb the emissions of methane, as one of the worst green house gases, has made extracting coalbed methane into an environmental necessity. Therefore, with coal deposits

currently identified in over 70 countries in the world and since all these deposits contain some coalbed methane, there is a huge potential of coalbed methane resources available all over the world.

Thus the significance of coalbed methane as a natural gas source was realised in the industry. Therefore its deposits have started to be utilised commercially in the last few years in numerous countries, including the USA, Canada, Australia and China. In the USA, as 2008 numbers show, coalbed methane accounted for 10% of its total natural gas production.[27] According to industry reports, coalbed methane is currently being actively produced in at least 17 countries, and this is only the tip of the iceberg, where this utilisation is still in its infancy.

However, an important question remains, which is whether it is easier and cheaper to convert coal directly into syngas, rather than extracting coalbed methane, especially with the development of various coal gasification technologies. The answer to this question is complex. It depends on many factors, including the comparable economics of the two processes, environmental legislation and political will. It is also essential to realise that syngas is not natural gas per se and thus it cannot substitute natural gas straight away in many appliances and power stations that have already been designed to use natural gas, consequently the cost of modifications required to all these equipment has to be included in the math.

Data availability on coalbed methane original gas-in-place (OGIP), resources and reserves are better than tight natural gas or shale gas data. Some sources, such as the BGR, publish data on country-by-country basis, alongside regional and global data. Table 10.11 summarises some of the latest world coalbed methane OGIP reported by different sources as well as the estimated resources and reserves calculated using conservative recovery factors. The data in the above table are calculated using the same methodology implemented in estimating tight natural

---

[27] Advanced Resources International, 2009, (http://www.adv-res.com).

gas and shale gas, though the recovery factors are different. The calculation uses the published data of OGIP while accounting for cumulative coalbed methane production from its deposits, which are estimated from all producing countries from the data presented primarily by the BGR, but are supplemented from other sources.[28,29,30]. The reserves are estimated using a conservative recovery factor of 40% which is an average of the often quoted recovery factors of between 20% and 60%. The resources are estimated using an ultimate recovery factor of 90%.[31] It is observed that there is a considerable disagreement between data sources; however, the data reported by the BGR are adopted here, as it is the only set of complete data available in the public domain.

Table 10.11: World coalbed methane reserves and resources

Data Source	Publication year	Reserves	Resources	Gas in Place
			trillion standard m^3	
BGR	2007	*101*	228	254
IFP	2005	*71*	161	180
Swindell	2008	*101*	228	254

Source: BGR (Energierohstoffe 2009 – Reserven, Ressourcen, Verfügbarkeit); IFP (Armelle Sanière, Gas Reserves, Discoveries and Production, Panorama 2006); Swindell (http://gswindell.com).
Note 1: Numbers in italic are calculated by the author.
Note 2: IFP numbers are average values from published range.

Based on the latest data published data by the BGR, an average OGIP of 253 trillion standard m^3 is used. Table 10.12 lists the top ten countries in the world, in terms of coalbed methane resources and the calculated reserves estimates. These countries account for almost 97.5% of the global estimated reserves and resources. The reserves data in the above table are calculated in a similar manner to the reserves calculations described in the preceding sections.

The table below mirrors the top ten countries in terms of coal reserves, as expected (see Section 11.1), we can see that no one

---

[28] Ibid.
[29] Navigant Consulting, North American natural gas assessment, 2008, (http://www.navigantconsulting.com).
[30] EIA, 2009, (http://www.eia.gov/dnav/ng/ng_prod_top.asp).
[31] BGR, Energierohstoffe 2009 – Reserven, Ressourcen, Verfügbarkeit.

country dominates this category. The leading country, Russia, accounts for 25.7% and 25.8% of the global estimated reserves and resources respectively, with coalbed methane being distributed worldwide. Note that only one OPEC member, Indonesia, is placed in the top ten countries, in terms of coalbed methane resources or reserves and it only accounts for 4.4% of the total reserves and resources. This is not surprising as it mirrors the coal reserves occurrence, which is more evenly distributed around the world; it also confirms that OPEC countries are extremely poor in terms of coal reserves.

Table 10.12: World coalbed methane reserves and resources – top ten countries (2007)

Rank	Country	Reserves billion standard m^3	Share %	Resources billion standard m^3	Share %	Gas in Place billion standard m^3	Share %
1	Russia	25976.5	25.82	58451.5	25.70	64946.5	25.69
2	Canada	20573.9	20.45	46355.2	20.38	51511.4	20.38
3	Ukraine	15635.8	15.54	35185.8	15.47	39095.8	15.47
4	China	13990.2	13.90	31490.2	13.85	34990.2	13.84
5	USA	9849.0	9.79	23061.0	10.14	25703.4	10.17
6	Australia	4782.5	4.75	10782.5	4.74	11982.5	4.74
7	Indonesia	4474.6	4.45	10067.9	4.43	11186.5	4.43
8	Turkey	1200.0	1.19	2700.0	1.19	3000.0	1.19
9	UK	919.3	0.91	2069.3	0.91	2299.3	0.91
10	Germany	696.5	0.69	1571.5	0.69	1746.5	0.69
	TOTAL	98098.3	97.49	221734.8	97.50	246462.1	97.50
	WORLD	100623.3		227421.8		252781.5	

Source: BGR (Energierohstoffe 2009 – Reserven, Ressourcen, Verfügbarkeit).
Note 1: Totals may not add up due rounding.

Table 10.13 shows the coalbed methane reported resources and estimated reserves in the Arab world, highlighting the Arab world's share to the world total reserves and resources. As shown in the table, there are no reported reserves or resources of coalbed methane in the Arab countries, which is expected since the Arab world is extremely poor in terms of coal. Based on the fact that coalbed methane exists with all coal deposits, we can guess that trivial deposits may exist in Egypt, Algeria and Morocco, where minor coal resources exist. However, the quantities are insignificant, so for the purposes of the analysis in this book they are ignored. It may therefore be deemed that the Arab world has no coalbed OGIP, resources or reserves.

Table 10.13: Coalbed methane reserves and resources in the Arab countries (2007)

Country	Reserves billion standard m^3	Rank	Share %	Resources billion standard m^3	Share %	Gas In Place billion standard m^3	Share %
Bahrain	0.0			0.0		0.0	
Iraq	0.0			0.0		0.0	
Jordan	0.0			0.0		0.0	
Kuwait	0.0			0.0		0.0	
Lebanon	0.0			0.0		0.0	
Oman	0.0			0.0		0.0	
Palestine	0.0			0.0		0.0	
Qatar	0.0			0.0		0.0	
Saudi Arabia	0.0			0.0		0.0	
Syria	0.0			0.0		0.0	
UAE	0.0			0.0		0.0	
Yemen	0.0			0.0		0.0	
Algeria	Note 1			Note 1		Note 1	
Comoros	0.0			0.0		0.0	
Djibouti	0.0			0.0		0.0	
Egypt	Note 1			Note 1		Note 1	
Libya	0.0			0.0		0.0	
Mauritania	0.0			0.0		0.0	
Morocco	Note 1			Note 1		Note 1	
Somalia	0.0			0.0		0.0	
Sudan	0.0			0.0		0.0	
Tunisia	0.0			0.0		0.0	
ARAB WORLD	0.0			0.0		0.0	

Source: BGR (Energierohstoffe 2009 – Reserven, Ressourcen, Verfügbarkeit).
Note 1: Reported unquantified deposits.

10.2.4 Methane Hydrates

Methane hydrates have enjoyed a lot of publicity recently, having been dubbed as the energy of the future. There have been massive amounts of unsubstantiated claims that these hydrates will displace all existing fossil fuels, with exaggerated estimates of their deposits, to levels exceeding all other fossil fuels tenfold. Chemically, many stable gas hydrates exist, however, the overwhelming majority of these hydrates are methane hydrates, which constitute over 99.9% of all hydrates, that contain up to

160 m³ methane (at atmospheric conditions) stored in each 1 m³ of hydrates.[32]

Methane hydrates can be classified into two main categories based on their location: permafrost and deep sea, with deposits from the former perceived to be easier to develop and extract than from the latter. However, the majority of deposits exist in the latter with estimates ranging between 99.5% and 85.7% of the total.[33,34] A useful resource triangle for methane hydrates is presented by Meggs[35] in which methane hydrate deposits are divided into several sub-categories that are listed in terms of increasing quantity and costs, decreasing quality, level of technical difficulty and ultimate recovery. The easiest sub-categories being hydrates in the Arctic permafrost followed by deposits in deepwater sandstone and then non-sandstone marine deposits. It can be seen that estimates of deposits quantities are over 100000 times for the more difficult sub-category compared to the easiest one.

Unlike tight natural gas, shale gas, and coalbed methane, there are no commercial processes to date that extract gas from methane hydrates, though pilot extraction is being tried and/or planned in Russia, USA, Canada, Germany, Japan, South Korea, India, and China, however no production is foreseen in the immediate future. Therefore all deposits of methane hydrates are treated only as prospective resources, so they are not included as reserves by any data source. Even if the technology is developed and the costs fall in the near future to render the processes feasible, the recovery factors will remain very low, thus triggering a minor reclassification, by placing only a small percentage of the methane hydrates resources in the reserves class.

Like data for other unconventional natural gas, gas-in-place data

---

[32] B.B. Rath, Methane Hydrates: An Abundance of Clean Energy?, MRS Bulletin, Volume 33, April 2008.
[33] BGR, Energierohstoffe 2009 – Reserven, Ressourcen, Verfügbarkeit.
[34] Gary Swindell, 2008, (http://gswindell.com).
[35] Tony Meggs, Natural gas supply – with a focus on the United States, MIT Energy Initiative, 2009.

for methane hydrates are erratic with no annually updated compilations available in the public domain on either a global or on a country-by-country basis. Furthermore, methane hydrates data vary enormously by orders of magnitude! Some of the estimates have no basis in fact, where estimators mistakenly believed that all ocean floors are covered with methane hydrates! Table 10.14 summarises the latest world methane hydrate OGIP as reported by different sources with the estimated resources calculated, based on conservative ultimate recovery factor of 1%.[36] For obvious reasons, the calculation accounts for no cumulative production from these deposits. The variation in available data is so extreme as to be completely unreliable for the purpose of future planning. In order to progress, it is crucial that further research is carried out if there is to be any sort of consensus on the numbers.

Table 10.14: World methane hydrates reserves and resources

Data Source	Publication year	Reserves	Resources	Gas in Place
			trillion standard m3	
BGR	2007	0	*605*	60500
Klauda & Sandler	2005	0	*1200*	120000
IFP	2005	0	*185*	18500
Rath	2008	0	*210*	21000
Swindell	2008	0	*10*	989

Source: BGR (Energierohstoffe 2009 – Reserven, Ressourcen, Verfügbarkeit); Klauda & Sandler (Jeffrey B Klauda & Stanley I Sandler, Global distribution of methane hydrate in ocean sediment, Energy and Fuels, Vol 19, p 459-470, 2005); IFP (Armelle Sanière, Gas Reserves, Discoveries and Production, Panorama 2006); B.B. Rath, Methane Hydrates: An Abundance of Clean Energy?, MRS Bulletin, Volume 33, April 2008; Swindell (http://gswindell.com).
Note 1: Numbers in italic are calculated by the author.
Note 2: IFP and BGR numbers are average values from published range.
Note 3: Swindell states that this is the lowest estimate.

Moreover, due to the lack of data, it is impossible to rank the top ten countries in the world in terms of methane hydrates resources quantitatively. However from the limited data it is possible to make an assessment with the help of the widely available maps in the public domain that show approximate locations of the deposits, thus it is possible to 'guestimate' which are the top five countries qualitatively, listed in Table 10.15.

---

[36] B.B. Rath, Methane Hydrates: An Abundance of Clean Energy?, MRS Bulletin, Volume 33, April 2008.

Table 10.15: World methane hydrates reserves and resources – top five countries (2007)

Country	Notes
Russia	Note 1
Canada	Note 1
USA	Note 1
Peru	Note 1
Australia	Note 1

Source: Jeffrey B Klauda & Stanley I Sandler, Global distribution of methane hydrate in ocean sediment, Energy and Fuels, Vol 19, p 459-470, 2005.
Note 1: Reported unquantified deposits.

Despite an extensive search for data, the BGR was the only data source that reports estimates of methane hydrates on a regional level; however, since there is no set of detailed data on a national level, it is not possible to detail these resources for the Arab world on a country-by-country basis. The regional estimates of BGR for the Middle East and Africa cannot be used as a basis, since it is derived by counting the occurrences of methane hydrates.[37] Instead, as it is the most detailed and up-to-date set of data, the study of Klauda and Sandler[38] which estimates methane hydrate volumes on a 1° latitude by 1° longitude global grid, is used to estimate the relative existence of methane hydrates per region. Methane hydrates resources estimation for the Arab world is shown in Table 10.16. In order to reach this estimation, the share of the Arab world is derived as a percentage of the total using the map presented in the above study, by estimating the Arab world's share of the resources shown to exist in their exclusive economic zones. It can be estimated that the Arab world's share of the methane hydrates is around 5.1% of the world's total, and using a conservative estimate of 605 trillion standard m^3 for the world resources it is assessed that the Arab world will be home to 31 trillion standard m^3 of methane

---

[37] Uwe Remme, Markus Blesl & Ulrich Fahl, Global Resources and energy trade: An overview for coal, natural gas, oil and uranium, IER, Universität Stuttgart, 2007.
[38] Jeffrey B Klauda & Stanley I Sandler, Global distribution of methane hydrate in ocean sediment, Energy and Fuels, Vol 19, p 459-470, 2005.

hydrates, which is a very modest quantity indeed, and thus if methane hydrates are ever to be developed and become a factor in the energy markets, the significance of the Arab world in natural gas markets will diminish drastically. Although the Arab countries should not be concerned about this issue in the short or medium terms as bringing methane hydrates into play is still a long way off.

Table 10.16: Methane hydrates reserves and resources in the Arab countries (2007)

	Original Gas in Place	Remaining Resources	Reserves	Share
	trillion standard m^3			%
Arab World	3082	31	0	5.1
World	60500	605	0	

Source: BGR (Energierohstoffe 2009 – Reserven, Ressourcen, Verfügbarkeit); Jeffrey B Klauda & Stanley I Sandler, Global distribution of methane hydrate in ocean sediment, Energy and Fuels, Vol 19, p 459-470, 2005
Note 1: Arab world's share is calculated by the author.

10.2.5 Geopressurised and Hydropressurised Gas

Contrasting other unconventional natural gas categories, geopressurised and hydropressurised gas (aka geopressured aquifer gas) deposits are unique, since not all data sources and industry experts consider them to be part of the natural gas endowment, with many treating this natural gas source as geothermal rather than fossil fuel.

It is universally accepted that methane is found, under very high pressure, in porous sand or silt layers located at great depths, usually at a depth over 3000 metres below the surface of the earth, often dissolved or dispersed in underground water aquifers, with the amount of methane increasing by increasing depth and salinity of the water aquifer. Several graphs are published in the public domain that detail these correlations, which estimate that on average, under normal pressure conditions, 1 m^3 of groundwater will contain between 0.5 and 5 m^3 of dissolved methane, with some zones contain as much as 90 m^3 methane per

1 m³ of water.[39]

Similar to methane hydrates, there are no commercial processes to date that extract gas from geopressurised or hydropressurised zones, though pilot extraction has been tried or is being planned in the USA, Canada, Italy, Australia, China and Brazil. Most of this extraction occurs in shallow aquifers and its main aim is often to extract rare minerals dissolved in water, rather than natural gas.

Therefore, with current recovery factors set to zero, all deposits of gas in geopressurised and hydropressurised zones are treated as resources but are not recognised as reserves by any data source. Even if we are to assume that technology will be developed in the near future to exploit these deposits or if the costs will fall accordingly to render the processes feasible, the recovery factors will remain extremely low due to geological and environmental aspects as well as due to the low density of dissolved methane in many groundwater aquifers. Thus merely a small percentage of the geopressurised and hydropressurised gas resources are envisaged as being reclassified into reserves in the medium time scale.

Furthermore, even more than with other unconventional natural gas categories, most data sources omit mentioning geopressurised and hydropressurised gas. Consequently, the few data estimates available are incomplete, inaccurate, inconsistent, conflicting and reported irregularly, with no updated annual compilation available in the public domain. Table 10.17 summarises the latest world geopressurised and hydropressurised gas-in-place as reported by few sources and the estimated resources calculated based on a conservative recovery factor. The data in the above table are calculated using all the data available, whether they are OGIP or resources data, and then back-calculate the other data. Obviously, since these deposits are still speculative with no commercial production, no cumulative production from these deposits is considered in the calculation. The calculations are

---

[39] BGR, Energierohstoffe 2009 – Reserven, Ressourcen, Verfügbarkeit.

performed using a conservative ultimate recovery factor of 3%.[40] Note that, with the exception of BGR data reported by Remme[41], who estimated (rather than measured) these deposits based on groundwater data, there is no other global estimate of these resources. The number quoted by BRG in its 2009 report[42] estimates a mammoth 10 million trillion standard m^3 of OGIP, which is incredibly high but is not used here in further calculations due to its questionable veracity.

Table 10.17.: World geopressurised and hydropressurised gas reserves and resources

Data Source	Publication year	Reserves	Resources	Gas in Place
			trillion standard m^3	
BGR	2007	0	*300000*	*10000000*
Remme	2007	0	663	22105
IFP	2005	0	*185*	18500
Swindell	2008	Note 3	Note 3	Note 3

Source: BGR (Energierohstoffe 2009 – Reserven, Ressourcen, Verfügbarkeit); Remme (Uwe Remme, Markus Blesl & Ulrich Fahl, Global Resources and energy trade: An overview for coal, natural gas, oil and uranium, IER, Universität Stuttgart, 2007); IFP (Armelle Sanière, Gas Reserves, Discoveries and Production, Panorama 2006); Swindell (http://gswindell.com).
Note 1: Numbers in italic are calculated by the author.
Note 2: IFP and BGR numbers are average values from published range.
Note 3: Swindell does not report this gas as part of his total world fossil fuel resources..

Due to the lack of data, it is not possible to rank the top ten countries in the world in terms of geopressurised and hydropressurised gas resources quantitatively. However from the vague data that is available and with the help of the groundwater maps available in the public domain, it is possible to 'guestimate' the top five countries, which are listed in Table 10.18.

Here also, only the BGR provides estimates of geopressurised and hydropressurised gas on a regional level. Since there is no set of detailed data on a national level, it is impossible to detail these resources for the Arab world on a country-by-country basis. Though using the regional estimates of BGR for the Middle East

---

[40] Uwe Remme, Markus Blesl & Ulrich Fahl, Global Resources and energy trade: An overview for coal, natural gas, oil and uranium, IER, Universität Stuttgart, 2007.
[41] Ibid.
[42] BGR, Energierohstoffe 2009 – Reserven, Ressourcen, Verfügbarkeit.

Table 10.18: World geopressurised and hydropressurised gas reserves and resources – top five countries (2007)

Country	Notes
Russia	Note 1, 2
Australia	Note 1, 2
USA	Note 1
Canada	Note 1
Brazil	Note 1

Source: Uwe Remme, Markus Blesl & Ulrich Fahl, Global Resources and energy trade: An overview for coal, natural gas, oil and uranium, IER, Universität Stuttgart, 2007.
Note 1: Reported unquantified deposits.
Note 2: according to Remme et al. (2007) has just over 10% of world's total.

Table 10.19: Geopressurised and hydropressurised gas reserves and resources in the Arab countries (2007)

	Original Gas In Place	Remaining Resources	Reserves	Share
	trillion standard m^3			%
Arab World	2349	70	0	10.6
World	22105	663	0	

Source: Uwe Remme, Markus Blesl & Ulrich Fahl, Global Resources and energy trade: An overview for coal, natural gas, oil and uranium, IER, Universität Stuttgart, 2007.
Note 1: Arab world's share is calculated by the author.

and Africa as a basis[43], geopressurised and hydropressurised gas resources estimation in the Arab world is shown in Table 10.19. The estimation methodology resembles the one implemented in calculating methane hydrates in the Arab world. The difference is, to use the total area, instead of the exclusive economic zones area. The share of the Middle East and Africa of the total world's resources is calculated by apportioning the Arab world a share of the geopressurised and hydropressurised gas resources proportional to the land areas of the Arab and non-Arab countries in the Middle East and Africa. This is a very rough estimate, and needs revising as soon as more data become available. It is estimated that the Arab world's share of the geopressurised and

---

[43] Ibid.

hydropressurised gas will be around 10.6% of the world's total. Using a conservative resources estimate of 663 trillion standard m^3 for the world, it is assessed that the Arab world will be home to 70 trillion standard m^3 of geopressurised and hydropressurised gas, which is a respectable quantity. Thus if these gas deposits are ever to be developed and become a factor in the energy markets, the significance of the Arab world in natural gas markets will be maintained, as its share of the world's gas will remain important.

10.2.6 Reliability of Unconventional Natural Gas Reserves and Resources Data

As with oil, there are no standards on reporting unconventional natural gas reserves and resources data. Here again, not only is the definition of what is unconventional is not agreed, but also unlike conventional natural gas reserves and resources data, almost all official numbers published by governments ignore unconventional natural gas. As seen in the previous section, most data were obtained from unofficial data sources such as the BGR, with even the WEC completely ignoring unconventional gas in its reports. The data presented here are based on rough estimations or are from scientific publications, neither of which is endorsed by any official organisations. None of the data is complete, as there are no detailed data reported for individual countries. Furthermore, no independent verification or audit has ever been done on a complete set of data. Many countries have never conducted a full assessment of their unconventional natural gas potential, especially in the Middle East and Russia, where the countries see no compelling reason to do so, while their conventional natural gas resources are in abundance. Thus one can state confidently that there are far more unconventional resources to be identified and the numbers reported to date only give a partial and incomplete picture, but with time, the blur in the picture will disappear as more data become available. As an example, some sources already report more tight natural gas and shale OGIP, in the USA alone, which exceeds BGR numbers in the whole of North America.[44]

---

[44] Advanced Resources International, 2009, (http://www.adv-res.com).

The following discussion is almost identical to unconventional oil reserves and resources estimates. A major cause of unreliability in the unconventional data is the persistent confusion in terms between resources and reserves and is therefore re-emphasized here. Refer to Section 6.3 for a comprehensive discussion, where examples were presented that highlight clearly the confusion and the wide spread mix-up between the two terms.

Also, the wide range of uncertainty in recovery factors used plays a large role in the reliability of the estimates. For example, a 1% increase in the methane hydrate deposits recovery factor from 1% to 2% will increase the world's resources by over 605 billion standard m^3. As stated earlier, in this book the practice has been always to adopt conservative data, and thus adopt the recovery factors that arrive at conservative estimates of reserves and resources.

Interestingly, until now, political conditions have had no significant effects in manipulating unconventional natural gas data and thus affecting its reliability, as many of the arguments regarding these reserves and resources are still in the scientific domain, and so did not enter into the political arena. This situation is expected to change soon, with more and more countries starting to uncover their unconventional natural gas potential, thus inevitably, political wrangling will come into play and start manipulating the published data.

Another significant source of data confusion is that some data sources report several unconventional categories together, such as the Petroleum Economist, who reported a combined resource of tight natural gas, shale gas and coalbed methane, without giving any details[45]; while others, such as the BGR, amalgamate tight gas and shale gas as one resource. This latter practice is inaccurate as, even though the recovery factors from the two resources can be close, their economics differ significantly. So,

---

[45] N J Watson, Unconventional gas could add 60-250% to global reserves, Petroleum Economist, 2009, (http://www.petroleum-economist.com/default.asp?page=14&PubID=46&ISS=25487&SID=722880).

even though this practice does not affect gas-in-place data, it has a determinant effect on estimating actual resources and reserves data, which leads to inconsistent estimates.

Finally, a major confusion which has already been highlighted earlier in this chapter is the inclusion of some conventional natural gas resources, such as Arctic or deepwater natural gas, in the unconventional natural gas data. As already stated in this book these resources and reserves are treated as conventional natural gas resources and reserves, and thus are included in conventional natural gas quantities.

10.3 Overall Natural Gas Reserves and Resources

Based on the data presented in the previous sections, the overall natural gas reserves and resources in the Arab world, plus their share to the world's total, can be estimated. The results of the estimates are shown in Table 10.20.

Table 10.20: Overall natural gas reserves and resources in the Arab world and their share to the world's total

		Conventional Natural Gas	Tight Natural Gas	Shale Gas	Coalbed Methane	Methane Hydrates	Geopressurised & Hydropressurised Gas	Total
		\multicolumn{7}{c}{billion standard m^3}							
Remaining Reserves	Arab world	53.2	3.3	10.1	0.0	0.0	0.0	66.6
	World	177.1	39.6	91.1	100.6	0.0	0.0	408.5
	Arab world Share (%)	30.0	8.4	11.1	0.0	N/A	N/A	16.3
Remaining Resources	Arab world	94.0	8.7	20.2	0.0	30.8	70.5	224.2
	World	441.1	106.9	182.4	227.4	605.0	663.2	2226.1
	Arab world Share (%)	21.3	8.1	11.1	0.0	5.1	10.6	10.1
Original Gas in Place	Arab world	111.3	16.7	50.6	0.0	3081.9	2349.1	5609.6
	World	590.1	207.8	456.4	252.8	60500.0	22105.2	84112.3
	Arab world Share (%)	18.9	8.0	11.1	0.0	5.1	10.6	6.7

Source: Calculated based on data in Sections 10.1 and 10.2.

It can be seen that the share of the Arab world, in terms of overall reserves and resources, decreases significantly and stands at approximately 16.3% and 10.1% of overall remaining reserves and resources respectively. As shown previously in Section 10.2, the Arab world is extremely poor in terms of coalbed methane resources and is relatively poor, in terms of tight gas resources and methane hydrates. However, looking at the numbers of the reserves illustrates that the Arab world remains very significant. Even though its share of the total natural gas reserves decreases when unconventional reserves are taken into consideration, it remains high, accounting for a sixth of all reserves and more significantly, its share comes mostly from the easier and more feasible types. One has to note that this percentage will inevitably change (most probably decrease) in the future, with better recovery factors for both conventional and unconventional resources with continuous exhaustion of conventional reserves. It is anticipated that the share of unconventional reserves will grow, leading to an eventual decline in the share of the Arab world of total natural gas reserves, though it will remain significantly higher than the Arab world's share of the total natural gas resources, and thus cementing the position of the Arab world (if taken as one entity) as a leading force of recoverable natural gas in the world. However, the decline in the Arab world's share may prompt the Arab countries to explore more unconventional deposits to defend their share, which may lead to an increase in the Arab world share, if more unconventional reserves are discovered.

10.4 Natural Gas Production

Unlike the well publicised pessimism of peak oil theory advocates, that warns of an imminent oil production peak, which will lead to a severe decline in oil production, very few voices are heard that warn of peak natural gas production, and their warnings of severe declines in gas production are, to date, mere whispers.

Natural gas was initially accidentally discovered while looking for oil and so it was often produced in oil fields. Initially natural gas was often seen as a nuisance and used to be flared. This attitude changed mainly after the Arab oil embargo in 1973, where the view regarding natural gas changed radically and it was viewed as a valuable energy source that can substitute oil in some uses. Furthermore, the drive towards greener policies and emissions reduction, promoted natural gas further as its emissions are lower than oil or coal.

Currently the majority of natural gas is produced onshore, with a share of approximately 74%, though offshore production is quickly catching up and accounted for 26% in 2006.[46]

The terminology of natural gas production is rather confusing with different data sources often reporting different quantity types, which many media outlets mistakenly use interchangeably. Three types of production terms are often used and must be differentiated: gross, marketed and dry production. *Gross production* is the full well stream produced from a gas well, excluding lease condensates. It includes natural gas plant liquids and all non-hydrocarbon gases. Parts of the gross production are reinjected into the gas wells to maintain their pressure or are vented or flared; however the majority of the gas, which is the remainder gas after removing most non-hydrocarbon gases in treating or processing operations, is labelled *marketed production*. Finally *dry production* is the marketed production less gas extraction loss, which includes gas lost in transmission and shrinkage, the latter is the volume of natural gas that is transformed into liquid products during processing, primarily at natural gas processing plants. To put these quantities in context, according to Cedigaz[47], 3.5% of gross production in 2008 was flared or vented, 11.4% was re-injected, while 5.6% was lost in transmission and shrinkage.

Nowadays the upward trend in natural gas production is irrefutable and is expected to continue as long as there is demand. This demand is projected to grow by all major forecasts. The demand is projected to reach 4330 and 4300 billion standard m^3 in 2030 by the EIA and IEA respectively.[48,49] This contradicts some peak oil theory advocates, who extended their estimates to predict also natural gas production peak and who estimate the

---

[46] Armelle Sanière, Gas Reserves, Discoveries and Production, Panorama 2006, IFP, 2006.
[47] Cedigaz, Natural Gas in the World, 2009 Esition.
[48] EIA, International Energy Outlook 2009.
[49] IEA, World Energy Outlook 2009.999999

world production will peak by 2020.[50]

Also, it is important to note that historically, no decline in the global natural gas production capacity or actual output has ever been recorded, with reported short intermittent declines only affecting actual production due to localised domestic economic factors, political factors or technical and environmental incidents.

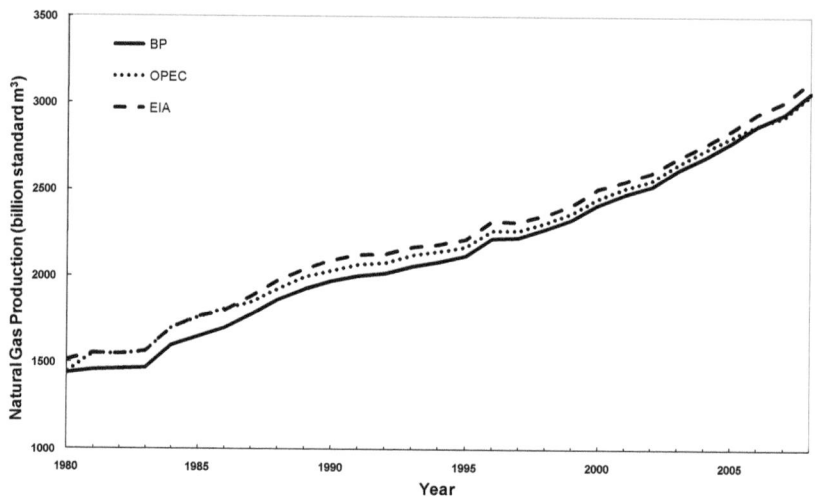

Figure 10.3: World natural gas production (1980-2008)

Source: EIA (http://www.eia.doe.gov/international); OPEC (Annual Statistical Bulletins 1999-2008); BP (Statistical Review of World Energy 2001-2009).

This indisputable upward trend to date is supported by all major data sources, who report that the world's natural gas production is continuing to increase. Figure 10.3 demonstrates this and shows the reported total natural gas production since 1980 as published by the EIA, BP, and OPEC, with all sources reporting an increase of approximately 107-113% in total natural gas production between 1980 and 2008. Total dry natural gas production numbers include both conventional and unconventional natural gas, but exclude NGL production, which is included within oil production. The numbers also exclude GTL production as counting it as an additional gas production will

---

[50] Bentley, R.W., Viewpoint - Global oil & gas depletion: an overview, Energy Policy 30: 189–205, 2002.

constitute double counting (see Section 10.4.1), and they also exclude other gaseous fuels, including biofuels (see Section 10.4.2). It can be seen that the differences in the reported data are insignificant and that the major data sources, not only agree on the upward trend, but also on the quantitative natural gas production. Unavoidably, there are slight differences and these differences can be attributed to different practices in data reporting (i.e. what is exactly included), with OPEC for example reporting marketed production, while the EIA reporting dry production. According to the EIA, the total natural gas production in 2008 is estimated to total 3136 billion standard m^3.

Similar to oil, reported natural gas production data are often more reliable and consistent than reported reserves and resources data. As explained earlier, this is largely due to the fact that the reported production data reflect actual tangible production, which is measurable and accounted for both physically and financially. Therefore it differs from estimates of reserves and resources, which uses a combination of mathematical models and physical facts to reach an assessment. Yet no one can claim that natural gas production data reporting is transparent. On the contrary, it is worryingly more confusing and non standard than its oil counterpart – though the confusion exists to a lesser extent than in reserves and recourses data. As an exercise, try picking a few natural gas production reports and compare the definition of production reported. Undoubtedly you will come across production numbers that refer to gross production, marketed production, dry production, production less flared and reinjected, etc. Furthermore, production numbers may include or exclude unconventional natural gas or include or exclude biogas and other gases. Moreover natural gas production is often reported in different units (e.g. m^3, ft^3, tonnes, boe, Btu, etc.). In this book, when referring to natural gas production, the term refers to dry production, includes conventional and unconventional gas, but excludes other gas production such as biogas.

Table 10.21 lists the leading ten countries in the world, in terms of dry natural gas production. These countries account for over 64.9% of the global production. An interesting point to note is that four of these countries are not in the top ten countries, in terms of natural gas reserves (refer to Table 10.1), which – similar to oil but contrary to the perceived wisdom by the public – illustrates that the level of natural gas production is not determined solely by the proved natural gas reserves, but depends on several factors including economics, politics and strategic considerations. Note that only four of the top ten countries are OPEC members.

Table 10.21: World total dry natural gas production – top ten countries (2008)

Rank	Country	Dry Natural Gas Production billion standard m^3	Share %
1	Russia	662.20	21.11
2	USA	582.22	18.56
3	Canada	170.94	5.45
4	Iran	116.30	3.71
5	Norway	99.20	3.16
6	Algeria	86.51	2.76
7	Netherlands	84.69	2.70
8	Saudi Arabia	80.44	2.56
9	Qatar	76.98	2.45
10	China	76.04	2.42
	TOTAL	2035.51	64.90
	WORLD	3136.24	

Source: EIA (http://www.eia.doe.gov/international).
Note 1: Totals may not add up due rounding.

The Arab world's overall dry natural gas production and its share to the world's total since 1980, are shown in Figure 10.4. The data shows that the dry natural gas production in the Arab countries increased steadily from 1980, almost quintupled by 2008, reaching almost 419 billion standard m^3. If taken as one entity, the Arab world will be placed at the third place of the top ten list in terms of dry natural gas production, with two thirds of the production belonging to the top placed Russia. Unlike the erratic trend observed in Arab oil production, the Arab natural gas production shows uninterrupted continuous increase, which suggests there is a disconnection between natural gas and political situation in the Arab world. This is actually untrue, since the main natural gas producers in the Arab world have been politically stable since 1980, avoiding the major political turmoil that engulfed the major Arab oil producing countries, such as the Iraq-Iran war, the Iraqi invasion of Kuwait, and the Anglo-American invasion of Iraq. Note that in Algeria, natural gas producing areas escaped the effects of the civil war. However even though production trends in countries that enjoy political stability, such as Norway or the UK, do not show production swings with production trends usually grow or decline gradually depending purely on economic and geologic factors. Production swings were witnessed in some countries due to political conditions, especially in Russia, whose short-lived 'dip' was due to the collapse of the Soviet Union and the turmoil that followed.

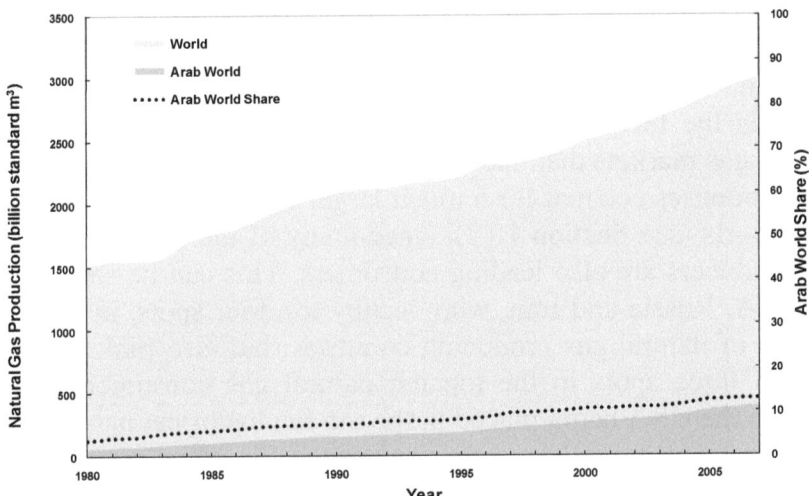

Figure 10.4: Arab world total dry natural gas production and its share to the world's total (1980-2008)

Source: EIA (http://www.eia.doe.gov/international).

The share of the Arab dry natural gas production, to the world total dry natural gas production, mirrored the Arab dry natural gas production trend qualitatively but not quantitatively. It can be seen that the share is in a continuous ascending trend, reaching approximately 13.4% in 2008, which emphasises the significance of the Arab world in terms of dry natural gas production, since its production is ranked third and accounts for almost a seventh of the world's total production.

In an identical position to oil, the fact that natural gas production in the Arab countries is not controlled by commercial companies (that are accountable for their share holders), but is instead totally controlled by national oil and gas companies (that are managed as governmental departments), resulted in considerably lower rates of investment in further production facilities. This hindered additional production to match the potential of the natural gas reserves, and thus resulted in the Arab world's share of production being only a third of its share of reserves. This is because the national oil and gas companies are treated as cash cows by the governments, who use their revenues for their political agenda rather than investing in additional production capacity.

The data discussed above does not tell the whole story. It conceals the fact that the Arab world is far more important to natural gas markets than the numbers suggest. This is because the Arab countries account for a much larger share of the natural gas net exports (see Section 10.7), since many of the leading natural gas producers are also leading consumers. This can be seen with the USA, Russia and Iran, who occupy top four spots, in the top ten list of natural gas producing countries, but also rank high in the top three spots in the top ten natural gas consumers. This leads to the USA being placed in the top ten list of net natural gas importers and drops Iran from the top ten list of net natural gas exporters. Several previous and current net natural gas exporting countries headed, and are heading, into becoming net importers, where the rapid increase in natural gas consumption is forcing them to cut their exports gradually until they turn eventually into net importers. Examples include the UK and UAE, who have turned already into a net natural gas importers, and China and Iran, who are heading that way! Refer to Sections 10.6 and 10.7 for more details.

The total dry natural gas production for all Arab countries in 2008, including details of production for each country, is reported in Table 10.22. A complete set of data from 1980 can be obtained

Table 10.22: Total dry natural gas production in the Arab countries (2008)

Country	Dry Natural Gas Production billion standard m^3	Rank	Share %
Bahrain	12.64	9	0.40
Iraq	1.88	12	0.06
Jordan	0.25	13	0.01
Kuwait	12.70	8	0.40
Lebanon	0.00		
Oman	24.00	6	0.77
Palestine	0.00		
Qatar	76.98	3	2.45
Saudi Arabia	80.44	2	2.56
Syria	6.04	10	0.19
UAE	50.24	4	1.60
Yemen	0.00		
Algeria	86.51	1	2.76
Comoros	0.00		
Djibouti	0.00		
Egypt	48.30	5	1.54
Libya	15.90	7	0.51
Mauritania	0.00		
Morocco	0.06	14	0.00
Somalia	0.00		
Sudan	0.00		
Tunisia	2.97	11	0.09
ARAB WORLD	418.91		13.36

Source: EIA (http://www.eia.doe.gov/international).
Note 1: Totals may not add up due rounding.

from *"http://www.2050consulting.com/books"*. From the above table we can detect that natural gas production processes are active in 14 Arab countries. We can classify these countries into five tiers, in terms of dry natural gas production: major, significant, medium, minor or no natural gas producing countries. The Arab countries that fall into this classification are as follows:
- Four countries belong to the first tier, with a total production exceeding 50 billion standard m^3 each, and a combined share of approximately 9.4% of the world's total dry natural gas production. These countries, in descending order of total dry natural gas production are:

Algeria, Saudi Arabia, Qatar and UAE, with the first three also ranked in the top ten list globally including top spot (see Table 10.21). Natural gas production plays a significant role in these countries respective economies, though it is ranked second as the main contributor to their GDPs, with the exception of Qatar where it is ranked first. All these countries are members of OPEC. Incidentally the production ranking does not reflect the reserves ranking, as already discussed earlier, however, with major projects underway in Qatar, its production is expected to soar and occupy the top ranking spot soon. An important fact to notice is that the UAE is a net natural gas importer, despite its huge production, while Saudi Arabia consumes all its production, leaving only Algeria and Qatar as net exporters.

- Five countries belong to the second tier, with a total production exceeding 10 billion but less than 50 billion standard m^3 each and a combined share of approximately 3.6% of the world's total dry natural gas production. These countries, in descending order, in terms of total dry natural gas production are: Egypt, Oman, Libya, Kuwait and Bahrain. Two countries, Libya and Kuwait, are OPEC members. Natural gas production plays a significant role in these countries respective economies, contributing substantially to their GDPs. However, the combined natural gas production from these countries plays an important role in the global natural gas market, especially in the LNG market, as will be explained in Section 10.7. An important fact to notice is that Kuwait and Bahrain consume all their production, leaving only Egypt, Oman, and Libya as net exporters.

- Two countries, Syria and Tunisia, belong to the third tier, with a total production exceeding 2.5 billion but less than 10 billion standard m^3 and a combined share of just under 0.3% of the world's total dry natural gas production. Natural gas contribution is significant to Syria's economy, whereas its contribution to Tunisia's GDP is minimal. Both countries are net importers.

- Four countries belong to the fourth tier with a total production less than 2.5 billion standard m^3 each and a combined share of less than 0.1% of the world's total dry natural gas production. These countries, in descending order, in terms of total dry natural gas production are: Iraq, Jordan, Morocco and Yemen. Only Iraq is a member of OPEC. Natural gas production plays a minor role in these countries respective economies, although with the development of production facilities in Iraq and Yemen, this is expected to change. It is worth noting that Yemen is absent from production numbers in 2008, as its commercial production only started in 2009, therefore it will feature in the tables soon and will then overtake Jordan and Morocco in terms of production. Two of these countries, Jordan and Morocco, are net natural gas importers (refer to Section 10.7). None of the above countries is an OPEC member.
- Finally, seven countries have no natural gas production operations. These countries are Lebanon, Palestine, Sudan, Mauritania, Djibouti, Somalia and Comoros. There are plans for production in Sudan, Mauritania and Palestine.

As with oil, a fundamental factor to be considered is to differentiate between natural gas production capacity and actual production. The latter is always reported. However, spare capacity is what really matters, as it plays a major role in the stability of the natural gas markets, it is vital in determining natural gas price and subsequently, the future of proposed production projects, hence providing additional future capacity!

However, spare capacity numbers are often hard to verify, they rely on political and economic factors, which affect their disclosure. The data presented earlier does not reveal the exact production spare capacity. However, in natural gas, the majority of this capacity is not controlled by one country, but is shared between different countries, with substantial spare capacity in Russia and Qatar. Therefore the effect of spare capacity in determining natural gas price is not as important as in the case of oil, and the price is more influenced by other market factors.

> ### The Middle East Perspective
>
> The Middle East region is the source of approximately 12.2% of the world's dry natural gas production, 69.3% of which are in its Arab countries. The region's countries occupy four places in the top ten list of leading natural gas producing countries. This indicates the significance of the region as a major natural gas producing area in the world, but also highlights that its production is far below its reserves share and thus suggests that the region has further potential in natural gas production capacity.
>
> Source: EIA, (http://www.eia.doe.gov/international), based on 2008 data.

> ### OPEC Perspective
>
> OPEC member states are the source of approximately 18.1% of the world's dry natural gas, 57.1% of which are in its Arab member states. The organisation's members occupy four places in the top ten list of leading natural gas producing countries. This illustrates the influence the organisation has in the natural gas markets. However, it also highlights that its production is far below its reserves share and thus suggests that the organisation has further potential in natural gas production capacity and a more significant role to play in shaping global energy policies and economics.
>
> Source: EIA, (http://www.eia.doe.gov/international), based on 2008 data.

10.4.1 Converted Natural Gas Production

Processes to convert produced natural gas to liquid fuel are technically possible and have been discussed in Section 2.8. These processes are secondary transformation processes and therefore their production numbers are already included in natural gas primary production quantities. So avoid double counting, if gas to liquid (GTL) is to be considered as part of oil production, a correction to natural gas production data has to be applied to exclude these values from natural gas production. Alternatively, if accounted for in natural gas production data, as is the usual practice, GTL production must be excluded from oil production data. In this book, the later approach is adopted.

GTL processes enjoyed publicity for the way in which refineries can convert some of their gaseous waste products into more valuable liquid fuels, and as an alternative way to develop stranded natural gas deposits. However, to date GTL processes have been confined to only three countries: South Africa, Malaysia and Qatar, with many other planned projects being shelved, such as in Algeria and Nigeria.[51] Currently Shell and Sasol run commercial facilities to produce liquid fuel from natural gas in Malaysia and South Africa respectively, with production capacity of only 40000 bpd. Both companies also constructed large facilities in Qatar, which have just started operations with production capacity of 175000 bpd. Accordingly it can be seen that the overall capacity of all GTL processes is trivial with less than 0.3% of the total oil production (see Section 9.4), and the projected increase in capacity is not expected to exceed 500000 bpd by 2030.[52] Consequently, we can conclude that GTL is unlikely to be a major constituent of the global energy market in the near future since its economics are very unfavourable, despite proven technology, with other alternatives such as LNG being more competitive.

On the other hand, as explained in Section 4.5, to date no commercial processes have been developed that convert natural gas into solid fuel.

10.4.2 Other Gaseous Fossil Fuels Production

As discussed in Section 3.9 earlier, gaseous fossil fuels are commonly derived from oil or coal using gasification or pyrolysis. However since these processes are secondary transformations of the oil or coal, where the original feedstocks of the oil or the coal have already been accounted for in the original oil or coal production numbers, the production of this 'secondary' gas is not included in any production numbers, though it is often included in consumption numbers as at that stage it is harder to distinguish the source of the gas.

---

[51] The National Petroleum Council (NPC), Working Document of the NPC Global Oil & Gas Study, Topical Paper #9, Gas to Liquids (GTL), 2007.
[52] Ibid.

An interesting point to consider in new coal gasification processes, where coal is to be gasified in-situ, coal is not produced physically per se, but it is processed and only syngas is produced. In this case the gas has to be accounted for. One option is to define this gas production number as a distinctive category since it is not strictly speaking natural gas, or alternatively, the other option is to calculate the amount of spent coal that has been used to generate the gas and add it, as part of the coal production. The latter approach is preferred as it keeps the coal and natural gas production numbers distinct, whereas the first option is statistically confusing and can lead to mistakes by placing different gases into one category.

10.5 Natural Gas Reserve to Production Ratio (R/P)

The reserve to production ratio (R/P) was introduced earlier in Section 9.5; please refer to that chapter as the discussion regarding its definition is not repeated here. We already demonstrated that the R/P ratio has defects and these defects are still applicable here. Using R/P as an indicator of how long a reserve will last is useless, since it neglects both production and reserve growth or reduction effects. In natural gas it neglects flared and vented gas as well. Therefore, it tells us nothing useful about either how many years of natural gas we have left or when we are likely to run out of natural gas. Furthermore, R/P ratios reported usually use only conventional natural gas reserves as their basis, thus ignoring the fact that some of the natural gas produced in the world is coming from unconventional resources.

Therefore, R/P ratios are not used as a credible analysis tool, they are only mentioned to complete the picture, while highlighting all their deficiencies that render them useless.

Figure 10.5 shows the world's R/P ratios since 1980, calculated from data obtained from the EIA. R/P calculations are based on conventional natural gas reserves only and exclude unconventional natural gas reserves; the natural gas production data on the other hand are based on dry natural gas production and thus exclude NGL production. The figure also shows the R/P ratio for the Arab world.

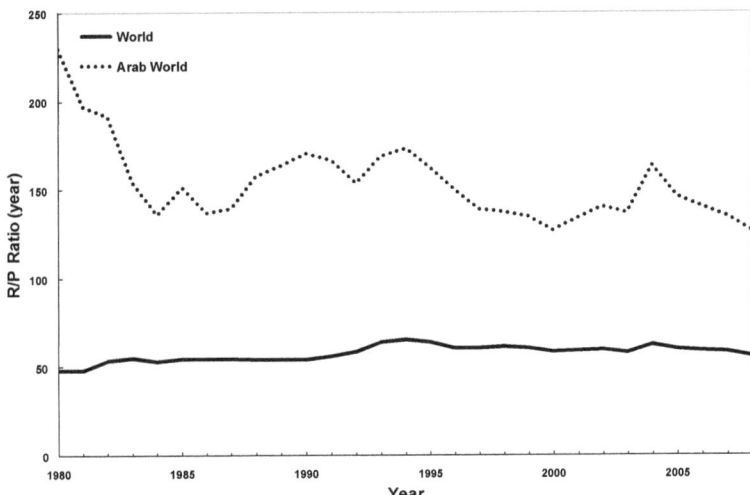

Figure 10.5: World natural gas reserves to production ratio (1980-2008)
Source: Calculated based on data in Sections 10.1 and 10.4.

The figure suggests that no imminent decline in natural gas production is envisaged, as the global ratio has been more or less stable since early 1990s and has increased by over 10 years since 1980, despite all natural gas utilisation. This seems very odd as this is very similar to the R/P of oil and, despite changes in both natural gas reserves and production, the ratio hardly changed. Is reverse engineering the reason for these similar figures? I am not suggesting this, but I leave it to the readers to draw their own conclusions.

Pay particular attention to the Arab world, the R/P ratio doubled. In fact the ratio fluctuated several times over 30 years, up or down, in a matter of few years, which does not reflect physical reality, but rather political and economic instabilities.

10.6 Natural Gas Consumption

As a result of calls for diversification in the energy markets, natural gas share of the total world energy consumption continues to grow, particularly in electricity generation sector where its share stands at a significant 25% (see Section 3.10). Whilst we keep hearing complaints from all sectors of society calling for less dependence on natural gas in the electricity sector, there is no indication that anyone is listening to these calls. Natural gas consumption continues to grow. Numerous power stations are being built as a fast and cheap way of adding electricity generation capacity and reduce emissions.

Natural gas consumption is often referred to in terms of dry consumption, which follows the definition used in dry gas production. Thus it excludes NGL, vented and flared gas as well as wet products separated or gas reinjected into the reservoirs. It accounts for all consumption utilised for fuel or chemical use. Obviously, consumption data includes natural gas produced from conventional and unconventional resources, as at this stage it is impossible to differentiate consumption origins.

Note that consumption quantities are calculated rather than a measured numbers. For practical reasons consumption excludes gas transformed into other fossil fuels (see Section 3.8), as this gas consumption is accounted for via the consumption of solid or liquid fuels, and there are no numbers released that distinguish this consumption. On the other hand consumption includes other fossil fuels transformed into gas and also biogas, as again there is no way to distinguish the source of gas once it is consumed. As already explained, this is in contrast to production where natural gas transformed into liquid or solid fuels is accounted for in natural gas production, whereas liquid or solid fuels transformed into gas are accounted for by their respective fossil fuels production.

Recent data has shown that growth in the world's natural gas consumption is accelerating. Figure 10.6 shows the reported natural gas consumption from different sources since 1980. It can be noted that, similar to trends observed in natural gas reserves or production data, the world's natural gas consumption shows a continuous increase since 1980, where between 1980 and 2008 consumption increased by around 110%, reaching just over 3145 billion standard m^3, according to the EIA. Reasons for this increase have already been mentioned and include energy diversification, by using natural gas as a substitute to other energy resources.

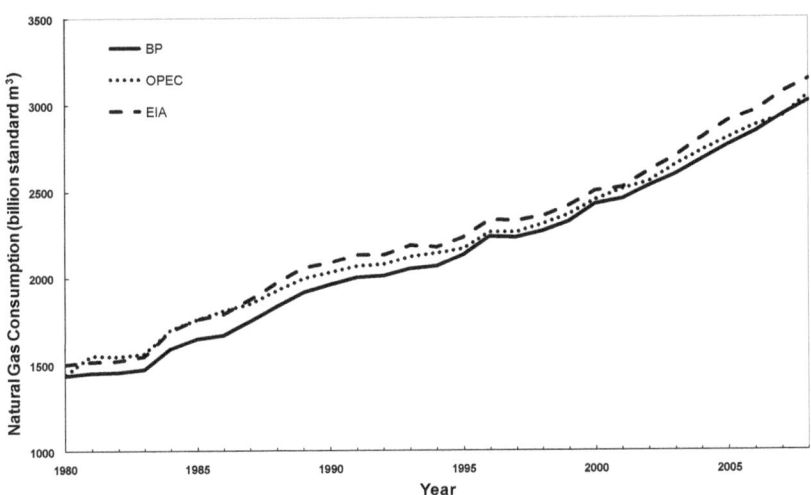

Figure 10.6: World natural gas consumption (1980-2008)
Source: EIA (http://www.eia.doe.gov/international); OPEC (Annual Statistical Bulletins 1999-2008); BP (Statistical Review of World Energy 2001-2009).

In the last 10 years Europe has become increasingly dependent on natural gas for its heating and electricity generation, this has been portrayed in the media as the main reason for the increase in the world's natural gas consumption. While it is a fact that consumption by European countries increased by 25% during that period, the whole continent combined, accounts for only 18.5% of the world's total consumption. The chief culprit, once again, in terms of natural gas consumption was, and undoubtedly still is the USA, which consumes almost 20.9% of the world's total, i.e. more than the whole of Europe.

Thus, with increasing public awareness regarding climate change and the politically stated need to reduce fossil fuels consumption, countries are coming up with different, creative, reporting methods to mask the true fossil fuel consumption picture portraying themselves in a better light to try and reduce public perception of their guilt and responsibility. This phenomenon has already been explained regarding oil in Section 9.7 and is re-emphasised here. As a result, (similar to measuring the oil consumption rate) measuring natural gas consumption rate has also become increasingly contentious issue on the global stage, as due to differences in economic and social circumstances, countries tend to prefer using different measures which portray them in a positive way. Thus they appear to be reducing their natural gas consumption rate, and cannot be accused of being the main offenders when policies to combat global climate change are discussed. Therefore rich countries, such as the USA, prefer measuring the natural gas consumption per GDP, while China and India favour measuring consumption per capita. Neither method gives a true indication of the actual natural gas consumption, as the numbers can be skewed to punish countries with small populations and reward rich countries with large populations.

Table 10.23: World natural gas consumption – top ten countries (2008)

Rank	Country	Dry Natural Gas Consumption billion standard m^3	Share %
1	USA	657.17	20.89
2	Russia	475.70	15.12
3	Iran	118.95	3.78
4	Japan	101.14	3.22
5	UK	95.94	3.05
6	Germany	95.79	3.05
7	Italy	84.88	2.70
8	Canada	82.93	2.64
9	Ukraine	80.80	2.57
10	Saudi Arabia	80.44	2.56
	TOTAL	1873.76	59.57
	WORLD	3145.42	

Source: EIA (http://www.eia.doe.gov/international).
Note 1: Totals may not add up due rounding.

Table 10.23 lists the top ten countries in the world, in terms of dry natural gas consumption. These ten countries combined account for approximately 59.6% of the world's total natural gas consumption. Unlike oil consumption, these countries are not all amongst the richest in the world in terms of total GDP. Poor countries appear on the list. The prime example is the Ukraine, which is very dependent on natural gas for its heating and industry. Countries with large populations inevitably have more needs, which lead to higher natural gas consumption. Surprisingly though, two OPEC countries are also on the list. These countries are Iran and Saudi Arabia. Saudi Arabia's natural gas consumption is disproportionally large, though this consumption is not for domestic use, but rather due to its usage as fuel in oil production operations. Note that Iran's natural gas sector is lagging behind and, until now is not even an exporter, despite massive reserves. This is due to international sanctions than hinder the sector's development.

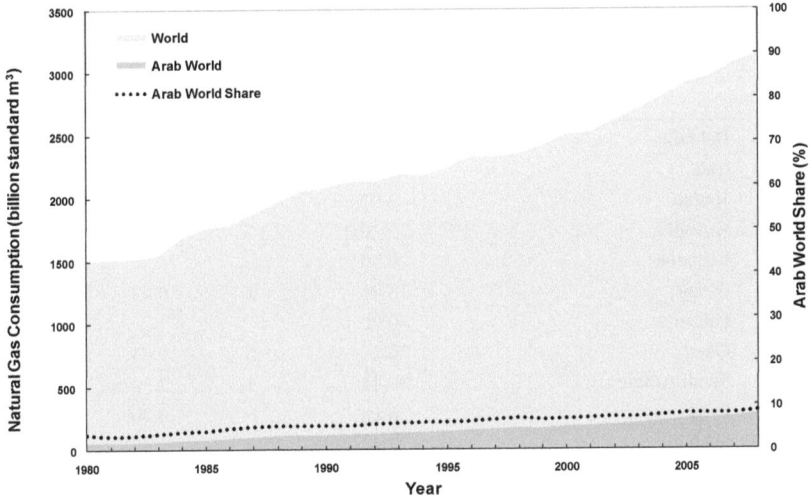

Figure 10.7: Arab world natural gas consumption and its share to the world's total (1980-2008)

Source: EIA (http://www.eia.doe.gov/international).

The Arab world's natural gas consumption and its share to the world's total, since 1980, are shown in Figure 10.7. The figure reveals that the Arab world's share of the world's total consumption is relatively small, barely exceeding 8% and reaching 8.9% in 2008, with a consumption of 278 billion standard m^3. Even if the Arab world is treated as one entity, it will be placed third in the top ten list, with just over a third of the USA consumption. Once again this reflects the relatively low state of development in the Arab countries, especially in terms of the industrial development, and the low usage of enhanced oil recovery methods (in comparison the USA) to maximise their oil production, which if it is to be performed widely, will result in substantial increases in natural gas consumption. Here one should not be fooled, as this relatively low natural gas consumption does not reflect environmentally friendly policies. Many Arab countries have very weak environmental protection legislation and the low consumption reflects a short-sighted view of still relying on oil rather than moving into another abundant resource with fewer emissions.

Table 10.24: Natural gas consumption in the Arab countries (2008)

Country	Dry Natural Gas Consumption billion standard m^3	Rank	Share %
Bahrain	12.64	8	0.40
Iraq	1.88	13	0.06
Jordan	2.97	12	0.09
Kuwait	12.70	7	0.40
Lebanon	0.00		
Oman	13.46	6	0.43
Palestine	0.00		
Qatar	20.20	5	0.64
Saudi Arabia	80.44	1	2.56
Syria	6.18	9	0.20
UAE	59.42	2	1.89
Yemen	0.00		
Algeria	26.84	4	0.85
Comoros	0.00		
Djibouti	0.00		
Egypt	31.38	3	1.00
Libya	5.50	10	0.17
Mauritania	0.00		
Morocco	0.56	14	0.02
Somalia	0.00		
Sudan	0.00		
Tunisia	4.22	11	0.13
ARAB WORLD	278.39		8.85

Source: EIA (http://www.eia.doe.gov/international).
Note 1: Totals may not add up due rounding.

The natural gas consumption for all Arab countries in 2008,

including details of the latest consumption estimates per country, is reported in Table 10.24. A complete set of data from 1980 can be found at *"http://www.2050consulting.com/books"*. The above table lists dry natural gas consumption in 22 Arab countries, where it can be seen it is only consumed in 14 of them. In terms of natural gas consumption, we can classify the countries into five tiers: major, significant, medium minor and non-natural gas consuming countries. The Arab countries fall into this classification as follows:

- Two countries belong to the first tier, with a total consumption exceeding 50 billion standard m^3 each, having a combined share of approximately 4.5% of the world's total consumption. These countries, in descending order of dry natural gas consumption are Saudi Arabia and UAE, with Saudi Arabia appearing in the top ten list of consumers (see Table 10.23). Both countries are members of OPEC. However both are not natural gas exporters and their large consumption mostly occurs in their oil production industries, with relatively small consumption used in other sectors, though the increasing domestic use in the UAE has recently turned the country into a net importer.

- Six countries belong to the second tier, with a total consumption exceeding 10 billion but less than 50 billion standard m^3 each, with a combined share of approximately 3.7% of the world's total consumption. These countries, in descending order of dry natural gas consumption are Egypt, Algeria, Qatar, Oman, Kuwait and Bahrain. Note that three of these countries are OPEC members. Interestingly, despite the relatively large natural gas consumption due to the large oil production and the need to use natural gas in the production process, the first four countries are net natural gas exporters. This reflects the fact that, for them, natural gas is a big revenue generator, and is still of very little use in other domestic usages such as electricity generation or heating.

- Four countries belong to the third tier, with a total consumption exceeding 2.5 billion but less than 10 billion standard m³ each, with a combined share of approximately 0.6% of the world's total consumption. These countries, in descending order of dry natural gas consumption are Syria, Libya, Tunisia and Jordan. Only Libya is an OPEC member and is a net natural gas exporter. The other three are net importers, using natural gas mostly for power generation and, increasingly, in other industrial areas.

- Two countries belong to the fourth tier, with a total consumption less than 2.5 billion standard m³ each, with a combined share of less than 0.1% of the world's dry natural gas consumption. These countries, in descending of dry natural gas consumption are Iraq and Morocco. Iraq is a member of OPEC, but despite its huge reserves is neither an exporter nor a big consumer, due to the poor state of oil producing facilities; the use of natural gas in enhanced oil recovery is minimal. Morocco on the other hand is a net importer, and has just started using natural gas in some industries. It is also planning to increase its usage in power generation.
- Finally, eight countries belong to the fifth tier, with no natural gas consumption. These countries are Lebanon, Palestine, Yemen, Mauritania, Sudan, Djibouti, Somalia and Comoros. Yemen recently turned into net gas exporter but still consumes no gas domestically! Some of the other countries are exploring using natural gas for electricity generation.

The Middle East Perspective

The Middle East region consumes approximately 10.5% of the world's dry natural gas consumption, only 63.6% of which are in the Arab countries of the Middle East. Two countries in the region occupy places in the top ten list of leading natural gas consuming countries including Iran in the third spot. This demonstrates that the Middle East as a region is still a relatively low natural gas consumer, and has great potential to export, though with Iran having big potential to consume more in EOR or fuel usage.

Source: EIA, (http://www.eia.doe.gov/international), based on 2008 data.

OPEC Perspective

OPEC member states consume approximately 12.8% of the world's dry natural gas consumption, 51.6% of which are in the Arab member states. Two countries in the organisation occupy places in the top ten list of leading natural gas consuming countries, with Iran in the third spot. This illustrates that the organisation's consumption is small, thus has great potential to export.

Source: EIA, (http://www.eia.doe.gov/international), based on 2008 data.

Table 10.25: Comparison of natural gas consumption data presentation in the Arab countries (2008)

Country	Dry Natural Gas Consumption standard m³/US$ GDP	Rank	Dry Natural Gas Consumption standard m³/capita	Rank
Bahrain	0.60	1	17368	2
Iraq	0.02	13	65	13
Jordan	0.14	8	468	9
Kuwait	0.08	11	4719	4
Lebanon	0.00		0	
Oman	0.22	3	3938	5
Palestine	0.00		0	
Qatar	0.20	4	24241	1
Saudi Arabia	0.17	6	2804	6
Syria	0.11	9	306	12
UAE	0.23	2	12384	3
Yemen	0.00		0	
Algeria	0.17	7	785	8
Comoros	0.00		0	
Djibouti	0.00		0	
Egypt	0.19	5	378	11
Libya	0.06	12	872	7
Mauritania	0.00		0	
Morocco	0.01	14	16	14
Somalia	0.00		0	
Sudan	0.00		0	
Tunisia	0.10	10	402	10
ARAB WORLD	0.15		788	

Source: EIA (http://www.eia.doe.gov/international); CIA – The World Factbook
Note 1: Totals may not add up due rounding.

As discussed above, inspecting the rate of consumption per GDP or per capita, leads to skewed results, as the increase in the oil price since 2006 results in an apparent disproportional reduction in natural gas consumption per GDP in the Arab countries. This

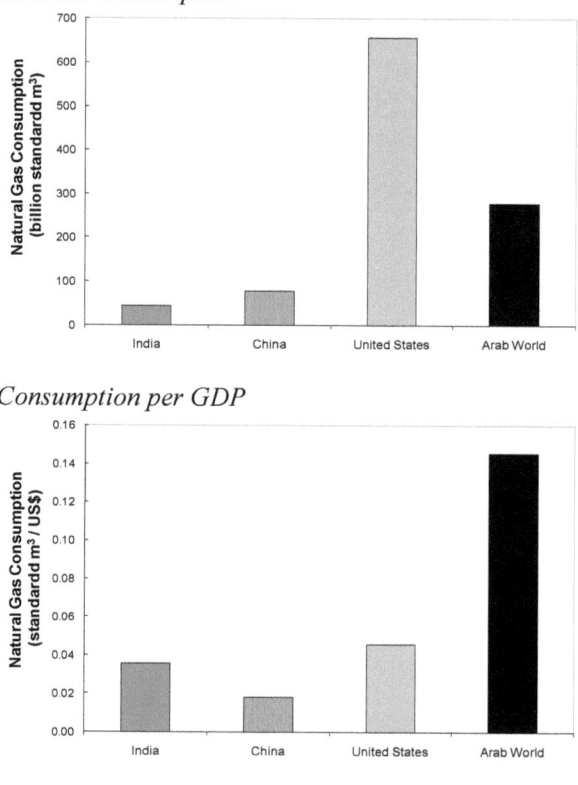

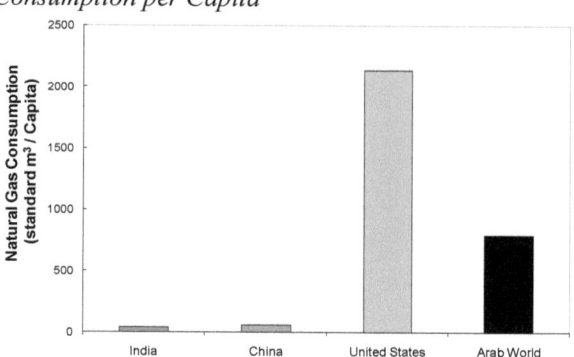

Figure 10.8: Comparison of natural gas consumption data presentation in the USA, China, India and the Arab world (2008)

Source: EIA (http://www.eia.doe.gov/international), CIA – The World Factbook

was masked even with the effects of the credit crunch of 2008-09. Similarly, the explosion in the population of the Arab countries results in a disproportional reduction in natural gas consumption per capita.

This is shown in Figure 10.8 and Table 10.25, demonstrating clearly how consumption data can be manipulated and presented as required! One can notice easily how the rankings just get muddled. This story is very similar to oil consumption story – refer to Section 9.7.

10.7 Natural Gas Trade
As is always the norm with commodities and repeating the story of the oil trend, the main natural gas producing countries are not the main natural gas consuming countries. Therefore, following the basic rules of supply and demand, natural gas is traded on the open market and is transported from producing regions to consuming regions, both internationally and within national borders.

Despite the growing trade in natural gas, it still lags way behind oil, which is the most traded commodity. This is mainly due to the limitations imposed by natural gas transport, which means that, until recently, several regional natural gas markets developed and operated semi-independently instead of being an integrated single global market, though the emergence of LNG trading has started changing this. Despite these limitations, natural gas trade is enormous. Using EIA data we can calculate that in the year 2008 just over 25% of total natural gas production was traded internationally.

Unlike most commodities, but similar to oil, the price of natural gas is not only determined by supply and demand forces, several other factors including political situations, weather conditions, and environmental laws play significant roles. Therefore predicting the price of natural gas is an extremely complicated issue and is outside the scope of this book. The price of natural gas and oil are not interlinked, since natural gas trade is usually locked into long term contracts, we can see that in the last two years, even though oil price recovered significantly after reaching its record price and dropping from there, the natural gas price failed to recover after its 2008 peak and this price is spiralling downwards.

Besides price, other factors affect natural gas trade. These include logistics, transportation and storage capacity. The costs of the latter two factors play an important role in defining trade patterns and trade partners. This has lead to the growth of LNG market as a convenient way of transporting natural gas to areas which were considered out of reach previously.

You may think this sounds familiar. It is true, but reporting natural gas trade numbers is confusing as there are no standards agreed, also due to political and commercial factors, not all the data is disclosed. The data can be reported in many different ways. The following are just few examples: in terms of volume, mass or energy; in terms of marketed gas, dry gas or LNG; detailing exports or imports or a net trade value.

Since any natural gas exported by a country is imported by another, theoretically speaking, the data for the world's exports and imports should be identical. However, a similar story to oil trade is repeated here, and data discrepancies often occur, with each data source using different methods to report trade data. To illustrate, note that here again BP, in its Statistical Review, reconciles the values of exports and imports by adding 'unidentified' quantity and thus reports identical exports and imports data. On the other hand, OPEC only reports exports data, while the EIA reports slight differences between exports and imports data, attributing this to various factors such as stock change and storage variations.

Figure 10.9 shows the reported natural gas exports and imports from different sources since 1980. As with the other data reported, it can be seen that all major sources estimates' do not differ significantly and the differences are often due to variations in compiling and reporting data, as explained earlier. Note that all sources agree that the net trade have massively increased, with no interruption, by over 300% between 1980 and 2008. According to the EIA's 2008 compiled data, the net exports and imports of dry natural gas stand at 788 and 797 billion standard m³ respectively.

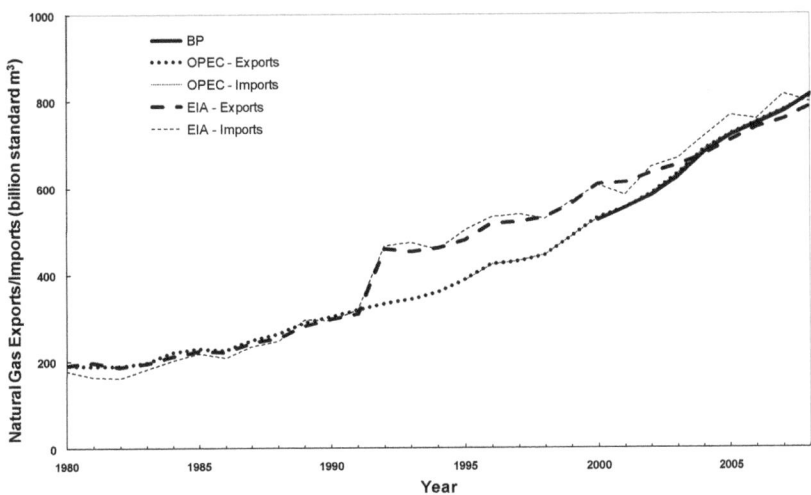

Figure 10.9: World natural gas trade (1980-2008)

Source: EIA (http://www.eia.doe.gov/international); OPEC (Annual Statistical Bulletins 1999-2008); BP (Statistical Review of World Energy 2001-2009).

Therefore, to simplify the matter, the same methodology used to present the oil trade is also used here. Therefore, in the remainder of this section, the natural gas trade data are reported in terms of an overall net value of either imports or exports, depending on the country. This is perceived to be the most relevant data that gives a full picture of a country's trade. This value is simply calculated as the difference between dry natural gas production (as defined in Section 10.4) and the dry natural gas consumption (as defined in Section 10.6), and as such it accounts for all traded natural gas whether in pipelines or LNG.

Table 10.26 lists the top ten countries in the world, in terms of net dry natural gas exports. These ten countries combined, account for a massive 83.4% of the world's total net dry natural gas exports. Unlike oil, this list is remarkably different to the world leading natural gas reserves countries list (see Table 10.1), with only four countries belonging to both lists. However, it is similar to oil, in terms of producing countries with six countries appear also in the top dry natural gas producing countries list (see Table 10.21). This situation is surprising as countries with large natural gas reserves are expected to be large producers, exporting their excess production, whereas, in the case of natural gas, big producers are big exporters. However countries with big reserves are not necessarily big producers or exporters. Four OPEC members appear in the list with the highest two occupying fourth and fifth spots.

Table 10.26: World net natural gas exports – top ten countries (2008)

Rank	Country	Net National Gas Trade (Exports) billion standard m^3	Share %
1	Russia	186.50	23.66
2	Norway	95.23	12.08
3	Canada	88.00	11.16
4	Algeria	59.67	7.57
5	Qatar	56.78	7.20
6	Turkmenistan	49.50	6.28
7	Netherlands	36.35	4.61
8	Indonesia	33.50	4.25
9	Malaysia	31.03	3.94
10	Nigeria	20.55	2.61
	TOTAL	657.11	83.37
	WORLD	788.22	

Source: Calculated based on data in Sections 10.4 and 10.6.
Note 1: Totals may not add up due rounding.

Table 10.27 lists the top ten countries in the world, in terms of net dry natural gas imports. These ten countries combined, account for a massive 71.5% of the world's total net dry natural gas imports. This number is noteworthy as it demonstrates that over two thirds of the world dry natural gas trade is dominated by ten partners! The net dry natural gas importing countries list shares six countries with the world leading dry natural gas consuming countries list (see Table 10.23). This situation is not surprising as countries with large dry natural gas consumption are expected to be large dry natural gas importers. The situation is further supported by the fact that, with the exception of the USA, none of the other countries on the top net importers are on the list of leading natural gas reserves countries. The USA is also amongst the top ten list of dry natural gas producers, however since its consumption exceeds its production markedly, it is a major net dry natural gas importer.

Table 10.27: World net natural gas imports – top ten countries (2008)

Rank	Country	Net Natural Gas Trade (Imports) billion std m3	Share %
1	Japan	95.78	12.01
2	Germany	79.43	9.96
3	Italy	75.63	9.48
4	USA	74.95	9.40
5	Ukraine	61.00	7.65
6	France	48.35	6.06
7	Spain	38.17	4.79
8	Turkey	36.16	4.54
9	South Korea	34.32	4.30
10	UK	26.04	3.27
	TOTAL	569.84	71.46
	WORLD	797.41	

Source: Calculated based on data in Sections 10.4 and 10.6.
Note 1: Totals may not add up due rounding.

Note that the numbers in the two tables above are for net dry natural gas exports and imports, defined effectively as the difference between total exports and total imports. Some countries can be exporting and importing simultaneously from different regions, which means that they can export from one region and import to another one within the same country, as is the case of the USA and the UK. In this section the matter of concern is whether a country is a net exporter or importer; the exact details of the trade are not pertinent to this book.

The Arab world dry natural gas exports and its share to the world's total since 1980 are shown in Figure 10.10. The figure reveals that the Arab world's share of the world's total has increased significantly from a tiny 1.4% in 1980 to just under 19.6% in 2008 and standing at just over 154.3 billion standard m³. If treated as a single entity, the Arab world would be placed second in the leading net dry natural gas exporters list, behind

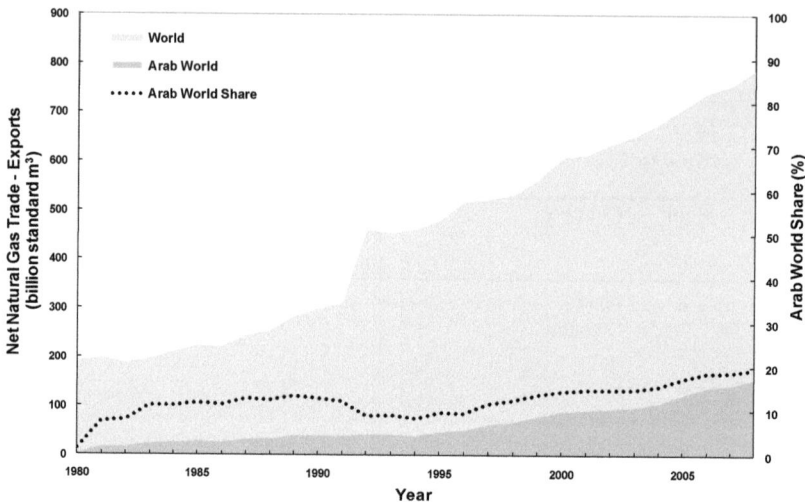

Figure 10.10: Arab world net natural gas exports and its share to the world's total (1980-2008)

Source: Calculated based on data in Sections 10.4 and 10.6.

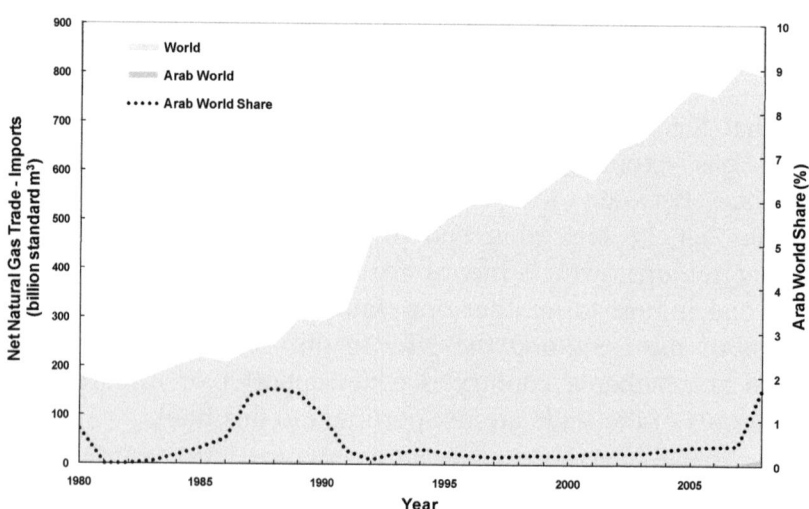

Figure 10.11: Arab world net natural gas imports and its share to the world's total (1980-2008)

Source: Calculated based on data in Sections 10.4 and 10.6.

world leader Russia, with its exports amounting to over 80% of the Russian exports. This confirms the Arab world's position as a major player in natural gas exporting world.

In a similar fashion, Figure 10.11 shows the Arab world net dry natural gas imports and its share of the world's total since 1980. The figure reveals that the Arab world's share of the world's total is insignificant; it remains below 2% of the world's total net imports recording slightly over 1.7% in 2008, where it stands at approximately 13.8 billion standard m^3 in 2008. If treated as a single entity, the Arab world still would not appear on the leading net dry natural gas importers list.

Both figures above treat intra-Arab trade as international trade since each Arab country reports its trade numbers separately. If the Arab world is taken as one entity then the intra trade has to be excluded and then a recalculation of the numbers shows that the Arab world will be a net exporter with a total global share of 18.1% in 2008, and obviously its net dry natural gas imports will become none.

The net dry natural gas exports for all Arab countries in 2008, including details of the latest estimates per country, are reported in Table 10.28. A complete set of data from 1980 can be found at "*http://www.2050consulting.com/books*". The above table shows that five Arab countries are net exporters. In terms of net dry natural gas exports, we can classify the countries into five tiers: major, significant, medium and minor dry natural gas exporting countries, with the fifth tier being those countries whose net dry natural gas exporting equals zero. Obviously some of these will be net importers, but others simply do not consume natural gas. The Arab countries fall into this classification as follows:
- Two countries belong to the first tier, with net exports exceeding 50 billion standard m^3 each, and a combined share of approximately 14.8% of the world's total dry natural gas exports. These countries, in descending order, in terms of dry natural gas exports are: Algeria and Qatar, both of which are ranked in the top ten list globally (see Table 10.27). Natural gas production plays a significant

role in these countries respective economies, where it is ranked as the main contributor to their GDPs. Both countries are members of OPEC. Incidentally the production ranking does not reflect the reserves ranking, however, with major projects underway in Qatar; its production is expected to soar.

- Three countries belong to the second tier, with total net dry natural gas exports exceeding 10 billion but less than 50 billion standard m^3 each, with a combined share of approximately 4.8% of the world's total dry natural gas exports. These countries, in descending order, in terms of dry natural gas exports are Egypt, Oman and Libya, with only the latter being an OPEC member. As Egypt and Oman are relatively poor, in terms of oil reserves and, with Egypt turning a net importer, natural gas is increasingly playing a significant role in their GDPs and both countries are major players in LNG markets. On the other hand, natural gas trade has a less important role in Libya, which relies heavily on oil exports.
- None of the Arab countries belong to the third tier, with total net dry natural gas exports exceeding 2.5 billion but less than 10 billion standard m^3.
- None of the Arab countries belong to the fourth tier, with net exports less than 2.5 billion standard m^3.
- Finally, the remaining seventeen countries belong to the fifth tier, with no net dry natural gas exports. These countries are Bahrain, Iraq, Jordan, Kuwait, Lebanon, Palestine, Saudi Arabia, Syria, UAE, Yemen, Mauritania, Morocco, Sudan, Tunisia, Djibouti, Somalia, and Comoros. Note however that Yemen started exporting natural gas as LNG in 2009, and thus its numbers will be updated from 2009 onwards, when it joined the club of exporting countries.

The net dry natural gas imports for all Arab countries in 2008, including details of the latest net dry natural gas imports estimates for each country, are reported in Table 10.29. A complete set of data from 1980 can be obtained from

Table 10.28: Net natural gas exports in the Arab countries (2008)

Country	Net National Gas Trade (Exports) billion standard m^3	Rank	Share %
Bahrain	0.00		
Iraq	0.00		
Jordan	0.00		
Kuwait	0.00		
Lebanon	0.00		
Oman	10.54	4	1.34
Palestine	0.00		
Qatar	56.78	2	7.20
Saudi Arabia	0.00		
Syria	0.00		
UAE	0.00		
Yemen	0.00		
Algeria	59.67	1	7.57
Comoros	0.00		
Djibouti	0.00		
Egypt	16.92	3	2.15
Libya	10.40	5	1.32
Mauritania	0.00		
Morocco	0.00		
Somalia	0.00		
Sudan	0.00		
Tunisia	0.00		
ARAB WORLD	154.31		19.58

Source: Calculated based on data in Sections 10.4 and 10.6.
Note 1: Totals may not add up due rounding.

"*http://www.2050consulting.com/books*". The table lists the imports of net dry natural gas in five Arab countries. In terms of net natural dry gas imports, we can classify the countries into five tiers: major, significant, medium and minor dry natural gas importing countries, with the fifth tier being those countries whose net dry natural gas importing equals zero. Obviously some of these will be net exporters, but others simply do not consume or produce natural gas. The Arab countries fall into this classification as follows:

- None of the Arab countries belong to the first tier, with a total net dry natural gas imports exceeding 50 billion

standard m^3.
- None of the Arab countries belong to the second tier, with a total net dry natural gas imports exceeding 10 billion but less than 50 billion standard m^3.

Table 10.29: Net natural gas imports in the Arab countries (2008)

Country	Net Natural Gas Trade (Imports) billion standard m^3	Rank	Share %
Bahrain	0.00		
Iraq	0.00		
Jordan	2.72	2	0.34
Kuwait	0.00		
Lebanon	0.00		
Oman	0.00		
Palestine	0.00		
Qatar	0.00		
Saudi Arabia	0.00		
Syria	0.14	5	0.02
UAE	9.18	1	1.15
Yemen	0.00		
Algeria	0.00		
Comoros	0.00		
Djibouti	0.00		
Egypt	0.00		
Libya	0.00		
Mauritania	0.00		
Morocco	0.50	4	0.06
Somalia	0.00		
Sudan	0.00		
Tunisia	1.25	3	0.16
ARAB WORLD	13.79		1.73

Source: Calculated based on data in Sections 10.4 and 10.6.
Note 1: Totals may not add up due rounding.

- Two countries belong to the third tier with a total net dry natural gas imports exceeding 2.5 billion but less than 10 billion standard m^3 each, with a combined share of approximately 1.5% of the world's total net dry natural gas imports. These countries, in descending order, in terms of total net dry natural gas imports are UAE and Jordan. Interestingly the UAE is a major producer,

however with its large consumption, it recently turned into a net importer, whereas Jordan has minimal natural gas production, and with its increasing reliance on natural gas for electricity generation, it is also a net importer.
- Three countries belong to the fourth tier, with a total net dry natural gas imports of less than 2.5 billion standard m^3 each, and a combined share of just over 0.2% of the world's total net dry natural gas imports. These countries, in descending order, in terms of total net dry natural gas imports are Tunisia, Morocco and Syria. As seen earlier, these three countries are natural gas producers; however, their gas demands exceed their production capacities and are therefore forced to import natural gas to plug the deficit.
- Finally the remaining seventeen countries belong to the fifth tier with no net dry natural gas imports. These countries are Bahrain, Iraq, Kuwait, Lebanon, Oman, Palestine, Qatar, Saudi Arabia, Yemen, Algeria, Egypt, Libya, Mauritania, Sudan, Djibouti, Somalia and Comoros.

The Middle East Perspective

The Middle East's share of the world's net natural gas exports is just over 8.5% of the total, 100% of which are from the Arab countries of the region. The region's countries are only represented by Qatar in the top ten list of leading net natural gas exporting countries, with Iran not even featuring in the exports list at all.

Predictably none of the Middle Eastern countries is in the list of leading net natural gas importers, and the region accounts for just over 1.8% of the world's net oil imports 82.9% of which are to the Arab countries.

The numbers illustrate the importance of the Middle East as the leading net exporting region in the world, with the situation very similar to the Arab world situation.

Source: Calculated based on data in Sections 9.4 and 9.7, *based on 2008 data.*

OPEC Perspective

OPEC member states are significant net natural gas exporters with a

> share of approximately 23.0% of the total, 70.1% of which are in its Arab member states. The organisation's members occupy four positions in the top ten list of leading net natural gas exporting countries, with Algeria occupying the highest rank at fourth.
>
> Predictably none of OPEC members is in the list of leading net natural gas importers, and the organisation accounts for only 1.7% of the world's natural gas imports, 68.9 of which is to its Arab member states. This surprising net overall natural gas import is due to UAE turning recently into a net importer.
>
> The numbers illustrate that OPEC can play a significant role in natural gas exporting markets, which further exceeds its production power.
>
> Source: Calculated based on data in Sections 9.4 and 9.7, based on 2008 data.

10.7.1 Natural Gas Transport

As already explained in Section 3.11, the majority of natural gas is transported using pipelines, though with the growth of LNG as a medium of natural gas trade, tankers are being increasingly used. In addition, a tiny proportion uses road transport (from Mexico to the USA).

Natural trade via pipelines accounted for over 72.2% of the world's total trade in 2008.[53] This mode of transport is the cheapest means of transporting natural gas, compared to using tankers for shorter distances, and has a distinct cost advantage for transporting gas up to 2000 km offshore and 3000 km onshore, though the distance can increase with increasing pipeline diameter and capacity. This advantage diminishes with increases in the transport distance required, and after a threshold, transporting gas using tankers has a cost advantage. This situation is different from oil transport, where tankers have the clear advantage. It reflects the high cost associated with the need to construct LNG export and import terminals, which affects adversely the cost of transport for shorter distances.

Natural gas pipelines are used both domestically and internationally The former transport natural gas from production areas to consumption areas by means of extensive national grids

---

[53] BP, Statistical Review of World Energy, 2009.

that consist of large network of pipelines criss-cross many countries. The latter transport natural gas across boundaries and seas between countries, where they can link two countries either directly or via transit countries.

Natural gas international pipelines are more widespread than oil pipelines, with significant networks connecting Russia with the former Soviet Union republics and with the European Union countries; Canada and the USA; Norway and the European Union as well as across South America. In the last few years several strategic natural gas pipelines have been constructed, mainly to export the Central Asian natural gas either westwards via Georgia to Turkey, eastwards to China, or southwards to Iran, as well as across Africa. In addition, numerous pipelines are on the drawing board all over the world, but mainly to connect Russia to East Asian countries; Nigeria to Europe via Algeria; Central Asia to Europe and Southeast Asian countries.

There are several natural gas international pipelines that connect the Arab world to other countries. Nowadays natural gas trade, via pipelines, is also the major mode of trade between the Arab world and its trading partners.

Currently several pipelines connect Algeria to Europe either directly under the sea or transiting via Tunisia or Morocco. These lines are: the Medgaz, connecting Algeria to Spain via underwater direct route, the Maghreb–Europe Gas Pipeline (MEG), connecting Algeria to Spain and Portugal via Morocco and the Trans-Mediterranean Pipeline connecting Algeria to Italy and Slovenia via Tunisia. In addition, a strategic trans-African pipeline is in the planning stages that will connect Nigeria to Algeria via Niger and then use current existing Algerian pipelines to transport gas to Europe. Also the GALSI pipeline will connect Algeria and Italy directly via submarine pipeline, with other Algerian pipelines to Europe also being planned. Other operating international pipelines are the Greenstream pipeline that connects Libya to Italy, and the Arish-Ashkelon that connects Egypt to Israel, while international pipelines are also proposed to connect Iran and Iraq, as well as Iraq and Turkey, so that Iraqi natural gas can be used as a potential source for the European Nabucco pipeline.

Furthermore, two major intra-Arab pipelines are operational: the Arab Gas Pipeline (AGP) and the Dolphin pipeline. The former connects Egypt to Jordan, Syria and Lebanon, with planned extensions to Turkey to connect to Nabucco, to Iraq, and to Cyprus. The latter connects Qatar to UAE and Oman, with planned extensions to Kuwait northwards that may also link to Bahrain, and Iraq to connect to Nabucco, and proposed extensions to Pakistan and India eastwards.

The practice of transporting natural gas in tankers as LNG has expanded rapidly in the last two decades. Even though the first LNG liquefaction terminal was inaugurated in Algeria in 1964, with the first shipments sent to the UK and France, the high costs of LNG terminals hindered the expansion globally. However, with increasing natural gas demand and prices, the natural gas market expanded rapidly,. In numbers, LNG trade did not exist till 1964, but increased steadily to reach 27.8% of the global natural gas trade in 2008[54], with LNG production levels increasing by seven fold since 1980.[55]

Obviously, the main advantage of transporting natural gas as LNG using tankers, over transporting the gas via pipelines, is that it allows for the globalisation of the natural gas markets rather than the regionalised markets that are the norm when trade is only conducted via pipelines. LNG made countries not limited to export or import natural gas based on distance constraints, but became rather free to source their gas from far away countries if the commercial conditions are acceptable. Furthermore LNG trading brought many previously stranded natural gas deposits into play and into the market. On the other hand, despite all its advantages, LNG has a major drawback, which is the relatively high cost. There are considerable initial costs associated with LNG trade as it requires building two terminals: liquefaction and

---

[54] BP, Statistical Review of World Energy, 2009.
[55] LNGpedia, 2009, (http://www.lngpedia.com/wp-content/html/Historical.html).

regasification terminals, even though the cost of transporting LNG in tankers is lower than pipelines for the same distance.[56]

Currently 16 countries have liquefaction LNG terminals, with the largest exporters being Qatar, Malaysia and Indonesia; while 19 countries have regasification terminals, with the largest importers being Japan, South Korea and Spain. The list of countries using this exclusive LNG club is expanding rapidly, with more countries joining every year as more LNG terminals springing up, transforming gas trade to become similar to crude oil trade. A complete transformation is envisaged to be gradual, as the majority of current LNG trade contracts are still long term and very little LNG is available on the spot market.

As already stated, the LNG industry started in Algeria, which inaugurated the first export terminal in 1964. Since then, Algeria has expanded its LNG capabilities and five more Arab countries joined the club, with Yemen the latest member joining in 2009. Other projects in the Arab countries are also being considered. The Arab countries accounted for 42.7% of the global LNG market.[57].

Thus far, regasification terminals exist only in Kuwait and the UAE, in the Arab world, and only a few are planned in the future. This lack of regasification activity is due to the fact that between the Arab countries, which are located in the middle of natural gas producing and exporting region, transporting natural gas via pipelines makes more financial sense.

---

[56] BGR, Energierohstoffe 2009 – Reserven, Ressourcen, Verfügbarkeit.
[57] EIA, (http://www.eia.doe.gov/international).

Chapter 11
COAL – GLOBAL AND ARAB PERSPECTIVE

For the last 200 years coal played a major role in human race development causing cultural changes and improving the standard of living. It was pivotal in providing the bulk of energy needed for the industrial revolution. Its role even extended to global politics, since coal was one of the main reasons that pushed towards establishing the EU, which was born as the successor of the European Coal and Steel Community. Even though coal is also a finite commodity, it differs from oil and natural gas fundamentally as due to the large volume of its endowment it is considered to be 'effectively' abundant and as such, it is widely accepted that its usage will end and be substituted by other alternative fuels before it ever runs out. Consequently debate analogues to peak oil (as already discussed in previous chapters) regarding peak coal is not really taken seriously and peak fossil fuel theorists are remarkably quiet, though some of them started making noises recently and it is expected that the time is approaching when they ignite this debate.

Thus, remembering that coal has been produced commercially for over 250 years and its production is yet to show any signs of decline, in this chapter, peak coal debate is not even entertained. This chapter aims to present a brief quantitative analysis of coal globally and in the Arab world, specifically in terms of four main aspects: reserves and resources, production consumption and trade.

Similar to Chapters 9 and 10, this chapter is structured systematically with each section starting by presenting the global standpoint of a certain aspect of coal. It is then followed by assessing and quantifying the contribution of the Arab world and then discussing the importance of the Arab world, in relation to that specific aspect. An overall analysis of the significance (or in this case irrelevance) of the Arab world's coal, along with its positioning into the overall energy picture globally, as well as any implications of this positioning are discussed in Chapter 12.

In a nutshell this chapter endeavours to answer three main questions:
- How much coal reserves and resources are there in the Arab world?
- How long will these coal reserves last?
- Will the Arab world's lack of coal affect its position as the world's main energy power broker?

After reading this chapter I once again encourage the readers to draw their own conclusions – with the help of the hints thrown along the way!

11.1 Conventional Coal Reserves and Resources

11.1.1 Conventional Coal Reserves

Compared to oil and natural gas data, reporting coal reserves data is even more erratic, on the one hand reserves estimates remain constant for many years without any updates, on the other hand they keep changing up and down depending on the price of coal with price changes forcing repeated evaluations of the reserves based on the economic conditions. The fact that reserves definitions for coal are not standardised between countries adds to the confusion. Obtaining coal reserves data has been harder than for other fossil fuels reserves data and finding updated time series also proved to be a mammoth task. For reasons unknown to the author, all data sources that report coal reserves break with the tradition practiced with other fossil fuels reporting. They report only the latest reserves data for a single year instead of the data for a time series that goes back several years. It is vital to bear in mind that all coal referred to in all sources is conventional coal, as the term unconventional coal is not customary and thus it is rarely in use since coal deposits are classified as either resources (some of which are reserves) or ignored totally (see Section 4.1). This position appears to be the consensus at the moment, although it is currently on its last legs, as it is being challenged, with the advancement of in-situ coal gasification, as we will discuss later.

Classifying coal into ranks has been introduced in Section 4.1. Since coal ranks differ significantly in many characteristics such as heating value, ash, moisture content and sulphur content, one expects that coal reserves should be reported per rank. This expectation is only partially met, and coal reserves are often reported in two main macro-categories rather than one overall reserve value. These two macro-categories are hard coals (which include anthracite and bituminous coal) and brown coals (which include sub-bituminous and lignite). This major division is deemed necessary by the industry as the two broad macro-categories differ significantly in terms of heating value and the applications for which they are used, although what is included in the macro-categories differ, so for example what the BGR reports as brown coal is different to what the IEA reports as brown coal. Historically reserves of hard coals have always been significantly higher than those of brown coals and due to their higher heating values, they have always been considered more valuable. However in this book we report coal reserves in terms of total coal as the exact definition of the ranks is not similar between countries, thus reporting the sum of all ranks minimises inaccuracies while at the same time simplifies the analysis, which is considered adequate for the message advocated by this book.

Not all major data sources report coal data. So, while for example the EIA, IEA and BP do, OPEC does not ignoring coal data entirely. When reporting reserves data, most sources report them in terms of recoverable reserves, which is a terminology analogous to proved reserves in oil and natural gas circuits. These reserves values are provided in various ways: overall coal amalgamating all ranks; divided into two or three macro-categories, each of which including few ranks or rarely detailed per rank.

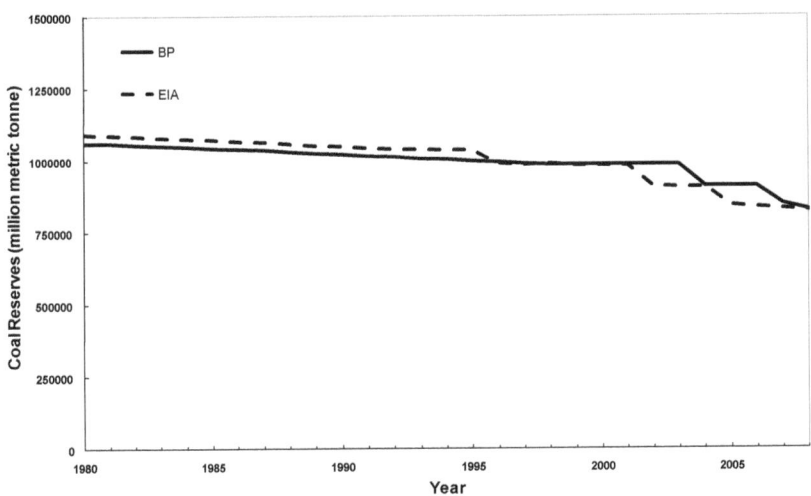

Figure 11.1: World conventional coal reserves (1980-2008)

Source: EIA (http://www.eia.doe.gov/international); BP (Statistical Review of World Energy 2001-2009).

Figure 11.1 shows the recoverable coal reserves since 1980, reported by the EIA and BP, with both sources reporting a modest decrease in recoverable reserves of approximately 20% between 1980 and 2009. Note that in this graph not all data points are reported explicitly, due to data unavailability. Therefore some data points are calculated by using as a reference point the data of the reserves available for a certain year. The reserves of the preceding or the following year are estimated by adding or subtracting the relevant coal production for the estimated year while assuming that no additional reserves are added to the reserves estimates through discovery or reclassification of resources. Examining the reserves data in the above figure shows that both data sources are in agreement that the world's coal reserves are declining, with insignificant quantitative estimates differences, which are unavoidable due to different practices in data reporting. The rather shy downward trend in reserves is evident and is even endorsed by the majority of media outlets. However both data sources and media agree that this reduction is due to many countries downgrading their reserves to resources because of economic conditions or environmental restrictions rather than deposits exhaustion. Interestingly one has to note that since they were first reported in 1930s, global coal reserves remained almost.

The data in Figure 11.1 show only conventional coal, which means coal assessed to be in recoverable deposits and excludes any other 'unconventional' coal. This is analogous to the situation for both oil and natural gas where unconventional reserves are excluded as well. It is observed that all major data sources. Using the data estimated from the EIA, the world's total recoverable coal reserves in 2009 stand at 825 billion metric tonnes, of which approximately 51% are hard coals and 49% are brown coals deposits. Graphite resources are not included and are considered trivial, accounting to only 71 million metric tonnes in 2009.[1]

Note that recoverable reserves estimates have fluctuated up and down frequently in the last few years. This is attributed to two main reasons, first, the continuous changes in coal prices, which continue to force the reclassification of marginally economical resources as reserves if the price increases and back into resources if the price decreases; and second, the tightening of the environmental legislation that restricts the usage of coal due to its high emissions, which forced many countries, especially in Europe, to abandon production from their reserves that turned uneconomical as a result of the imposed controls, thus forcing the deposits to be reclassified as resources.

Table 11.1 lists the recoverable coal reserves in the leading ten countries in the world, which demonstrates that they account for almost 92.5% of the global recoverable reserves. These countries are often the major coal producers and consumers, as we will see in the following sections. This highlights a fundamental difference between coal on the one hand, and oil and natural gas on the other hand, where countries rich with oil and natural gas reserves are not necessarily the leading producers or consumers. On the contrary, the norm is that producing countries and

---

[1] USGS, Mineral Commodity Summaries, Graphite (Natural), January 2010.

consuming countries are different, which creates the massive trade volumes in oil and natural gas markets. Also note that none of these top ten countries in the above table is an OPEC member. This demonstrates that the organisation is not important in terms of coal reserves and it has a weak position in coal markets, where it cannot rely on its reserves to manipulate and influence the prices, though it can still try to influence energy markets as a whole by manipulating all and gas prices, which will have an indirect effect on coal price.

Table 11.1: World conventional coal reserves – top ten countries (2009)

Rank	Country	Total Coal Reserves	Share
		million metric tonne	%
1	USA	236140.31	28.63
2	Russia	156113.29	18.92
3	China	107093.94	12.98
4	Australia	75445.36	9.15
5	India	55049.47	6.67
6	South Africa	47272.06	5.73
7	Ukraine	33692.77	4.08
8	Kazakhstan	30997.25	3.76
9	Serbia	13885.00	1.68
10	Poland	7058.58	0.86
	TOTAL	762748.04	92.46
	WORLD	824934.72	

Source: EIA (http://www.eia.doe.gov/international).
Note 1: Totals may not add up due rounding.

Figure 11.2 shows the Arab world total recoverable coal reserves and their share to the world's total since 1980. The data shows that the Arab world's reserves are minute and are estimated at approximately 80 million metric tonnes in 2009. If taken as a single entity, the Arab world will not feature in the top of the top ten list in terms of recoverable coal reserves – far from it in fact. The data thus reveals the irrelevance of the Arab world in terms of recoverable coal reserves, where its share of the world's total stands at a mere 0.01% in 2009. This share is a far cry from the Arab world's oil and natural gas massive shares and shows that, in terms of recoverable coal reserves, the Arab world is not even a player but rather a disinterested bystander in the coal market, where the reserves are more evenly distributed globally.

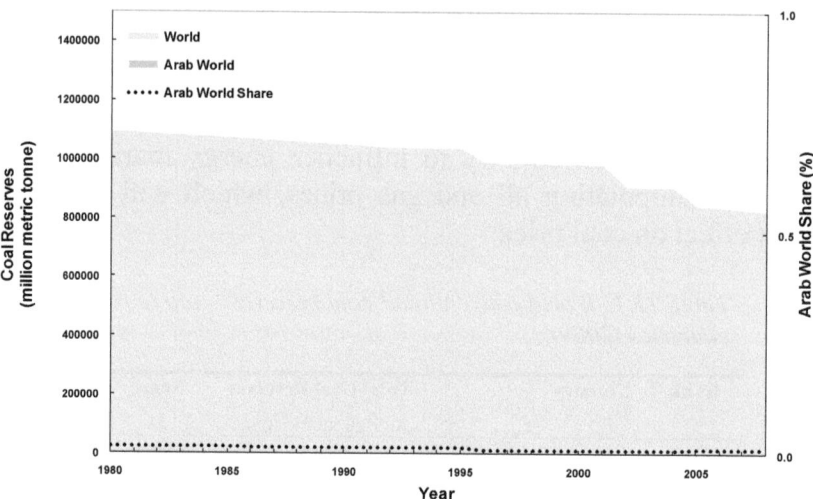

Figure 11.2: Arab world conventional proved coal and its share to the world's total (1980-2009)

Source: EIA (http://www.eia.doe.gov/international).

Similar to oil and natural gas, coal reserves are only one part of the equation that forms the coal market. Other parts include actual coal production and consumption and to a lesser extent, coal export and import facilities, particularly the geographical locations of both the deposits and the trading partners. In this market, global economic conditions and political conditions play only minor roles, in comparison to oil and natural gas. One point that deserves mentioning here is that, despite its weak influence, OPEC has a larger role in coal markets that is bigger than its reserves suggest. This is because Indonesia (an Open member, though with suspended membership) is a major coal exporter, as its production of coal is large while its consumption of coal is small. On the other hand, the fact is that Arab world's weight in the coal market is minimal as can be deduced from its deposits figures.

The total Arab world recoverable coal reserves in 2009, including details of the latest reserves estimates for all Arab countries, is reported in Table 11.2. A complete set of data from 1980 can be

Table 11.2: Conventional coal reserves in the Arab countries (2009)

Country	Total Coal Reserves	Rank	Share
	million metric tonne		%
Bahrain	0.00		
Iraq	0.00		
Jordan	0.00		
Kuwait	0.00		
Lebanon	0.00		
Oman	0.00		
Palestine	0.00		
Qatar	0.00		
Saudi Arabia	0.00		
Syria	0.00		
UAE	0.00		
Yemen	0.00		
Algeria	59.00	1	0.01
Comoros	0.00		
Djibouti	0.00		
Egypt	20.93	2	0.00
Libya	0.00		
Mauritania	0.00		
Morocco	0.00	3	0.00
Somalia	0.00		
Sudan	0.00		
Tunisia	0.00		
ARAB WORLD	79.92		0.01

Source: EIA (http://www.eia.doe.gov/international).
Note 1: Totals may not add up due rounding.
Note 2: Morocco's reserves are negligible.

obtained from *"http://www.2050consulting.com/books"*. The table above shows that recoverable coal reserves exist in only three Arab countries. We can classify countries into five tiers in terms of recoverable coal reserves, countries with major, significant, medium, minor or no reserves. The Arab countries fall into this classification as follows:

- None of the Arab countries belongs to the first tier, with recoverable coal reserves exceeding 10 billion metric tonnes.
- None of the Arab countries belongs to the second tier, with coal reserves exceeding 1 billion but less than 10

billion metric tonnes.
- None of the Arab countries belongs to the third tier, with recoverable coal reserves exceeding 250 million but less than 1 billion metric tonnes.
- Three countries belong to the fourth tier, with recoverable coal reserves less than 250 million metric tonnes each, and a combined share of less than 0.01% of the world's total coal reserves. These countries, in descending order in terms of recoverable coal reserves, are Algeria, Egypt and Morocco. None are ranked in the top ten list globally (see Table 11.1). Ironically, Algeria's tiny reserves are the largest reserves of any Arab country, however as an OPEC member rich in both oil and natural gas, it ceased developing its coal reserves. The other two countries are both energy hungry as a result they have developed and exploited their reserves. Currently only Egypt has any production activities, though they do not satisfy its consumption so it is forced to import its remaining coal needs. Historically both Algeria and Morocco mined some coal deposits. While Algeria decided to stop coal production in 1999 despite reserves remaining in place, Morocco's mines were commercially exhausted and ceased to produce in 2002. Some 'unfeasible' reserves are obviously still there, and are reported by the BGR though not by EIA.
- Finally 19 countries belong to the fifth tier with no recoverable coal reserves. These countries are Bahrain, Iraq, Jordan, Kuwait, Lebanon, Oman, Palestine, Qatar, Saudi Arabia, Syria, UAE, Yemen, Libya, Mauritania, Sudan, Tunisia, Djibouti, Somalia, and Comoros.

The Middle East Perspective

The Middle East region is home to a minute 0.2% of the world's recoverable coal reserves, all of which are in Iran with none in its Arab countries. This illustrates the unimportance of the region as a coal source globally.

Source: EIA, (http://www.eia.doe.gov/international), based on 2009 data.

OPEC Perspective

OPEC member states are home to approximately 0.7% of the world's recoverable coal reserves. The majority of the reserves are in Indonesia, with a tiny 1% of them in the organisation's Arab member states – namely Algeria. The organisation's members do not feature in the top ten list of recoverable coal reserves countries. This illustrates the weak position the organisation has in the coal markets, and demonstrates the limits of its role in an integrated energy market.

Source: EIA, (http://www.eia.doe.gov/international), based on 2009 data.

11.1.2 Conventional Coal Resources

Analogous to oil and natural gas resources data (see Sections 9.1.2 and 10.1.2), coal resources data are also not readily available. However in the case of coal the situation can be fairly described as dire. Therefore coal resources data need to be estimated using the latest available set of statistics. To achieve this, a comparable method, resembling the method implemented in attaining an estimate of conventional oil and natural gas resources, is carried out here to estimate conventional coal resources on a global level. Remember that data sources use different definitions of reserves and resources for coal, on the one hand, and oil and natural gas on the other hand. As a result all statistical input data has to be examined carefully to ensure choosing the relevant numbers and avoiding wrong numbers that may cause over- or under-estimation. The resultant estimation is clearly only a conservative approximation.

The methodology implemented was outlined earlier in Sections 9.1.2 and 10.1.2 by using the available data from different sources to calculate resources quantities. Here the methodology is applied with a few necessary modifications that are needed, due to the differences in the definitions, along with the availability of the reported reserves and resources data between coal and other fossil fuels.

The original coal-in-place (OCIP) first needs to be evaluated as using its value, other quantities can be evaluated based on the recovery factors proposed. This calculation assumes uniform recovery factors for all global coal resources, but this is not true as each coal field has a unique recovery factor and recovery factors differ significantly between mining methods, with higher factors for surface mining in comparison to underground mining. Nevertheless the use of a uniform recovery factor is justified to achieve approximate values, in the absence of detailed data for every single coal field and mine in the world.

As already discussed, the BGR[2] publishes what it labels 'resources estimates' on a country-by-country basis for fossil fuels, though its definition of resources differs between coal on the one hand and oil and natural gas on the other hand. These resources are in fact remaining coal-in-place, excluding recoverable reserves and cumulative production, thus they exclude remaining coal in exhausted mines (i.e. coal remaining after production). The sum of these resources, the recoverable reserves, the remaining coal in exhausted mines and the cumulative production leads to the original coal-in-place estimated quantity.

To perform this OCIP calculation, three quantities are already available in the public domain. The first quantity is the remaining coal-in-place data excluding recoverable reserves (labelled resources in BGR reports) and it is obtained from BGR[3]. Cumulative production data are obtained from several sources to cover the historic spectrum[4,5,6], and finally, recoverable reserves data are obtained from BGR[7] and EIA[8]. In a few instances one of the two quantities, either the 'resources (as labelled by the BGR)' or the reserves data, were estimated from each other if one of them was not available. The estimation was carried out based on uniform recovery factors that are outlined below with reserves data from both sources were compared and reconciled. Then the only quantity that needed estimating was the remaining coal in

---

[2] BGR, Energierohstoffe 2009 – Reserven, Ressourcen, Verfügbarkeit.
[3] BGR, Energierohstoffe 2009 – Reserven, Ressourcen, Verfügbarkeit.
[4] M K Hubbert, The world evolving energy system, Am. J. Phys. Vol 49, 11, p 1007-29, 1981.
[5] BGR, Energierohstoffe 2009 – Reserven, Ressourcen, Verfügbarkeit.
[6] EIA, (http://www.eia.doe.gov/international). Updated from 2008 numbers.
[7] BGR, Energierohstoffe 2009 – Reserven, Ressourcen, Verfügbarkeit.
[8] EIA, (http://www.eia.doe.gov/international).

exhausted mines, which has to be evaluated prior to performing the full OCIP summation. To evaluate this remaining coal in place, an average recovery factor based on the USA mining history is used, assuming recovery factors of 90% from surface mining and 50% from underground mining. Although nowadays 60% of the coal is mined underground, while 40% is extracted via surface mining[9,10,11], historically most coal was mined underground. Thus the average recovery factor used here is calculated based on the current 60:40 split, which means we are underestimating the remaining coal in disused mines, but this is done to stay conservative. Therefore the average recovery factor calculated is just over 69%. After reconciling the data for all countries, a global OCIP value is obtained (as is shown in Table 11.3), and is then used to back calculate all other quantities.

Table 11.3: World and Arab world conventional coal in place and coal resources (2009)

Quantity (million metric tonne)	World	Arab World	Share
	Estimated	Estimated	%
Original Coal in Place (OCIP)	22707581	584	0.0
Remaining Coal in Place (RCIP)	22389984	550	0.0
Ultimate Coal Recoverable Resources (UCRR)	6812274	175	0.0
Remaining Coal Recoverable Resources (RCRR)	6494677	142	0.0
Remaining Coal Resources Excluding Proved Reserves	5669743	62	0.0
	Reported	Reported	
Remaining Coal Proved Reserves	824935	80	0.0
Cumulative Production	317597	33	0.0

Source: EIA (http://www.eia.doe.gov/international); BGR (Energierohstoffe 2009 – Reserven, Ressourcen, Verfügbarkeit).
Note 1: All share values for the Arab world are below 0.05%.

The total recoverable coal resource is then estimated, which is considered to be the ultimate coal recoverable resources (UCRR). An ultimate recovery factor of 30% is used. This factor is based on calculating the current USA recovery factor from the EIA

---

[9] BGS, (http://www.bgs.ac.uk/mineralsuk/statistics/home.html).
[10] WC, (http://www.worldcoal.org/coal/coal-mining).
[11] EIA, http://www.eia.doe.gov/glossary/glossary_r.htm).

data, and including additional recovery by multiplying this recovery factor by 4, to account for the increase of coal recoverability due to in-situ gasification.[12,13] Based on the reconciled OCIP data calculated above, the UCRR can be calculated by multiplying the OCIP with the ultimate recovery factor.

As with other fossil fuels, the above estimated OCIP and UCRR numbers are calculated, not only considering deposits from producing and future coal fields, but also taking into account coal remaining in place in the fields that have ceased production, but still contain significant deposits that can be produced in the future if conditions change such as higher price, advances in technology, or using different production method (for example in-situ gasification instead of mining). The inclusion of the latter is important in order to achieve more accurate coal resources estimation and unfortunately is often overlooked by many data sources. Furthermore, all the above values include only recognised conventional coal deposits but exclude other coal occurrences such as coal in thin seams, coal with high ash content and scattered coal. Some of these may need to be re-evaluated with the imminent use of in-situ coal gasification, as until now there are no estimates of these occurrences whatsoever.

Thus, using the reconciled OCIP of 22708 billion metric tonnes, the UCRR is calculated to be 6812 billion metric tonnes by multiplying the OCIP estimates by 30%, ultimate recovery factor. This is significantly higher than the figure of 3400 billion metric tonnes estimated by the Intergovernmental Panel on Climate Change (IPCC)[14], who did not take in-situ coal gasification into consideration. By subtracting the cumulative production from the OCIP, the remaining coal-in-place (RCIP) can be calculated as 22390 billion metric tonnes. Also by subtracting the cumulative production from the UCRR, the remaining coal recoverable resources (RCRR) can be calculated, and is shown to be 6495 billion metric tonnes.

---

[12] EIA, (http://tonto.eia.doe.gov/energyexplained/index.cfm?page=coal_reserves).
[13] Clean Coal, (http://www.cleancoalucg.com/ucg_world_coal.aspx?id=68).
[14] SourceWatch, (http://www.sourcewatch.org/index.php?title=Coal_reserves).

Table 11.4: *Conventional coal in place and coal resources in the Arab countries (2009)*

Country	Original Coal in Place (OCIP)	Remaining Coal in Place (RCIP)	Ultimate Coal Recoverable Resources (UCRR)	Remaining Coal Recoverable Resources (RCRR)	Remaining Coal Resources Excluding Proved Reserves	Remaining Coal Proved Reserves	Cumulative Production
			million metric tonne				
Bahrain	0.00	0.00	0.00	0.00	0.00	0.00	0.00
Iraq	0.00	0.00	0.00	0.00	0.00	0.00	0.00
Jordan	0.00	0.00	0.00	0.00	0.00	0.00	0.00
Kuwait	0.00	0.00	0.00	0.00	0.00	0.00	0.00
Lebanon	0.00	0.00	0.00	0.00	0.00	0.00	0.00
Oman	0.00	0.00	0.00	0.00	0.00	0.00	0.00
Palestine	0.00	0.00	0.00	0.00	0.00	0.00	0.00
Qatar	0.00	0.00	0.00	0.00	0.00	0.00	0.00
Saudi Arabia	0.00	0.00	0.00	0.00	0.00	0.00	0.00
Syria	0.00	0.00	0.00	0.00	0.00	0.00	0.00
UAE	0.00	0.00	0.00	0.00	0.00	0.00	0.00
Yemen	0.00	0.00	0.00	0.00	0.00	0.00	0.00
Algeria	223.50	223.18	67.05	66.73	7.73	59.00	0.32
Comoros	0.00	0.00	0.00	0.00	0.00	0.00	0.00
Djibouti	0.00	0.00	0.00	0.00	0.00	0.00	0.00
Egypt	187.45	187.11	56.24	55.90	34.97	20.93	0.34
Libya	0.00	0.00	0.00	0.00	0.00	0.00	0.00
Mauritania	0.00	0.00	0.00	0.00	0.00	0.00	0.00
Morocco	172.76	140.13	51.83	19.19	19.20	0.00	32.64
Somalia	0.00	0.00	0.00	0.00	0.00	0.00	0.00
Sudan	0.00	0.00	0.00	0.00	0.00	0.00	0.00
Tunisia	0.00	0.00	0.00	0.00	0.00	0.00	0.00
ARAB WORLD	583.72	550.42	175.12	141.82	61.90	79.92	33.30

Source: EIA (http://www.eia.doe.gov/international), BGR (Energierohstoffe 2009 – Reserven, Ressourcen, Verfügbarkeit).

The same methodology described above is used to estimate the OCIP, RCIP, UCRR and RCRR in the Arab world. The results are shown in Tables 11.3 and 11.4. The Arab world's share of RCRR hardly differs from its share of reserves remaining below 0.1%, which confirms that the Arab world has no role to play in current or future coal markets. Also it illustrates that if coal usage becomes more prominent in the energy mix, the Arab world's role will decrease significantly. As expected though, the ranking of the Arab countries, in terms of coal resources is identical to their ranking in terms of conventional coal reserves. Immaterial in both!

A reminder that will not go away, but on the contrary will gain more prominence and that will alter the recovery factor in the coal resources, is that some coal resources will be utilised to produce energy using in-situ coal gasification. This technology will add additional recoverable reserves quantities not only from the conventional resources already identified, but also from some occurrences which are considered now to be outside any recognised resources. So to re-iterate, note that in-situ gasification will inevitably convert some of the conventional coal resources into reserves, as the technology will bring them firmly into a feasible possibility. Similarly, it will bring some of the coal occurrences which are not currently classified, even as resources, into the reserves and resources categories for the same reason. According to WEC[15] the additional reserves are 565 billion metric tonnes, which are assumed to be added to all countries with coal deposits but not on an equal basis, as geological factors will determine the exact allocation. Clean Coal Ltd[16], a coal consultancy, estimates that additional recoverable reserves from the USA alone will be 1000 billion metric tonnes.

11.1.3 Reliability of Conventional Coal Reserves and Resources Data

Following a now familiar practice, the reliability of conventional coal reserves data has been questioned by experts though, together with mass media and data sources they accept the enormity of the reserves and thus use the published data without fuss, still they question the lack of transparency in the provided data and the lack of updates. This situation is similar to natural gas and is equally strange when we consider that many of the arguments put forward by peak oil theory advocates to doubt conventional oil data, such as that most data reported have no independent auditing procedures and that the data seem to stay

---

[15] World Energy Council, 2007 Survey of Energy Resources.
[16] Clean Coal, (http://www.cleancoalucg.com/ucg_world_coal.aspx?id=68).

constant for years before having a sudden jump up or down are also true in the coal case. Still, no one seems to suggest that coal reserves data are inflated or exaggerated. Some plausible explanations are offered for this:
1. The majority of coal reserves discoveries are ancient and their estimations are well documented. As the deposits are physically more continuous and closer to the earth's surface, mining companies know that they can exploit deposits if there is demand. They see no need to search for additional deposits, unless it becomes necessary.
2. There is no talk of an impending coal peak, and few experts (with the exception of hard core alarmists) are predicting any shortage in production soon, therefore even alarmists are not focusing on the subject yet.
3. OPEC is not involved in coal and thus there is no coal production quota system that influences coal markets and gives rise to accusations of manipulating the data.
4. More coal deposits are located in areas where multinational mining companies operate and it is widely agreed that their numbers are more reliable, since they are regulated by international financial standards and are responsible to shareholders rather than governments.

It is important to note that some factors put forward to question the reliability of conventional oil or natural gas reserves and resources data, are also applicable in the case of coal. Amongst these factors are:
1. The multinational mining companies deliberately withheld data regarding part of the discovered recoverable reserves so that they can choose the best time of announcing them to the financial markets to maximise their impact on the share prices.
2. As explained in Chapter 6, it is easy to prove that coal reserves estimates can vary higher or lower, without the modification of any technical or physical parameters, just by changing economic factors, because this will mean that the appreciation of what is 'reasonable', when defining reserves, has changed.
3. Due to its abundance, many companies have no incentives

to invest in coal exploration activities. This is because they have enough reserves to maintain their production for years beyond the technical design life of the projects and so any money spent on additional exploration activities will have no return and will be deemed wasted.

In order to lessen doubts in the reported data, it is universally accepted that there is still huge potential for production from current coal fields. Thus there is no argument here regarding non-discovery of new major fields and no one bothers to suggest that the only major reason for increasing coal reserves is discovery. Hence by developing technology, alternative production methods, e.g. in-situ coal gasification, enhancements can be made to coal exploitation methods (analogous to enhanced oil and natural gas recovery methods). This will improve recovery factors, and improve feasibility of current recovery methods, which has the potential to increase coal reserves considerably.

An interesting note worth mentioning here is that some peak oil theory advocates have a tendency to discredit almost all data sources, as we already seen with both oil and natural gas data. In the case of coal this criticism reaches a new low, where for example Laherrère[17] cynically questions 'what he termed' the massive discrepancy between coal reserves data for Europe that are published by the BGR for 1993 and 1997, he uses this as a pad to launch a vicious attack discrediting all data. It is hard to believe that he did not notice that the data in that case was related to Russia (which is a transcontinental country) and was assigned to Europe rather than Asia for 1993 then, in the next estimate in 1997, it was reassigned back to Asia, so in fact there was no discrepancy. Malicious or incompetent, you decide.

Nevertheless, it is only fair to consider some limitations of coal reserves data, which are in line with criticisms of the conventional oil and natural gas reserves data. The most important criticism is that many data sources, including the EIA,

---

[17] Jean Laherrère, Uncertainty of data and forecasts for fossil fuels, Universidad de Castilla-La Mancha, 2007.

BP and IEA, publish data supplied by the governmental agencies without any independent verification. In addition, some odd and unexplainable discrepancies in coal reserves data are evident, which are often attributed to inaccurate or out-of-date data reporting For example the BGR[18] report that several former Soviet republics have no coal reserves when they already produce coal, as published by the same data sources who credit them with zero reserves. The same also is witnessed with data for France and Niger.

11.2 Unconventional Coal Reserves and Resources

A fundamental difference between coal and other fossil fuels is that there is no concept of unconventional coal. Coal occurrences are considered either resources, if they fulfil certain criteria (see Section 4.1), or not classified as resources at all. Examples of the these latter occurrences include deposits with ash content exceeding 50%, which are sometimes called *coal shale*, thin coal seams below certain thickness or isolated deposits.

With the development of in-situ gasification, the above attitude has been changing. The process can exploit some of the coal occurrences, turning them into recoverable resources or even recoverable reserves if the economics are right. Therefore the advancement of in-situ gasification is inevitably leading to the introduction of the term 'unconventional coal'.

Yet, to date, since the term unconventional coal is rarely used, there are no data available that evaluate its deposits in the public domain. Furthermore, even if some data will start to become available soon, estimating these reserves and resources will be significantly more difficult compared to the difficulties associated with estimating unconventional oil or natural gas reserves and resources (in Sections 9.2 and 10.2). It will face the usual hurdles including resource classification, definition, characterisation, and where to place the limits to differentiate conventional from non conventional.

---

[18] BGR, Energierohstoffe 2009 – Reserven, Ressourcen, Verfügbarkeit.

We must note here that, similar to both oil and natural gas, Arctic and Antarctic coal is classified as a conventional deposit. Whereas, in contrast to both oil and natural gas, deepwater coal is not even assessed in detail and the majority of its deposits are still considered inaccessible by traditional mining processes, unless undersea coal deposits are located very close to the dry land where mines can start onshore and extend under the sea to extract some of that coal. In-situ coal gasification will bring some of this coal into play.

Following from the discussion above, in this book, we follow the accepted convention, and exclude unconventional coal occurrences from the fossil fuel endowment. In the future, some of these disregarded coal occurrences may be even reclassified as resources. In the case of in-situ gasification, the coal resources may be classified as conventional if the gasification process is taking place in current conventional deposits, or they may have their own category if the process is taking place in unclassified occurrences.

However, to complement the picture of solid fossil fuels, peat reserves and resources are discussed in the next section as an analogous unconventional coal, even though, as already mentioned in Section 4.2, peat is strictly not coal but a precursor of it. It is included here since it can be used as a fuel in a similar way to coal and as fuel, its production and consumption data are often included within coal data implicitly. Thus, in this book, a proportion of peat resources are treated as potential resources and reserves, but rather than amalgamating them with conventional resources and reserves, they are labelled as 'unconventional resources and reserves'. The reserves are estimated based on conservative recovery factors values and the estimation methodology is similar to the one used for both unconventional oil and natural gas.

11.2.1 Peat
As already explained, peat is not coal, but a precursor that is widespread in wet areas and marshes. Dry, its heating value is comparable to lignite, therefore its data are included here for

completion since, theoretically it can be used as coal and transformed into other fossil fuels, even converted to liquid as coal-to-liquid (CTL).[19] This means it has to be included in fossil fuels analysis to plug the gaps and complete the picture. For more details on peat refer to Section 4.2.

Peat deposits cover approximately 2% of the world's land, with an estimated resource in place, of approximately 2 trillion metric tonnes.[20] Peat deposits are found in many places around the world, notably in Russia, Canada, USA, Finland, Sweden and Ireland.

Table 11.5: World peat reserves and resources

Data Source	Publication year	Reserves	Resources	Peat in Place
			billion metric tonne	
WEC	2007	*16*	*160*	1600
USGS	2000	20	200	2000
MII	2009	20	200	2000

Source: World Energy Council, 2007 Survey of Energy Resources; USGS (http://minerals.usgs.gov/minerals/pubs/mcs/2000/mcs2000.pdf); MII (http://www.mii.org/Minerals/photopeat.html).
Note 1: Numbers in italic are calculated by the author.
Note 2: WEC numbers in cubic metres, converted to tonnes assuming average dry peat density of 400 kg/m3 (http://www.simetric.co.uk/si_materials.htm)

Several assessments of world peat resources exist, most of which quantify it in terms of area rather than mass or energy. Table 11.5 summarises some of the latest estimates of the world's original peat-in-place (OPIP) as reported by different sources, though the data are converted to mass units to allow easier comparison to coal. The estimated resources and reserves are calculated using conservative recovery factors that take into account the potential of using peat as fuel only. These recovery factors are estimated from the data published by the WEC.[21] The data in the above table are calculated using the published data of OPIP and ignoring cumulative peat production for fuel from its deposits. This is justified, as the production is minute and thus falls within

---

[19] World Energy Council, 2007 Survey of Energy Resources.
[20] USGS, Mineral Commodity Summaries, Peat, February 2010.
[21] World Energy Council, 2007 Survey of Energy Resources.

the accuracy of the calculation. The reserves are estimated using a conservative recovery factor of 0.1%, which is very low since the economics of using peat as fuel are currently unfavourable. The resources are estimated using an ultimate recovery factor of 10%, which is significantly lower than coal but still is considered generous when one bears in mind that peat usage, as fuel, is in decline. Technically this decline can be reversed at any time since there are no constraints, with the exception of economics and political will.

Historically peat was an important energy source, where it once accounted for over 40% of the Soviet Union's electricity needs.[22] However its usage is in terminal decline and currently, commercial production from peat deposits for fuel takes place mostly in Europe, mainly in Finland and Ireland, though other countries use small quantities, including Belarus, Ukraine and Russia. Outside Europe minute usage is recorded in other places, e.g. Burundi and China.

Table 11.6: World peat reserves and resources – top ten countries (2007)

Rank	Country	Reserves billion metric tonne	Share %	Resources billion metric tonne	Share %	Peat In Place billion metric tonne	Share %
1	Canada	820.4	41.02	82042.5	41.02	820425.1	41.02
2	Russia	418.6	20.93	41858.4	20.93	418584.3	20.93
3	Indonesia	199.0	9.95	19897.5	9.95	198974.9	9.95
4	USA	157.7	7.89	15770.6	7.89	157706.0	7.89
5	Finland	65.6	3.28	6558.8	3.28	65588.0	3.28
6	Sweden	47.2	2.36	4716.4	2.36	47164.4	2.36
7	Malaysia	18.7	0.93	1868.9	0.93	18688.9	0.93
8	Belarus	17.7	0.88	1766.5	0.88	17664.6	0.88
9	Norway	17.5	0.87	1746.6	0.87	17465.6	0.87
10	UK	14.2	0.71	1419.4	0.71	14193.5	0.71
	TOTAL	1776.5	88.82	177645.5	88.82	1776455.4	88.82
	WORLD	2000.0		200000.0		2000000.0	

Source: World Energy Council, 2007 Survey of Energy Resources.
Note 1: Totals may not add up due rounding.

---

[22] Ibid.

Using the data from WEC[23], a derived list of the top ten countries, in terms of peat reserves and resources, is shown in Table 11.6. These countries account for almost 88.8% of the global estimated reserves and resources. The data in the above table is calculated using the same methodology described in deriving Table 11.5 above with a uniform ultimate recovery factor used for all countries.

Table 11.7: Peat reserves and resources in the Arab countries (2007)

Country	Reserves billion metric tonne	Rank	Share %	Resources billion metric tonne	Share %	Peat In Place billion metric tonne	Share %
Bahrain	0.0			0.0		0.0	
Iraq	13.2	1	0.66	1319.1	0.66	13191.3	0.66
Jordan	0.0			0.0		0.0	
Kuwait	0.0			0.0		0.0	
Lebanon	0.0			0.0		0.0	
Oman	0.0			0.0		0.0	
Palestine	0.0			0.0		0.0	
Qatar	0.0			0.0		0.0	
Saudi Arabia	0.0			0.0		0.0	
Syria	0.0			0.0		0.0	
UAE	0.0			0.0		0.0	
Yemen	0.0			0.0		0.0	
Algeria	0.2	4	0.01	16.2	0.01	162.1	0.01
Comoros	0.0			0.0		0.0	
Djibouti	0.0			0.0		0.0	
Egypt	0.3	3	0.02	33.9	0.02	339.0	0.02
Libya	0.0			0.0		0.0	
Mauritania	0.0			0.0		0.0	
Morocco	0.0			0.0		0.0	
Somalia	0.0			0.0		0.0	
Sudan	0.7	2	0.04	73.7	0.04	736.9	0.04
Tunisia	0.0	5	0.00	0.7	0.00	7.4	0.00
ARAB WORLD	14.4		0.72	1443.7	0.72	14436.7	0.72

Source: World Energy Council, 2007 Survey of Energy Resources.
Note 1: Totals may not add up due rounding.

Unlike conventional coal, it is interesting to note that a few countries, all located near the Arctic dominate peat resources and reserves. Note that one of the top ten countries, in terms of peat

---

[23] Ibid.

resources and reserves is a member of OPEC, and it accounts for just under 10% of the total peat resources and reserves.

Table 11.7 shows peat estimated resources and reserves in the Arab world, highlighting the Arab world's share of the world total resources and reserves. Using the methodology outlined above, the detailed country-by-country data in the table is estimated. It is interesting to note that the resources of peat are minor in the Arab world, where its share is a tiny 0.7% of the reserves and resources. This is expected as the Arab world is located in an arid zone and thus the area of wet lands it possesses is minimal. Interestingly peat resources in the Arab world exceed coal resources estimates, which mean in few million years the Arab world may possess more coal!

The table above suggests that although peat resources exist in five Arab countries, minute reserves exist, since peat is not used as fuel in any Arab country. Therefore even though similar tier classification to coal reserves can be applied, it is meaningless in this instance to categorise the Arab countries in terms of peat reserves.

11.2.2 Reliability of Unconventional Coal Reserves and Resources Data

We established already that the term 'unconventional coal' is not is use and therefore no such coal data are reported by any data source or published in the public domain, with the term occurrences sometimes used to describe some of these deposits. As a result any data sought is extremely difficult to locate, since one has to scan scientific papers and reports to collect scattered bits of information which are often incomplete, out of date, conflicting, and, to put it more bluntly, inadequate.

However with the new interest of in-situ coal gasification, one can confidently expect that more data will become available. Some of the so-called occurrences will be finally termed as unconventional coal, which means that there will be more unconventional resources to be identified and eventually the numbers will be reported and made available.

However, for the time being, the data reliability issue is of no real concern as the major arguments are centred on the identifying, characterising and classifying of the other coal occurrences into resources first.

The issue regarding the reliability of peat data is different when compared to data reliability of other unconventional fossil fuels. Peat data are often available since peat deposits are often identified and reported, however the data reported are regularly in poor format. Since peat is not used as fuel and, even if used as fuel, some countries classify it as renewable, such as Finland. Peat's data reporting is generally not manipulated or dressed up by governments as they appear to have no incentives to do so. It follows that peat data are not compiled with the same rigour of other fossil fuels and the reporting standards are significantly less strict. The data is often reported in area units, which is more suited to the horticultural uses of peat. Furthermore, the data are usually incomplete, inconsistent and out of date, with many data sources still quoting data estimated in 1990s.

11.3 Overall Coal Reserves and Resources

Table 11.8: Overall coal reserves and resources in the Arab world and their share to the world's total

		Conventional Coal	Peat	Total
		million metric tonne		
Remaining Reserves	Arab world	80	14	94
	World	824935	2000	826935
	Arab world Share (%)	*0.0*	*0.7*	*0.01*
Remaining Resources	Arab world	142	1444	1585
	World	6494677	200000	6694677
	Arab world Share (%)	*0.0*	*0.7*	*0.02*
Original Oil in Place	Arab world	584	14437	15020
	World	22707581	2000000	24707581
	Arab world Share (%)	*0.0*	*0.7*	*0.1*

Source: Calculated based on data in Sections 11.1 and 11.2.

Based on the data presented in the previous sections, the overall coal reserves and resources in the Arab world and their share to the world's total can be estimated. The results of the estimates are shown in Table 11.8.

It can be seen that the share of the Arab world, in terms of overall reserves and resources, is minimal and stands at less than 0.1% of overall remaining coal reserves and resources. This does not change the fact that the Arab world is irrelevant in terms of coal, and with these low numbers, Arab countries have no current incentives whatsoever to invest in exploration for any coal resources.

11.4 Coal Production
With abundant coal resources and reserves identified worldwide, the overwhelming majority of energy experts agree that there is no forthcoming peak coal point for at least a century to come. However, a few dissenting pessimists argue that peak coal is closer than we think, though their arguments advocating it are feeble at best and are not taken seriously, even by media sensationalists.

Historically coal production goes back to the 1700s, when the UK was the world leading producer. From then on, production increased rapidly with coal dominating energy markets throughout the 1800s and the first half of the 1900s. Its share peaked in 1910s when it was the source of over 70% of the world's total energy.[24] However, with the rapid emergence of oil as the fossil fuel of choice, coal was dethroned from its dominant position. It declined rapidly and was overtaken by oil in 1960s as the most dominant fuel, with coal's share deteriorating to levels similar to 1850s. By 1970, despite its share decline, coal regained some ground recently and cemented its place as an indispensible fossil fuel in the energy mix. Crucially one has to recognise that the decline in coal share, after its peak, was never as a result of running out, but rather as a result of the promotion of oil as a substitute fuel. Furthermore, the drive towards environmentally 'green' policies and emissions reduction is discouraging coal

---

[24] M K Hubbert, The world evolving energy system, Am. J. Phys. Vol 49, 11, p 1007-29, 1981.

usage in many places, as coal's emissions are higher than those from oil or natural gas.

At present almost all coal is produced by mining onshore, using either surface mining (40%) or underground mining (60%), with a tiny fraction produced from offshore mines, though until recently these offshore deposits were still accessed from onshore mines. Recent developments initiated offshore mining in Japan and China and also included investigating producing coal in-situ by gasification, which will bring a substantial amount of offshore coal deposits into play.

The quality of coal produced differs significantly between mines and unlike oil, there are no coals that are used as specific benchmarks to standardise the industry. Generally speaking, coal production is reported in two main macro-categories; hard coals (which include anthracite and bituminous coal) and brown coals (which include sub-bituminous coal and lignite). As explained earlier, this major division is deemed necessary by the industry as these two broad macro-categories differ significantly in terms of heating value and the applications for which coal is used. Historically production of hard coals has always been significantly higher than brown coals; this reflects both their higher reserves and their higher heating values, which render them the more valuable.

Despite losing its status as the dominant fuel, coal production always showed an increasing trend, though there were numerous short term declines in between. The recurring upward trend in coal production to date is supported by all major data sources, who report that the world's coal production is continuing to increase. Figure 11.3 demonstrates this and shows the total coal production since 1980 as published by the EIA, and BP; here again OPEC data are missing as it reports no coal production data. The data in the figure shows an increase of approximately 74-77% in total coal production between 1980 and 2008. It can be seen that the difference in the reported data is inconsequential, and that the major data sources agree on the upward trend both quantitatively and qualitatively, with the slight differences attributed to different practices in data reporting (i.e. what is exactly included). According to the EIA, the total coal production in 2008 is estimated to total 6597 million metric tonnes.

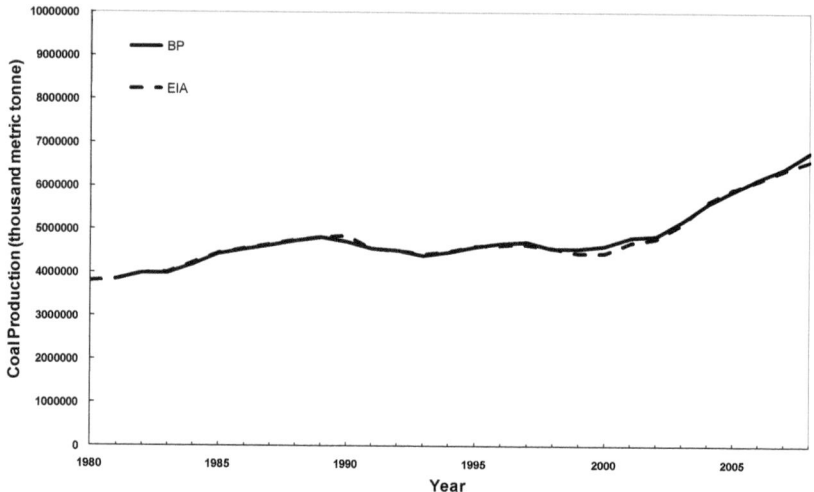

Figure 11.3: World total primary coal production (1980-2008)

Source: EIA (http://www.eia.doe.gov/international); BP (Statistical Review of World Energy 2001-2009).

Regardless of the environmental concerns the continuous upward trend in coal production shows no signs of slowing down, and is expected to continue as long as there is demand, which is projected to grow as reported in all major forecasts. The demand is projected to reach 9170 million metric tonnes and 8784 million metric tonnes by the EIA and IEA respectively in 2030.[25,26] This contradicts some peak oil theory advocates who extended their estimates to predict also coal production peak, they who estimate the world production will peak by 2050.[27]

Total coal production numbers include all ranks of coal, as for this book the details of the production per rank are of little interest. Though to illustrate, according to the IEA[28] production

---

[25] EIA, International Energy Outlook 2009.
[26] IEA, World Energy Outlook 2009.
[27] Jean Laherrère, Uncertainty of data and forecasts for fossil fuels, Universidad de Castilla-La Mancha, 2007.
[28] IEA, 2007, (http://www.iea.org/stats/coaldata.asp?country_code=29).

was split as follows: 1.2% anthracite, 75.0% bituminous coal, 9.6% sub-bituminous and 14.0% lignite. Production data include production from shale gas (whose production is just under 20 million metric tonnes or 0.3% of total coal production[29] and is included with lignite). The data also include peat (whose production of just under 13 million metric tonnes or 0.2% of total coal production) and also any other solid fossil fuels if they are burnt directly for fuel.

However, production numbers exclude solid biofuels. Furthermore, graphite production is also excluded from coal production data, since it is not treated as fuel and its production is extremely tiny. To illustrate, in 2008, world graphite production was 2.1 million metric tonnes[30], which is equivalent to less than 0.1% of total coal production, if it is to be included.

CTL production should be treated with caution to avoid double counting, so coal converted to liquid is counted once as primary coal production. The situation regarding coal-to-gas (CTG) is a bit more involved, as coal gasified in gasifiers should not be counted twice, since the original coal produced is accounted for. The situation regarding gasification in-situ is more tricky, as the coal consumed to produce the syngas needs to be counted in primary coal production (as explained above), whereas at the same time, the syngas produced must be excluded from natural gas production numbers as strictly speaking this syngas is not natural gas and its production is already previously accounted for as coal. To date no convention has been agreed and it is feared that some may add the syngas wrongly, as additional natural gas production. The in-situ technology is still in its infancy and it is expected that it will be a short matter of time before this type of coal production is included within the production numbers.

Similar to natural gas, the coal produced usually requires minimal processing before being used as fuel. Consequently, all the discussion so far relates to primary coal production, since most

---

[29] Jean Laherrère, Review on Oil shale data, 2005,
(http://www.hubbertpeak.com/laherrere/OilShaleReview200509.pdf).
[30] BGS, World Mineral Production 2004-2008.

production data are for coal that is mined and then used directly. However (as discussed in Section 4.4) some coal is further processed and transformed into other forms that suit specific industries and is then labelled secondary coal, i.e. coke, briquettes and patent fuels. Thus, analogous to oil refining, the production of this secondary coal should not be added to the production numbers as this will constitute double counting since the coal transformed has already been accounted for in the original production data.

Equally, similar to oil and natural gas, reported coal production data are often more reliable and consistent than reported reserves and resources data. The explanation here is similar to previous explanations and is largely due to the fact that the reported production data reflect actual tangible production, which is measurable and accounted for both physically and financially. Therefore it differs from estimates of reserves and resources, which both use a combination of mathematical models and physical facts as well as outdated data to produce assessments. Nevertheless, coal production data also share many bad attributes of oil and natural gas production data. They are also far from being transparent, they are confusing, with different standards applied to report production and with no rules agreed to standardise what is included. In addition coal production is often reported in different units (e.g. tonnes, boe, Btu, etc.). In this book, when referred to coal production, the term applies to coal produced from all ranks, as primary production. It excludes other solid biofuels production. Also to date, production data exclude CTG produced directly from in-situ coal gasification since it is not yet produced on a commercial scale, though this needs addressing as, when this commercial production commences coal gasified will be treated as primary production.

Table 11.9 lists the leading ten countries in the world, in terms of total coal production. These countries account for over 88.7% of the global production. In this case this list is very similar to the list of top ten countries, in terms of reserves, sharing seven countries (refer to Table 11.1). This, unlike oil and natural gas situation, agrees with the perceived wisdom by the public, and stems from the fact that countries produce coal if they have it, otherwise they will use alternative fossil fuels, or gradually turn into net importers to plug the gaps in their needs. Examples of the

Table 11.9: World total primary coal production – top ten countries (2008)

Rank	Country	Total Coal Production (Primary)	Share
		thousand metric tonne	%
1	China	2583647.11	39.17
2	USA	1062751.38	16.11
3	India	515574.02	7.82
4	Australia	397806.02	6.03
5	Russia	323126.01	4.90
6	Indonesia	284159.01	4.31
7	South Africa	235502.01	3.57
8	Germany	194456.01	2.95
9	Poland	143228.01	2.17
10	Kazakhstan	108688.00	1.65
	TOTAL	5848937.59	88.67
	WORLD	6596523.85	

Source: EIA (http://www.eia.doe.gov/international).
Note 1: Totals may not add to rounding.

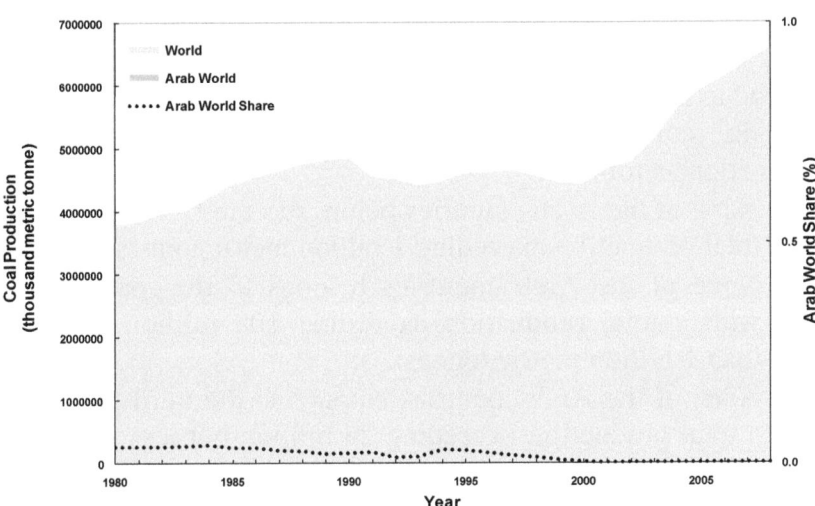

Figure 11.4: Arab world total primary coal production and its share to the world's total (1980-2008)

Source: EIA (http://www.eia.doe.gov/international).

later countries include the UK and, most recently, China, which is also heading this way. Refer to Sections 11.6 and 11.7 for more details. An interesting point to note is that even here one member of OPEC, Indonesia, occupies a spot in the list of top ten producing countries, the sixth.

The Arab world's overall total coal production and its share to the world's total since 1980 are shown in Figure 11.4. The data shows that total coal production in the Arab countries has always been minimal, never exceeding 0.1% of the total world production. In fact, the production decreased steadily from 1980 to a low of less than 0.01% of the total world production in 2008, which translates to a mere 24 thousand metric tonnes. If taken as a single entity, the Arab world situation will not place it anywhere on the list of top producing countries.

The total coal production for all Arab countries in 2008, including details of production for each country, is reported in Table 11.10. A complete set of data from 1980 can be found at "*http://www.2050consulting.com/books*". The above table shows that total coal production is currently only active in one Arab country, though coal reserves have been identified and even were mined in other North African locations in the past. Generally, we can classify coal producing countries into five tiers in terms of total coal production: major, significant, medium, minor or non- coal producing countries. The Arab countries fall into this classification as follows:
- None of the Arab countries belongs to the first tier, with a total production exceeding 1 billion metric tonnes.
- None of the Arab countries belongs to the second tier, with a total production exceeding 100 million but less than 1 billion metric tonnes.
- None of the Arab countries belongs to the third tier, with a total production exceeding 25 million but less than 100 million metric tonnes.
- Only one country, Egypt, belongs to the fourth tier, with a total production less than 25 million metric tonnes, with a share of less than 0.1% of the world's total coal

Table 11.10: Total primary coal production in the Arab countries (2008)

Country	Total Coal Production (Primary) thousand metric tonne	Rank	Share %
Bahrain	0.00		
Iraq	0.00		
Jordan	0.00		
Kuwait	0.00		
Lebanon	0.00		
Oman	0.00		
Palestine	0.00		
Qatar	0.00		
Saudi Arabia	0.00		
Syria	0.00		
UAE	0.00		
Yemen	0.00		
Algeria	0.00		
Comoros	0.00		
Djibouti	0.00		
Egypt	24.00	1	0.00
Libya	0.00		
Mauritania	0.00		
Morocco	0.00		
Somalia	0.00		
Sudan	0.00		
Tunisia	0.00		
ARAB WORLD	24.00		0.00

Source: EIA (http://www.eia.doe.gov/international).
Note 1: Totals may not add up due to rounding.

production. However, Egypt's production does not satisfy its consumption and it is forced to import its remaining coal needs. Historically, Morocco and Algeria mined some coal deposits in the past, but the former's mines were commercially exhausted and ceased to produce since 2002, while the latter decided to stop coal production in 1999.

- Finally 21 countries belong to the fifth tier, with no coal production. These countries are Bahrain, Iraq, Jordan, Kuwait, Lebanon, Oman, Palestine, Qatar, Saudi Arabia, Syria, UAE, Yemen, Algeria, Libya, Mauritania,

Morocco, Sudan, Tunisia, Djibouti, Somalia and Comoros.

Unlike oil and natural gas, spare production capacity is not the trump card in determining coal prices, as most coal consuming countries are self sufficient. Besides, due to the nature of mining industry, production capacity is much easier to expand if needed.

> **The Middle East Perspective**
>
> *The Middle East region is the source of less than 0.1% of the world's total coal production, all of which is in Iran, and none in the region's Arab countries. This indicates the irrelevance of the region as a coal producing area in the world.*
>
> Source: EIA, (http://www.eia.doe.gov/international), based on 2008 data.

> **OPEC Perspective**
>
> *OPEC member states are the source of approximately 4.5% of the world's total coal production, none of this production is from OPEC's Arab member states. The organisation's members occupy one place in the top ten list of leading coal producing countries. This illustrates that the influence of the organisation is minimal on coal markets, and highlights the dependence of its countries on oil and natural gas.*
>
> Source: EIA, (http://www.eia.doe.gov/international), based on 2008 data.

11.4.1 Converted Coal Production

Processes to convert produced coal to liquid fuel are technically possible and have been discussed in Section 2.8. These processes are secondary transformation processes and therefore their production numbers are already included in coal primary production quantities. Therefore, to avoid double counting, if coal-to-liquid (CTL) is considered as part of oil production, a correction to coal production data must be applied to exclude these values from coal production. Alternatively, if accounted for in primary coal production data, CTL production must be excluded from oil production data. In this book, the latter approach is adopted, though many sources mistakenly include it in oil production without deducting it from coal production.

CTL processes prospered during the Second World War, when Germany used them to produce the bulk of its oil needs and where coal was converted into more valuable liquid fuels. Later they prospered in South Africa under the apartheid regime and continue to be significant there, accounting for almost a 20% of the country's needs, with a production capacity of 150000 bpd.[31] Until recently, the technology was experiencing a revival with China investing heavily in new facilities, with a current capacity of 64000 rising to 364000 bpd, and with more expansion on the horizon. However, concerns over water supply and other environmental impacts have initiated a rethink, since the environmental impacts of CTL processes are more taxing and they may hinder further development. As a result, currently CTL processes have been confined to only two countries: South Africa run by Sasol and China run by Shenhua, though with many other countries are exploring the possibilities including the USA, Indonesia, the Philippines and Australia.[32] Accordingly, it can be seen that the overall capacity of all CTL processes is trivial, with less than 250000 bpd, accounting to less than 0.3% of the total oil production (see Section 9.4) and the projected increase in capacity is not expected to exceed 1 million bpd by 2030.[33] Consequently, we can conclude that CTL is unlikely to be a major constituent of the global energy market in the near future. Its economics are very unfavourable, despite proven technology, though it has more chance to prosper than oil shale, since the yield of oil from coal is higher than that of oil shale[34], and thus it has to be ranked higher, in terms of development and utilisation.

On the other hand, coal-to-gas (CTG) production is gaining both popularity and momentum as a cleaner alternative to coal burning and its usage is being encouraged by many countries. To date, all coal gasification is conducted by using mined coal in industrial gasifiers, thus it is considered a form of secondary

---

[31] EIA, International Energy Outlook 2009.
[32] The National Petroleum Council (NPC), Working Document of the NPC Global Oil & Gas Study, Topical Paper #18, Coal to Liquids and Gas, 2007.
[33] EIA, International Energy Outlook 2009.
[34] Jean Laherrère, Review on Oil shale data, 2005,
(http://www.hubbertpeak.com/laherrere/OilShaleReview200509.pdf).

transformation, where its production is already accounted for in the original coal mined. As mentioned previously, in-situ gasification is undergoing extensive research in many countries; its potential is enormous and may change the industry, though to date no commercially functioning projects are operating.

An interesting point to reiterate though is that in these new in-situ coal gasification processes, coal is not produced per se, only syngas is produced. In this case, this gas has to be accounted for and included as a distinct category since it is not, strictly speaking, natural gas. Or preferably, we can calculate the amount of spent coal that has been used to generate the gas and add it as part of the coal production. Refer to Section 10.4.2 for more details.

11.4.2 Other Solid Fossil Fuels Production
As discussed in Section 4.5 earlier, solid fossil fuels are sometimes derived from oil, though usually this is done in refineries where petroleum coke is produced from refined oil (see Section 9.6). However since this process is also a secondary transformation process, where the oil production has already been accounted for in the original oil production numbers, the production of this 'secondary' solid fuel is not included in any coal production numbers, though it is often included in consumption numbers as, at that stage, it is harder to distinguish the source of the coal.

On the other hand, as explained in Section 4.5, no commercial processes have been developed that convert natural gas into solid fossil fuels.

11.4.3 Secondary Coal Production
As already explained in Section 4.4 primary coal can be further processed and transformed to enhance its quality, thereby improving its heating value, reducing its ash content and eliminating smoke generation while burning. The resultant transformed manufactured fuels can be solid, liquid or gaseous and are called *secondary coal*. CTL and CTG were discussed in the preceding two sections. The discussion here is limited to

secondary coal, in the form of solid fuels. These latter transformation processes are analogous to oil refining since their aim is to produce more convenient and superior products to be used as fuel.

Coal transformation technologies and details of secondary coal production are outside the scope of this book. However, it is necessary to briefly visit the latter subject to remove any ambiguities or misunderstandings that occur in mass media outlets when they report issues concerning coal production.

A common mistake, often made in many reports, is that secondary coal production is added to the primary production to reach a grand total. This should never be performed as it constitutes double counting, so is fundamentally wrong. By definition, coke, briquettes and patent fuels are secondary products. As such, they are transformations of the 'same' produced primary coal, therefore their production numbers should not be added to coal production numbers, since they are implicitly accounted for.

Table 11.11 World coal secondary production

Secondary Coal Form	Production	Share Relative to Primary Coal
	thousand metric tonne	%
Coke	554415	8.8
Patent Fuel	148270	2.4
Brown Coals & Peat Briquettes	10778	0.2
Hard Coals Briquettes	14994	0.2
Coal Tar	8980	0.1

Source: IEA (http://www.iea.org/stats/coaldata.asp?COUNTRY_CODE=29); EIA (http://www.eia.doe.gov/international).
Note 1: Data are for 2007.

One main difference from oil refining though is that the percentage of coal transformed into secondary coal is considerably lower than its oil counterpart. So to put it in context, according to the EIA numbers of 2008, the production and share of secondary coal is shown in Table 11.11, where it can be seen that all forms combined accounted for only 11.8% of the primary coal production, where 8.8% of the total is produced as metallurgical coke, 2.4% as patent fuel, and a tiny 0.4% as briquettes.

No secondary coal production takes place in the Arab world.

11.5 Coal Reserve to Production Ratio (R/P)

The reserve to production ratio (R/P) was introduced in Sections 9.5, and then discussed for oil in that section and for natural gas in Section 10.5; the discussion regarding its definition, merits and defects is applicable to coal and not repeated here. Consequently, drawing on the conclusions from previous discussion, in this chapter R/P ratios are also not used as a credible analysis tool and are only mentioned to complete the picture.

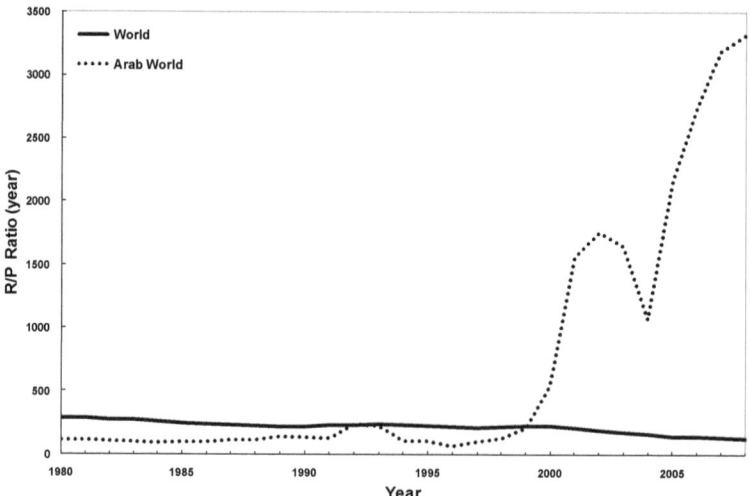

Figure 11.5: World coal reserves to production ratio (1980-2008)
Source: Calculated based on data in Sections 11.1 and 11.4.

Figure 11.5 shows the world's coal R/P ratios since 1980, calculated from data obtained from the EIA. R/P calculations are performed, based on total combined coal reserves (i.e. the summation of reserves of all coal ranks); similarly the coal production data are based on total combined primary coal production numbers. Both reserves and production data exclude peat reserves and production. Figure 11.5 also shows the R/P ratio for the Arab world.

The figure suggests that coal supplies are abundant with no signs of even a very slow decline envisaged. The global ratio is still over 100 years, despite changes in reserves and production, which demonstrates again the misleading nature of this ratio.

Regarding the Arab world, the ratio multiplied! This demonstrates, clearly, the misleading information that can be obtained from this ratio, since the Arab world has hardly any coal reserves or production, thus this figure and its enormity is simply meaningless.

11.6 Coal Consumption

Despite the bad press, the ardent campaigns orchestrated by environmental groups and lobbyists everywhere highlighting pollution issues particularly high CO_2 emissions caused by coal utilisation and calling for cuts in its consumption; the reality is that coal's share of the total world energy consumption continues to grow. Nowadays it is the predominant player in the power generation sector, where its share stands at a massive 40% (see Section 4.6). So, despite the complaints heard from all sectors of society, calling for less dependence on coal in the electricity generation sector, there is no indication that anyone in the electric sector or governments are listening. This is leading to ever increasing consumption with numerous power stations being built, especially in power hungry China and India, as a fast and cheap way of adding much needed electricity generation capacity regardless of additional emissions.

Coal consumption is often referred to in terms of 'total consumption', which includes the four coal ranks. It accounts for all consumption utilised for fuel or chemical use. Coal consumption data include both primary and secondary coal, as at this stage, it is impossible to differentiate consumption origins, though one has to be careful not to double count secondary coal, if it is already accounted for in primary coal consumption.

As established already, consumption is a calculated number, thus, for practical reasons, it excludes coal transformed into other gaseous or liquid fossil fuels (see Sections 2.8 and 3.9) as this coal consumption is accounted for via the consumption of gaseous or liquid fuels. There are no numbers released that distinguish this consumption. On the other hand, consumption includes fossil fuels transformed to solid fossil fuels, such as biochar, or burnt as solid fuel, such as peat or oil shale, as again, there is no way to distinguish the source of coal once it is consumed. To emphasise, this is in contrast to production, where coal transformed into gas or liquid is accounted for in original coal production, liquid fuels transformed into solid are accounted for by liquid fossil fuel production.

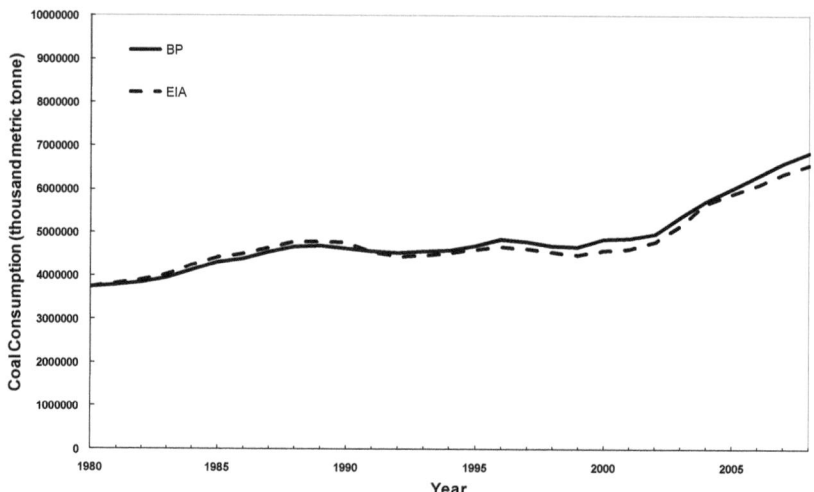

Figure 11.6: World coal consumption (1980-2008)

Source: EIA (http://www.eia.doe.gov/international); BP (Statistical Review of World Energy 2001-2009).

Despite numerous short peaks and valleys, recent data shows that the growth in the world's coal consumption is accelerating. Figure 11.6 shows the total coal consumption from different sources since 1980. Unlike reserves trends, but similar to trends observed in coal production, the world's coal consumption shows a solidly visible increase since 1980, where between 1980 and 2008 consumption jumped by around 75-82%, reaching just over 6566 million metric tonnes according to the EIA. Reasons for this increase have already been mentioned and are due to the rapid expansion in coal driven electricity generation capacities for many developing countries especially China and India, as well as the continuous reliance of the USA on coal as its major source for electricity.

In the last 10 years China and India have become increasingly dependent on coal for their power generation and as such, the familiar story of those countries being portrayed by the media as the villains responsible for the increase in the world's coal consumption. In fact, here the accusation is partially true, since the consumption of both countries, almost doubled during that period and both countries, combined, account for 47.9, almost half the world's total consumption. However, the USA may be regarded as one of the main culprits once again, as it consumes almost 15.5% of the world's total, surpassing the whole of Europe.

With the increasingly prominent public campaigns advocating reducing coal consumption to cut emissions and slow the rapid climate change, countries are manipulating their reporting methods. They are coming up with many creative solutions to mask their true fossil fuel consumption image and to portray themselves in a better light to reduce their guilt and responsibility. This has already been explained for oil and natural gas in Sections 9.7 and 10.6. It is further re-emphasised here. In the case of coal, countries are also advocating the use of carbon capture and storage (CCS), as a technical solution that can mitigate the effects of continuous coal consumption. They are therefore introducing legislation to make its inclusion mandatory in all future coal power stations, with plans to apply it retrospectively to existing stations. While this can be a temporary solution, its effectiveness has not yet been proven, and its economic feasibility is poor. As a result, analogous to measuring oil and natural gas consumption rates, measurement of the coal consumption rate is also a contentious issue on the global stage. Due to differences in economic and social circumstances, countries use different measurement rates, which depict them in a way in which they appear to be reducing their coal consumption rate, thus avoiding being accused of being the main offenders when policies to combat global climate change are discussed. Therefore, as shown earlier, rich countries such as the USA prefer measuring their coal consumption per GDP, while China and India favour measuring consumption per capita. Both methods do not give a true indication of actual coal consumption, as the numbers can be skewed to punish countries with small populations and reward rich countries.

Table 11.12: World coal consumption – top ten countries (2008)

Rank	Country	Total Coal Consumption	Share
		thousand metric tonne	%
1	China	2566893.11	39.09
2	USA	1017602.18	15.50
3	India	578350.03	8.81
4	Germany	244842.01	3.73
5	Russia	244654.01	3.73
6	Japan	185047.01	2.82
7	South Africa	175680.01	2.68
8	Australia	145617.01	2.22
9	Poland	135473.01	2.06
10	South Korea	102370.00	1.56
	TOTAL	5396528.38	82.18
	WORLD	6566391.46	

Source: EIA (http://www.eia.doe.gov/international).
Note 1: Totals may not add up due to rounding.

Table 11.12 lists the top ten countries in the world, in terms of total coal consumption. These ten countries, combined, account for approximately 82.2% of the world's total coal consumption. Similar to natural gas, but not oil, consumption these countries are not all amongst the richest in the world, in terms of total GDP, with countries like India and Poland featuring in the list, who are dependent on coal for their power generation and industry. Still, countries with large populations who are coal consumers inevitably have more needs, which lead to higher coal consumption. Logically, none of OPEC countries are on the list.

The Arab world's total coal consumption and its share to the world's total since 1980 are shown in Figure 11.7. The figure reveals that the Arab world's share of the world's total consumption never exceeded 0.2%, with only 0.1% in 2008 - a consumption of just over 7.9 million metric tonnes. Even if treated as a single entity, it ranks very low in terms of coal consumption, though it highlights the full dependence of the Arab world on oil and natural gas. It reflects the fact that the Arab countries have hardly any steel industry, which also relies heavily on coal.

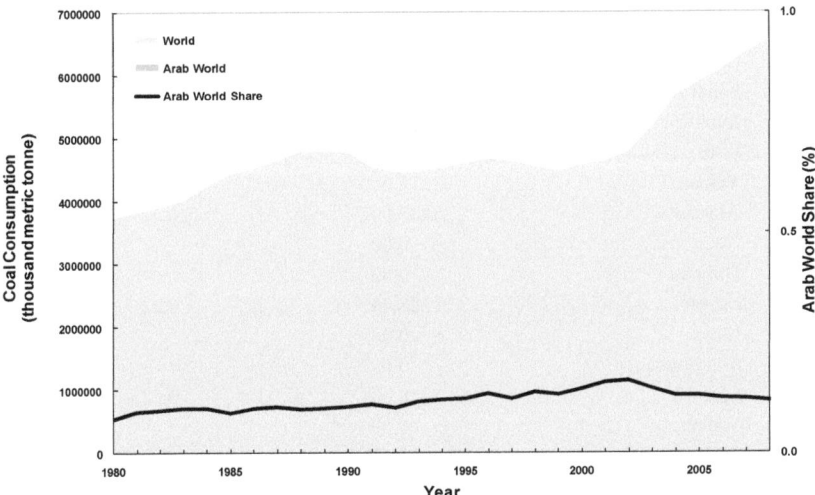

Figure 11.7: Arab world coal consumption and its share to the world's total (1980-2008)

Source: EIA (http://www.eia.doe.gov/international).

The total coal consumption for all Arab countries in 2008, including details of the latest consumption estimates per country, is reported in Table 11.13. A complete set of data from 1980 can be found at *"http://www.2050consulting.com/books"*. The table lists total coal consumption in 22 Arab countries, it can be seen that it is only consumed in six of them. In terms of coal consumption, we can classify the countries into five tiers: major, significant, medium, minor and non-coal consuming countries. The Arab countries fall into this classification as follows:
- None of the Arab countries belongs to the first tier, with a total consumption exceeding 1 billion metric tonnes.

Table 11.13: Coal consumption in the Arab countries (2008)

Country	Total Coal Consumption thousand metric tonne	Rank	Share %
Bahrain	0.00		
Iraq	0.00		
Jordan	0.00		
Kuwait	0.00		
Lebanon	216.00	4	0.00
Oman	0.00		
Palestine	0.00		
Qatar	0.00		
Saudi Arabia	2.00	6	0.00
Syria	4.00	5	0.00
UAE	0.00		
Yemen	0.00		
Algeria	1135.00	3	0.02
Comoros	0.00		
Djibouti	0.00		
Egypt	1208.00	2	0.02
Libya	0.00		
Mauritania	0.00		
Morocco	5218.00	1	0.08
Somalia	0.00		
Sudan	0.00		
Tunisia	0.00		
ARAB WORLD	7783.00		0.12

Source: EIA (http://www.eia.doe.gov/international).
Note 1: Totals may not add up due to rounding.

- None of the Arab countries belongs to the second tier, with a total consumption exceeding 100 million but less than 1 billion metric tonnes.
- None of the Arab countries belongs to the third tier, with a total consumption exceeding 25 million but less than 100 million metric tonnes.
- Six countries belong to the fourth tier, with a total consumption less than 25 million metric tonnes each and a combined share of just over 0.1% of the world's total coal consumption. These countries in descending order in terms of total coal consumption are Morocco, Egypt,

Algeria, Lebanon, Syria and Saudi Arabia. Morocco and Lebanon use coal to generate electricity, while Egypt and Algeria utilise it in their steel industries and, with the exception of some production in Egypt, all coal consumed is imported. Algeria is a member of OPEC, but its coal consumption is for industrial purposes in steel production, as already explained.

- Finally 16 countries belong to the fifth tier with no coal consumption. These countries are Bahrain, Iraq, Jordan, Kuwait, Oman, Palestine, Qatar, UAE, Yemen, Libya, Mauritania, Sudan, Tunisia, Djibouti, Somalia and Comoros.

The Middle East Perspective

The Middle East region consumes approximately 0.2% of the world's total coal consumption, only a minimal 1.5% of which are in the Arab countries of the Middle East, with the rest in Iran and Israel. This demonstrates that the Middle East as a region hardly registers on the radar in terms of coal consumption.

Source: EIA, (http://www.eia.doe.gov/international), based on 2008 data.

OPEC Perspective

OPEC member states consume approximately 1.3% of the world's total coal consumption, a trivial 1.4% of which are in the Arab member states, with the majority of OPEC's consumption in Indonesia. This illustrates that coal is very insignificant for the organisation's member states, which all depend on oil and natural gas, and demonstrates that OPEC can play no role in coal markets.

Source: EIA, (http://www.eia.doe.gov/international), based on 2008 data.

As discussed above, inspecting the rate of consumption, per GDP or per capita, leads to much skewed results. The increase in the oil price since 2006 led to rapid increases in the Arab countries GDPs, while the credit crunch of 2008 that affected many countries, clouded the matters further. Also the population explosion of the Arab countries results in a disproportional reductions in the perceived consumption per capita. Figure 11.8

and Table 11.14 demonstrate clearly how consumption data for coal can be manipulated and presented. This story is very similar

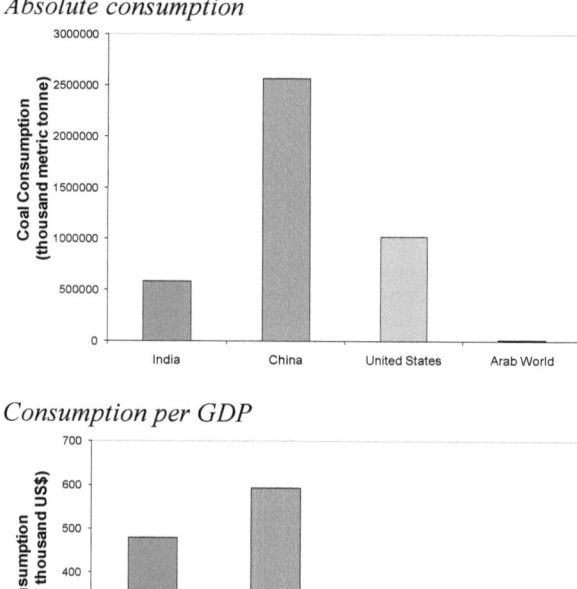

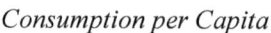

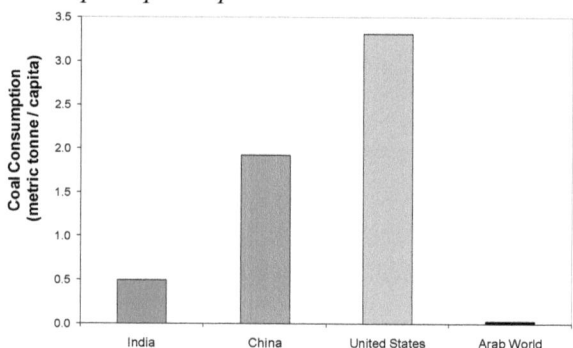

Figure 11.8: Comparison of coal consumption data presentation in the USA, China, India and the Arab world (2008)

Source: EIA (http://www.eia.doe.gov/international), CIA – The World Factbook

to oil and natural gas consumption stories – refer to Sections 9.7 and 10.6.

Table 11.14: Comparison of coal consumption data presentation in the Arab countries (2008)

Country	Total Coal Consumption metric tonne/thousand US$ GDP	Rank	Total Coal Consumption metric tonne/capita	Rank
Bahrain	0.0		0.00	
Iraq	0.0		0.00	
Jordan	0.0		0.00	
Kuwait	0.0		0.00	
Lebanon	7.4	3	0.05	2
Oman	0.0		0.00	
Palestine	0.0		0.00	
Qatar	0.0		0.00	
Saudi Arabia	0.0	6	0.00	6
Syria	0.1	5	0.00	5
UAE	0.0		0.00	
Yemen	0.0		0.00	
Algeria	7.1	4	0.03	3
Comoros	0.0		0.00	
Djibouti	0.0		0.00	
Egypt	7.4	2	0.01	4
Libya	0.0		0.00	
Mauritania	0.0		0.00	
Morocco	58.7	1	0.15	1
Somalia	0.0		0.00	
Sudan	0.0		0.00	
Tunisia	0.0		0.00	
ARAB WORLD	4.1		0.02	

Source: EIA (http://www.eia.doe.gov/international); CIA – The World Factbook
Note 1: Totals may not add to rounding.

11.7 Coal Trade

Contrary to the norm with most commodities and certainly in contrast to oil and natural gas, the majority of coal's main producing countries are also the main coal consuming countries. The main reason for this is the relative difficulty of coal transport compared to oil or natural gas transport. Therefore international coal trade volume is substantially smaller than oil or natural gas trade, has less reach and fewer countries are engaged in it. According to EIA data, in 2008, international coal trade accounted for only 12 or 15% of total coal produced relative to net or absolute trade numbers respectively, with the traded coal is predominantly of the hard variety.

It follows that coal behaves like most commodities, with its price determined by market forces following supply and demand equations and affected less by political factors. However recently, stringent environmental laws, restricting coal usage and making CCS mandatory, started having an impact on coal prices. With all these factors, predicting coal price remains a difficult issue and is outside the scope of this book. Note though, that the price of coal is independent of the prices of both oil and natural gas and since coal trade is usually locked into long term contracts, its price fluctuations are less than oil, as it is more resistant to market speculation.

In addition to price, other factors affect coal trade, such as logistics and transportation, with the cost of transport playing an important role in defining trading partners, leading to regional markets rather than a global market.

Over the course of this book you may have realised that reporting oil and natural gas trade data is confusing. I would suggest coal data reporting is even more confusing with no standards agreed – not even definitions of what to report are consistent, with nations reporting data using a different basis and reporting data for different coal ranks, with the definitions of the ranks being also inconsistent. Furthermore, political and commercial factors mean that not all the data is disclosed.

Additionally, the data can be reported in many different ways, e.g. in terms of mass or energy; for total coal, for hard and brown coal or for lignite, bituminous and anthracite or for primary and secondary coal. All these quantities can be reported in terms of exports or imports or a net trade value. Another complication that is foreseen is the reporting of gasified coal data, whether it will be in terms of syngas, natural gas or coal consumed. To date, no commercial syngas from coal is traded internationally, but with this becoming inevitable in the near future, this needs to be addressed. Note that in this book, coal trade is reported in terms of total coal using mass units.

Since any coal exported by a country is imported by another, theoretically the data for the world's exports and imports should be identical. However, a similar story to the oil and natural gas trade emerges here. Data discrepancies occur, with each data source using different methods to report trade data. So, as explained earlier, each data source follows a different approach, for example, the EIA reports slight differences between exports and imports data and attributes this to various factors such as stock change and storage inventory changes. While, as you are aware by now, OPEC does not address coal data in its reports.

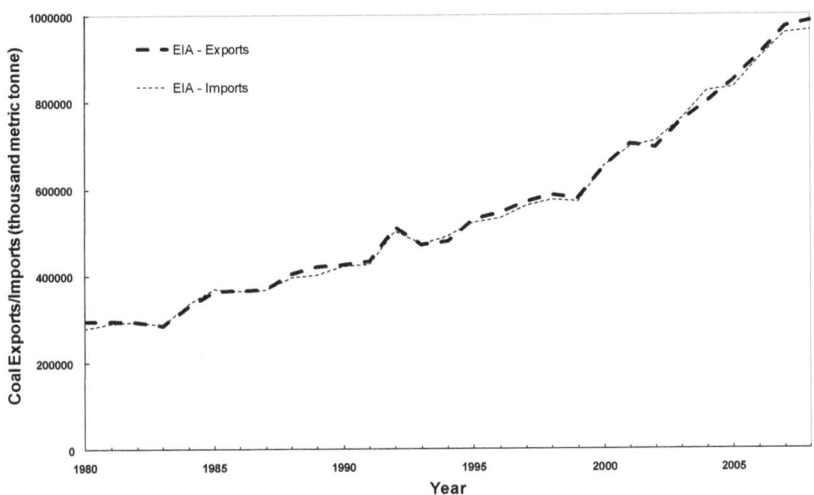

Figure 11.9: World coal trade (1980-2008)
Source: EIA (http://www.eia.doe.gov/international).

Interestingly, note that BP also does not also report any coal trade data, even though it reports oil and natural gas trade data, so Figure 11.9 shows the reported coal exports and imports from only the EIA since 1980. It can be seen that the net trade has fluctuated, but with an evident rising trend and increased by over 230% between 1980 and 2008. According to the EIA's 2008 compiled data the net export and imports of total coal stand at 821 and 791 million metric tonnes respectively.

Therefore, to simplify the presentation, the same methodology used to present oil and natural gas trade is also used here. In the remainder of this section, the coal trade data are reported in terms of an overall net value of either imports or exports, depending on the country, as it is perceived that this is the most relevant data that gives the full picture of a country's trade. This value is simply calculated as the difference between total coal production, as defined in Section 11.4, and total coal consumption, as defined in Section 11.6, thus it implicitly accounts for trade either as primary or secondary coal.

Table 11.15: World net coal exports – top ten countries (2008)

Rank	Country	Net Coal Trade (Exports)	Share
		thousand metric tonne	%
1	Australia	252189.01	30.71
2	Indonesia	206957.01	25.20
3	Russia	78472.00	9.56
4	Colombia	73908.00	9.00
5	South Africa	59822.00	7.29
6	USA	45149.20	5.50
7	Kazakhstan	26216.00	3.19
8	Vietnam	20037.00	2.44
9	China	16754.00	2.04
10	Canada	11519.00	1.40
	TOTAL	791023.23	96.33
	WORLD	821159.24	

Source: Calculated based on data in Sections 11.4 and 11.6.
Note 1: Totals may not add up due to rounding.

Table 11.15 lists the top ten countries in the world in terms of net total coal exports. These ten countries, combined, account for a massive 96.3% of the world's net total coal exports. This list is different to the world leading coal reserves countries list (see Table 11.1), with only six countries belonging to both lists. It is closer to the list of leading coal producing countries, with seven countries appearing in both lists (see Table 11.9). This situation is not surprising as countries with large coal reserves are also large producers and consumers. With coal trade being much smaller in size, compared to both oil and natural gas, many countries consume what they produce and therefore do not produce excessively for exporting purposes, as in many cases no buyers can be found. Interestingly, nominally OPEC member, Indonesia, is second in the top ten exporting countries list. Even though it is a modest producer, its consumption is very low as it relies more on oil and natural gas and therefore has a relatively high excess coal available for export.

Table 11.16: World net coal imports – top ten countries (2008)

Rank	Country	Net Coal Trade (Imports)	Share
		thousand metric tonne	%
1	Japan	185047.01	23.39
2	South Korea	99597.00	12.59
3	Taiwan	62955.02	7.96
4	India	62776.00	7.94
5	Germany	50386.00	6.37
6	UK	47145.00	5.96
7	Italy	25410.00	3.21
8	Turkey	19704.00	2.49
9	Brazil	19203.00	2.43
10	France	19160.00	2.42
	TOTAL	591383.04	74.76
	WORLD	791026.85	

Source: Calculated based on data in Sections 11.4 and 11.6.
Note 1: Totals may not add up due to rounding.

Table 11.16 lists the top ten countries in the world in terms of net total coal imports. These ten countries, combined, account for a massive 74.8% of the world's net total coal imports. This number is noteworthy as it demonstrates that over three quarters of the world coal trade is dominated by ten partners! The net total coal importing countries list shares four countries with the world leading total coal consuming countries list (see Table 11.12). This situation is not surprising as countries with large coal consumption are expected to be producers and therefore large consumers will only import if their production falls short.

Note that the numbers in the two tables above are for net total coal exports and imports, defined effectively as the difference between total exports and total imports, as some countries can be exporting and importing simultaneously from different regions. This means that some countries can export from one region and import to another one within the same country, as is the case of the USA, Canada, Russia and China. In this section the matter of concern is whether a country is a net exporter or importer; the exact details of the trade are not of interest in this book.

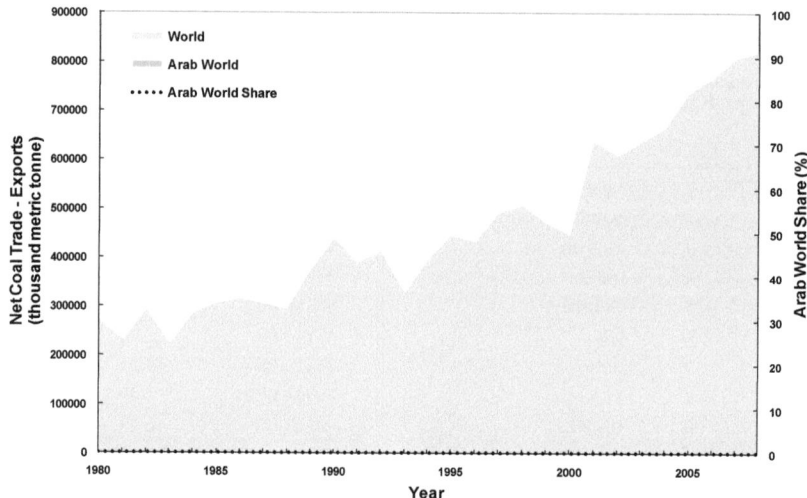

Figure 11.10: Arab world net coal exports and its share to the world's total (1980-2008)

Source: Calculated based on data in Sections 11.4 and 11.6.

The Arab world net total coal exports and its share to the world's total since 1980 are shown in Figure 11.10. The figure reveals that the Arab world plays no role whatsoever in coal exporting markets, with its share of the world's total standing at 0%.

In a similar fashion, Figure 11.11 shows the Arab world's net total coal imports and its share to the world's total since 1980. The figure reveals that the Arab world's share of the world's total is trivial, hovering just over 1% of the world's total net imports and recording 1%, with approximately 8.5 million metric tonnes in 2008. If treated as a single entity, the Arab world will not appear in the leading net total coal importers list.

Note that with no Arab exports, obviously there is no coal intra-Arab trade reported as international trade.

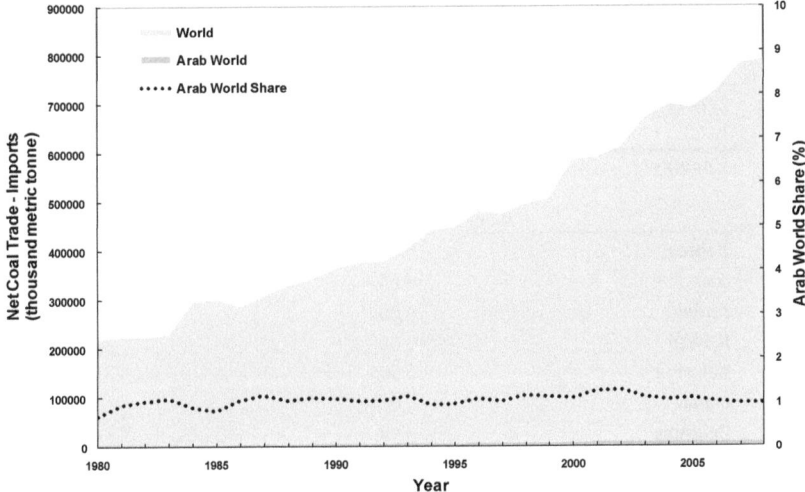

Figure 11.11: Arab world net coal imports and its share to the world's total (1980-2008)
Source: Calculated based on data in Sections 11.4 and 11.6.

The net total coal exports for all Arab countries in 2008, including details of the latest estimates per country, are reported in Table 11.17. A complete set of data from 1980 can be obtained from "*http://www.2050consulting.com/books*". The table shows that no Arab countries are net coal exporters. In terms of net total coal exports, we can classify the countries into five tiers: major, significant, medium and minor net total coal exporting countries, with the fifth tier being countries whose net total coal exporting equals zero. Obviously some of these will be net importers, but others simply do not consume coal. To be consistent with oil and natural gas methodology, the tiers are quantitatively defined using the same limits that are used in defining production and consumption tiers. This leads to an odd situation, where no countries in the world belong to the first tier. This situation is a significant indication of the small size of the total international coal trade. The Arab countries fall into this classification as follows:

- None of the Arab countries belongs to the first four tiers, with a net total coal exports less than 25 million metric tonnes.
- All 22 countries belong to the fifth tier with no net total coal exports.

Table 11.17: Net coal exports in the Arab countries (2008)

Country	Net Coal Trade (Exports) thousand metric tonne	Rank	Share %
Bahrain	0.00		
Iraq	0.00		
Jordan	0.00		
Kuwait	0.00		
Lebanon	0.00		
Oman	0.00		
Palestine	0.00		
Qatar	0.00		
Saudi Arabia	0.00		
Syria	0.00		
UAE	0.00		
Yemen	0.00		
Algeria	0.00		
Comoros	0.00		
Djibouti	0.00		
Egypt	0.00		
Libya	0.00		
Mauritania	0.00		
Morocco	0.00		
Somalia	0.00		
Sudan	0.00		
Tunisia	0.00		
ARAB WORLD	0.00		

Source: Calculated based on data in Sections 11.4 and 11.6.
Note 1: Totals may not add up due to rounding.

Similarly, the net total coal imports for all Arab countries in 2008, including details of the latest estimates for each country, are reported in Table 11.18. A complete set of data from 1980 can be obtained from "http://www.2050consulting.com/books". The table lists net total coal imports in six Arab countries. In terms of net total coal imports, we can classify the countries into five tiers:

Table 11.18: Net coal imports in the Arab countries (2008)

Country	Net Coal Trade (Imports) thousand metric tonne	Rank	Share %
Bahrain	0.00		
Iraq	0.00		
Jordan	0.00		
Kuwait	0.00		
Lebanon	216.00	4	0.03
Oman	0.00		
Palestine	0.00		
Qatar	0.00		
Saudi Arabia	2.00	6	0.00
Syria	4.00	5	0.00
UAE	0.00		
Yemen	0.00		
Algeria	1135.00	3	0.14
Comoros	0.00		
Djibouti	0.00		
Egypt	1184.00	2	0.15
Libya	0.00		
Mauritania	0.00		
Morocco	5218.00	1	0.66
Somalia	0.00		
Sudan	0.00		
Tunisia	0.00		
ARAB WORLD	7759.00		0.98

Source: Calculated based on data in Sections 11.4 and 11.6.
Note 1: Totals may not add up due to rounding.

major, significant, medium and minor net total coal importing countries, with the fifth tier being countries whose net total coal importing equals zero, obviously some of these will be net exporters, but others simply do not consume or produce coal. Similar to the approach adopted for coal exports, and to be consistent with oil and natural gas methodology, the tiers are quantitatively defined using the same limits that are used in defining production and consumption tiers. Obviously, this will replicate the odd situation observed in the exports case, where no countries in the world belonging to the first tier which, as already explained, is a significant indication of the small size of the total

international coal trade. The Arab countries fall into this classification as follows:
- None of the Arab countries belongs to the top three tiers with a net total coal imports exceeding 25 million metric tonnes.
- Six countries belong to the fourth tier with a net total coal imports less than 25 million metric tonnes each and a combined share of just under 1.0% of the world's net total coal imports. These countries in descending order in terms of net total coal imports are Morocco, Egypt, Algeria, Lebanon, Syria and Saudi Arabia. With the exception of Egypt, that produces a fraction of its consumption, the other five countries import all their coal requirements, either as primary or as secondary coal.
- Finally 16 countries belong to the fifth tier with no net total coal imports. These countries are Bahrain, Iraq, Jordan, Kuwait, Oman, Palestine, Qatar, UAE, Yemen, Libya, Mauritania, Sudan, Tunisia, Djibouti, Somalia, and Comoros.

The Middle East Perspective

The Middle East's share of the world's net total coal exports is zero, i.e. none of the countries in the region exports any coal.

On the other hand, as expected the region is very insignificant in terms of net total coal imports and accounts for only 1.7% of the world's net total coal imports, 1.6% of which are to the Arab countries, while the overwhelming majority of the rest is imported by Israel.

The numbers confirm that the Middle East is totally insignificant in terms of and has no influence on coal trade.

Source: Calculated based on data in Sections 9.4 and 9.7, *based on 2008 data.*

OPEC Perspective

Unexpectedly, OPEC member states are significant net total coal exporters with a share of approximately 26.0% of the total, none of which is from its Arab member states. The organisation's member, Indonesia, occupies the second position in the top ten list of leading net total coal exporting countries.

> *Predictably none of OPEC members is in the list of leading net total coal importers, and the organisation accounts for less than 0.2% of the world's net total coal imports in 2008, 84.2% of which is to its Arab member states. This apparently surprising net total coal import is mainly due to Algeria.*
>
> *The numbers illustrate that OPEC is a major player in coal exports markets despite being a minor producer.*
>
> ---
> Source: Calculated based on data in Sections 9.4 and 9.7, *based on 2008 data.*

11.7.1 Coal Transport

As already explained in Section 4.7, almost all internationally traded coal is transported using ships or barges, with a very small percentage using rail or road and obviously none using pipelines.

Ships or barges are the preferred mode of transport, since they provide the cheapest means of transporting coal. For transporting for the same distance, they are cheaper than transporting oil or natural gas, on the basis of equal energy units transported.[35] However, despite the ease of transporting coal via ships or barges, the coal market never emulated oil or LNG markets and remains stubbornly regional and fragmented.

When coal gasification becomes mature and commercially viable, it is envisaged that the resultant syngas will be transported using pipelines, analogous to natural gas pipelines via national and international grids.

---

[35] BGR, Energierohstoffe 2009 – Reserven, Ressourcen, Verfügbarkeit.

Chapter 12
FOSSIL FUELS - OVERALL GLOBAL AND ARAB PERSPECTIVES

In the previous Chapters 9, 10 and 11, we explored in detail the three types of fossil fuels oil, natural gas and coal, by analysing them separately in terms of four aspects: reserves and resources, production, consumption and trade. The next step is to look at the bigger picture and examine the combined fossil fuels in terms of the same four aspects.

Some may question the reason for amalgamating fossil fuels together, since an overwhelming proportion of the supply-demand chain of each fossil fuel type is independent. However even though this is true at present, it is technically possible to convert fossil fuels from one type to another and these developing conversion processes are becoming more feasible.

As with the previous three chapters, this chapter is also structured systematically. Each section starts by presenting the global standpoint of the amalgamated fossil fuels of each of the four aspects, followed by an overall assessment of the Arab world's position in relation to the particular aspect.

12.1 Interrelation of Fossil Fuels Analysis
As already discussed in detail in various parts of this book, it is technically possible, and increasingly feasible, to convert a fossil fuel from one type to the other. For detailed discussions readers are referred to Sections 2.8, 3.9, 4.5, 9.4.1, 9.4.2, 10.4.1, 10.4.2, 11.4.1 and 11.4.2. To give a comprehensive, but still, a simple description of the interrelation between the three main types of fossil fuels, one can inspect Figure 12.1, which demonstrates the

Fossil Fuels - Overall Global and Arab Perspectives

complex relationships between oil, natural gas, coal and peat, starting from their existence in the reservoirs, fields or bogs, through their production and supply, to their consumption. The figure is self explanatory and as they often say, a picture tells a thousand words.

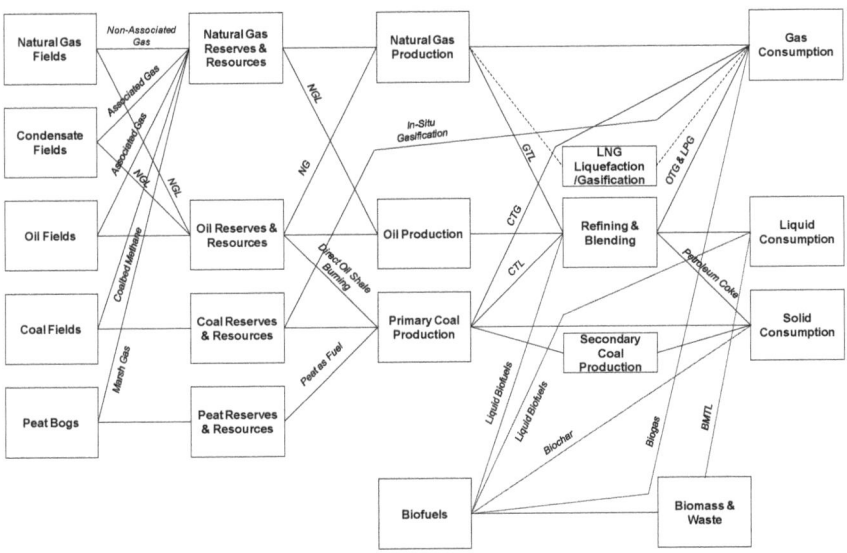

Figure 12.1: Interrelation between fossil fuels

12.1.1 Conversion Factors

Having established the inevitable correlations, there is a need to use a common measure to express all fossil fuels in some form of energy units so that we can compare them together. Several units exist such as mile oil, barrel of oil equivalent (boe) etc. Here we choose 'boe', which is based on the approximate energy released by burning one barrel of crude oil, used widely by oil and gas companies, as it is easier to understand, and with oil being the more dominant seems more appropriate to use.

Here again the conversion units are not universally agreed and each data source uses slightly different conversion factors that are dependent on specific fossil fuels characteristics such as density or heating value. In this book, uniformed conversion factors are used and are noted in Table 12.1 below.

Table 12.1: Uniformed Conversion Factors

Fuel Type	Unit	Conversion Factor boe	Mathematical Operation
Oil	barrel	1	Multiplication
Natural Gas	standard m^3	6.6 x 1E-3	Multiplication
Coal	metric tonne	3.5393 x 1E-3	Multiplication

Source: Calculated by using the average yearly data of coal in tonnes and tonne oil equivalent (toe) as reported by BP, Statistical Review of World Energy 2009).

Note 1: To calculate the coal conversion factor, coal production numbers in the above source reported in metric tonnes are divided over the corresponding numbers in the same table reported in boe. Subsequently to get a yearly mean value, the divisions' results are after that all are averaged. The average value used here is 2.071. Then to convert toe to boe multiply by 7.33.

12.2 Overall Conventional Fossil Fuels Reserves and Resources

12.2.1 Overall Conventional Fossil Fuels Reserves

We discussed the conventional reserves of oil, natural gas and coal in detail in Sections 9.1.1, 10.1.1 and 11.1.1. Therefore there is no need to repeat these facts and interested readers are advised to revisit the relevant sections for the relevant data. Here we perform a simple exercise of amalgamating the conventional reserves of the three types of fossil fuels in a standardised single unit, boe. This allows us to create an encompassing picture quantifying the overall conventional fossil fuels reserves, identify the major world players and determine the share of the Arab world in this overall picture, assessing its importance.

The world's overall conventional fossil fuels reserves are shown in Table 12.2. Here, it can be observed that oil is not the dominant fossil fuel, in terms of conventional reserves, with coal accounting for over 55.5% of the total reserves and oil relegated to second place with just over 22.2%, with the rest being natural gas, which accounts for almost as much as oil with just under 22.2%. Even though this may come as no surprise to the experts in the field, it may catch many people unaware, wondering, what is the fuss about oil then?

Table 12.2: World overall conventional proved fossil fuels reserves by type (2009)

Fuel Type	Reserves billion boe	Share %
Oil	1169.5	22.24
Natural Gas	1168.9	22.23
Coal	2919.7	55.53
TOTAL	5258.1	100.00

Source: Calculated based on data in Chapters 9, 10 and 11.
Note 1: Oil is reported in billion barrel.
Note 2: Totals may not add up due rounding.

Table 12.3: World overall conventional proved fossil fuels reserves – top ten countries (2009)

Rank	Country	Total Crude Oil Reserves billion barrel	Total Natural Gas Reserves billion boe	Total Coal Reserves billion boe	Total Fossil Fuels Reserves billion boe	Share %
1	Russia	60.00	313.97	552.54	926.51	17.62
2	USA	21.32	44.43	835.78	901.53	17.15
3	China	16.00	14.95	379.04	409.99	7.80
4	Iran	136.15	185.32	4.89	326.36	6.21
5	Saudi Arabia	266.71	48.31	0.00	315.02	5.99
6	Australia	1.50	5.61	267.03	274.13	5.21
7	India	5.62	7.09	194.84	207.56	3.95
8	Qatar	15.21	166.70	0.00	181.91	3.46
9	Kazakhstan	30.00	15.89	109.71	155.60	2.96
10	UAE	97.80	40.07	0.00	137.87	2.62
	TOTAL	650.31	842.33	2343.83	3836.47	72.96
	WORLD				5258.12	

Source: Calculated based on data in Chapters 9, 10 and 11.
Note 1: Totals may not add up due to rounding.

Table 12.3 shows the list of the top ten countries, in terms of overall conventional fossil fuels reserves, where they account for a massive 73.0% of the global proved reserves. The table makes interesting reading, as it highlights important facts that contradict the public perception regarding energy markets. Three main observations deserve emphasizing:
- The world's major energy heavy-weights, in term of conventional reserves, are not in the Middle East region

or members of OPEC, but are in fact the world's major political and military powers, with Russia edging the USA marginally to occupy the top spot, while China occupies the third spot.
- Four Middle Eastern countries, that are all members of OPEC, feature in the top ten list though their combined conventional reserves share, only barely, exceeds that of Russia or the USA by a tiny margin.
- Iran emerges as the top OPEC and Middle Eastern country, displacing Saudi Arabia.

In fact the Middle East's total share, in terms of overall conventional fossil fuels reserves, is only 23.5%, while that of OPEC is 29.8%, both significantly less than their massive shares in terms of conventional oil reserves often publicised of over two thirds and over four fifths of the world's total respectively, and notably less than their conventional natural gas reserves.

The Arab world's overall conventional fossil fuels reserves and their share to the world's total are shown in Table 12.4. The table also includes details of the overall fossil fuels reserves of all Arab countries. 19 Arab countries have fossil fuels reserves, however their rankings are changed significantly in comparison to their oil reserves' rankings. Even though Saudi Arabia maintained the top spot, Qatar is now a distant second overtaking oil heavy weights Iraq, Kuwait and the UAE.

Examining the above table and comparing it to Table 12.3 show that, if taken as a single entity, the Arab world will be placed at the top of the leading ten countries list in terms of overall fossil fuels reserves, with a share totalling 19.5% of the world's overall proved conventional reserves. The data reveals that the Arab world remains a major player in terms of overall fossil fuels reserves, though it is now only one of three players of equal strength. It is evident that the Arab word's power is significantly diminished, is far from the sheer domination it enjoys in terms of conventional oil reserves and has far less than its influence, in terms of conventional natural gas reserves.

Table 12.4: Overall conventional proved fossil fuels reserves in the Arab countries (2009)

Country	Total Crude Oil Reserves billion barrel	Total Natural Gas Reserves billion boe	Total Coal Reserves billion boe	Total Fossil Fuels Reserves billion boe	Rank	Share %
Bahrain	0.12	0.61	0.00	0.73	14	0.01
Iraq	115.00	20.92	0.00	135.92	4	2.58
Jordan	0.00	0.04	0.00	0.04	17	0.00
Kuwait	104.00	11.84	0.00	115.84	5	2.20
Lebanon	0.00	0.00	0.00	0.00		
Oman	5.50	5.61	0.00	11.11	9	0.21
Palestine	0.00	0.26	0.00	0.26	16	0.01
Qatar	15.21	166.70	0.00	181.91	2	3.46
Saudi Arabia	266.71	48.31	0.00	315.02	1	5.99
Syria	2.50	1.59	0.00	4.09	12	0.08
UAE	97.80	40.07	0.00	137.87	3	2.62
Yemen	3.00	3.16	0.00	6.16	10	0.12
Algeria	12.20	29.72	0.21	42.12	7	0.80
Comoros	0.00	0.00	0.00	0.00		
Djibouti	0.00	0.00	0.00	0.00		
Egypt	3.70	10.93	0.07	14.71	8	0.28
Libya	43.66	10.16	0.00	53.82	6	1.02
Mauritania	0.10	0.19	0.00	0.29	15	0.01
Morocco	0.00	0.01	0.00	0.01	19	0.00
Somalia	0.00	0.04	0.00	0.04	18	0.00
Sudan	5.00	0.56	0.00	5.56	11	0.11
Tunisia	0.43	0.43	0.00	0.85	13	0.02
ARAB WORLD	674.93	351.13	0.28	1026.35		19.52

Source: Calculated based on data in Chapters 9, 10 and 11.
Note 1: Totals may not add up due to rounding.

12.2.2 Overall Conventional Fossil Fuels Resources

Following on from the previous section and using the conventional recourses data reported in Sections 9.1.2, 10.1.2 and 11.1.2, the world's overall conventional fossil fuels resources are shown in Table 12.5. It can be seen that, similar to the reserves situation, oil is not the dominant fossil fuel in terms of conventional resources, with coal accounting for over 75.7% of the total resources, while oil is relegated to second place with almost 14.7% and natural gas is significantly less, accounting for only 9.6%. This illustrates clearly that oil and natural gas resources combined do not even constitute a quarter of the

Table 12.5: World overall conventional fossil fuels resources by type (2009)

Fuel Type	Resources billion boe	Share %
Oil	4456.6	14.68
Natural Gas	2911.5	9.59
Coal	22987.0	75.73
TOTAL	30355.1	100.00

Source: Calculated based on data in Chapters 9, 10 and 11.
Note 1: Oil is reported in billion barrel.
Note 2: Totals may not add up due to rounding.

world's total fossil fuels remaining resources, with coal accounting for an overwhelming three quarters. This inevitably triggers the simple question of why are we all obsessed with oil and natural gas, if coal as an alternative fossil fuel is so abundant. Several explanations may be offered. For example, some cynics may not believe the coal reserves data; others may suggest environmental concerns as the reason to exclude coal, while the 'conspiracy theory' obsessed may point the finger at hidden agendas. The truth is that the facts speak for themselves. It is obvious that we have plenty of the 'stuff', i.e. fossil fuel, remaining, and thus the above question is left open for now. Chapter 14 will offer some hints, but readers are invited to draw their own conclusions.

Table 12.6 summarises the Arab world's overall conventional fossil fuels resources and their share of the world's total are also shown. The table includes estimates of the remaining conventional resources and original conventional resources in place. It shows that, if taken as one entity, the Arab world's share of the total remaining conventional fossil fuels resources will diminish to just over 8.9%, which is significantly less than its share of the remaining fossil fuel reserves. The data thus reveals the Arab world's status as a major player, in terms of overall conventional fossil fuels resources, is unsustainable, and that its influence will wane in the future.

Table 12.6: World and Arab world overall conventional fossil fuels in place and overall conventional fossil fuels resources (2009)

			Oil	Natural Gas	Coal	Total
Remaining Reserves	Arab world	billion barrel	674.9	350.9	0.3	1026
	World	billion barrel	1169.5	1168.9	2919.7	5258
	Arab world Share	%	57.71	30.02	0.01	19.51
Remaining Resources	Arab world	billion boe	2077.1	620.4	0.5	2698
	World	billion boe	4456.6	2911.5	22987.0	30355
	Arab world Share	%	46.61	21.31	0.00	8.89
Remaining in Place	Arab world	billion boe	3099.0	693.9	1.9	3795
	World	billion boe	6854.9	3301.0	79246.1	89402
	Arab world Share	%	45.21	21.02	0.00	4.24
Original in Place	Arab world	billion boe	3406.1	734.7	2.1	4143
	World	billion boe	7994.4	3894.4	80370.1	92259
	Arab world Share	%	42.61	18.86	0.00	4.49

Source: Calculated based on data in Chapters 9, 10 and 11.
Note 1: Totals may not add up due to rounding.

12.3 Overall Unconventional Fossil Fuels Reserves and Resources

Repeating the same methodology in Sections 12.1 and 12.2 and using the unconventional reserves and recourses data reported in Sections 9.2, 10.2 and 11.2, the world's overall unconventional fossil fuels reserves and resources are shown in Tables 12.7, and 12.8. Here the situation differs from conventional reserves and

Table 12.7: World overall unconventional proved fossil fuels reserves by type (2009)

Fuel Type	Reserves billion boe	Share %
Oil	625.0	28.95
Natural Gas	1527.0	70.72
Coal	7.1	0.33
TOTAL	2159.0	100.00

Source: Calculated based on data in Chapters 9, 10 and 11.
Note 1: Oil is reported in billion barrel.
Note 2: Totals may not add up due to rounding.

Table 12.8: World overall unconventional fossil fuels resources by type (2009)

Fuel Type	Resources billion boe	Share %
Oil	2680.1	17.67
Natural Gas	11780.6	77.66
Coal	707.9	4.67
TOTAL	15168.6	100.00

Source: Calculated based on data in Chapters 9, 10 and 11.
Note 1: Oil is reported in billion barrel.
Note 2: Totals may not add up due to rounding.

resources. Even though oil is still not the dominant fossil fuel, with its share only 29.0% and 17.7% of the total unconventional reserves and resources respectively, natural gas replaces coal as the dominant fossil fuel, with a massive 70.7% and 77.7% of the total unconventional reserves and resources respectively. Coal is relegated to a distant third, accounting for a tiny 0.3% and 4.7% of the total unconventional reserves and resources respectively. This illustrates clearly that unconventional oil and natural gas resources combined, are extremely important and have to be taken into account when planning future oil and natural gas production scenarios. The above numbers suggest that there is a potential glut in natural gas rather than a peak and that, future low prices, rather than the scarcity of resources, may play a bigger role in constraining future supplies, as it will influence decisions on future investment.

Table 12.9 summarises the Arab world's overall unconventional fossil fuels reserves and resources and their share to the world's total are shown there. The table also includes estimates of the remaining and original resources in place. The table shows that, if taken as one entity, the Arab world's share of the total remaining unconventional fossil fuels reserves and resources is an insignificant 4.2% and 5.9% respectively, which is a far cry from the misleading two thirds 'scary number' we are always faced with in all media reports.

Table 12.9: World and Arab world overall unconventional fossil fuels in place and overall unconventional fossil fuels resources (2009)

			Oil	Natural Gas	Coal	Total
Remaining Reserves	Arab world	billion barrel	1.0	88.8	0.1	90
	World	billion barrel	625.0	1527.0	7.1	2159
	Arab world Share	%	0.16	5.82	0.72	4.16
Remaining Resources	Arab world	billion boe	28.1	859.4	5.1	893
	World	billion boe	2680.1	11780.6	707.9	15169
	Arab world Share	%	1.05	7.29	0.72	5.88
Remaining in Place	Arab world	billion boe	93.8	36288.5	51.1	36433
	World	billion boe	8984.6	551224.1	7078.7	567287
	Arab world Share	%	1.04	6.58	0.72	6.42
Original in Place	Arab world	billion boe	93.8	36288.5	51.1	36433
	World	billion boe	9006.1	551246.9	7078.7	567332
	Arab world Share	%	1.04	6.58	0.72	6.42

Source: Calculated based on data in Chapters 9, 10 and 11.
Note 1: Totals may not add up due to rounding.

12.4 Overall Totalised Fossil Fuels Reserves and Resources

To summarise, Tables 12.10 to 12.12 present the overall fossil fuels reserves and resources in the world and the Arab world's share of these reserves and resources. From these summary tables we can conclude that:

Table 12.10: World overall proved fossil fuels reserves by type (2009)

Fuel Type	Reserves billion boe	Share %
Oil	1794.5	24.19
Natural Gas	2695.8	36.35
Coal	2926.8	39.46
TOTAL	7417.2	100.00

Source: Calculated based on data in Chapters 9, 10 and 11.
Note 1: Oil is reported in billion barrel.
Note 2: Totals may not add up due to rounding.

- In terms of reserves, coal constitutes the majority of fossil fuels with a share of 39.5%, followed by natural gas with

Table 12.11: World overall fossil fuels resources by type (2009)

Fuel Type	Resources billion boe	Share %
Oil	7136.7	15.68
Natural Gas	14692.1	32.27
Coal	23694.8	52.05
TOTAL	*45523.6*	*100.00*

Source: Calculated based on data in Chapters 9, 10 and 11.
Note 1: Oil is reported in billion barrel.
Note 2: Totals may not add up due to rounding.

36.4%, while oil is 24.2%.
- In terms of resources, coal constitutes the biggest majority of fossil fuels with a share of 52.1%, followed by natural gas with 32.3%, while oil is 15.7%.
- The Arab world's share of the world's total fossil fuels is rather small and stands at 15.0% and 7.9% of the total remaining reserves and resources respectively.

Table 12.12: World and Arab world overall fossil fuels in place and overall fossil fuels resources (2009)

			Oil	Natural Gas	Coal	Total
Remaining Reserves	Arab world	billion barrel	675.9	439.7	0.3	1116
	World	billion barrel	1794.5	2695.8	2926.8	7417
	Arab world Share	%	37.67	16.31	0.01	15.04
Remaining Resources	Arab world	billion boe	2105.3	1479.8	5.6	3591
	World	billion boe	7136.7	14692.1	23694.8	45524
	Arab world Share	%	29.50	10.07	0.02	7.89
Remaining in Place	Arab world	billion boe	3192.7	36982.4	53.0	40228
	World	billion boe	15839.5	554525.0	86324.8	656689
	Arab world Share	%	20.16	6.67	0.06	6.13
Original in Place	Arab world	billion boe	3499.8	37023.2	53.2	40576
	World	billion boe	17000.5	555141.3	87448.9	659591
	Arab world Share	%	20.59	6.67	0.06	6.15

Source: Calculated based on data in Chapters 9, 10 and 11.
Note 1: Totals may not add up due to rounding.

Thus to re-iterate, please read the above three bullet points and ask yourselves why are we concerned about oil in general, and the Arab countries in particular.

12.5 Overall Fossil Fuels Production

A similar amalgamation exercise to the one explained in the previous sections while analysing fossil fuels reserves and resources is performed here to obtain the overall fossil fuels production, expressed in terms of boe, using the data previously presented in Sections 9.4, 10.4 and 11.4.

Table 12.13: World overall fossil fuels production by type (2008)

Fuel Type	Production thousand boepd	Share %
Oil	83835.1	40.99
Natural Gas	56710.0	27.73
Coal	63965.6	31.28
TOTAL	204510.7	100.00

Source: Calculated based on data in Chapters 9, 10 and 11.
Note 1: Oil is reported in thousand bpd.
Note 2: Totals may not add up due to rounding.

The world's overall fossil fuels production is shown in Table 12.13, where it can be observed that oil contributes the largest proportion of the overall fossil fuels production accounting for approximately 41.0% of the total production, though it is by no means the dominant fossil fuel. Coal is still a significant contributor, remaining in second place, despite all the calls to reduce its production to cut carbon dioxide emissions, where it currently accounts for approximately 31.3% of the total production, while the third remainder production is constituted of natural gas, with 27.7% of the total production.

Examining Table 12.13 highlights the dependence of the world on oil to provide the biggest proportion of its energy needs, which does not seem to be waning in the near future. This oil dependence is fuelling the development of alternative methods to produce more oil. These include EOR methods to extract more oil from known reserves, development of unconventional oil deposits, expansion of NGLs production, as well as the extensive research conducted to commercially develop conversion processes of CTL and GTL as substitutes to dwindling conventional oil production. In addition the majority of biofuels produced are in the form of liquid fuels that are geared towards being compatible with oil, to elongate the oil age.

Table 12.14: World overall fossil fuels production – top ten countries (2008)

Rank	Country	Total Oil Production thousand bpd	Total Natural Gas Dry Production thousand boepd	Total Coal Primary Production thousand boepd	Total Fossil Fuels Production thousand boepd	Share %
1	China	3973.13	1374.97	25053.26	30401.35	14.87
2	USA	7707.24	10527.76	10305.35	28540.35	13.96
3	Russia	9789.76	11974.03	3133.31	24897.10	12.17
4	Saudi Arabia	10701.12	1454.53	0.00	12155.65	5.94
5	Canada	3353.09	3090.90	660.41	7104.40	3.47
6	India	883.51	582.25	4999.45	6465.20	3.16
7	Iran	4149.30	2102.96	15.13	6267.39	3.06
8	Australia	585.35	817.66	3857.47	5260.48	2.57
9	Indonesia	1052.29	1265.75	2755.45	5073.49	2.48
10	Norway	2465.96	1793.75	31.83	4291.53	2.10
	TOTAL	44660.74	34984.56	50811.65	130456.95	63.79
	WORLD				204510.65	

Source: Calculated based on data in Chapters 9, 10 and 11.
Note 1: Totals may not add up due to rounding.

Table 12.14 shows the list of the top ten countries, in terms of overall fossil fuels production. These countries account for almost 63.8% of the global fossil fuels production. The table makes fascinating reading as it highlights that, similar to the conventional reserves situation, the world's major energy heavyweights, in terms of production, are also not located in the Middle East nor are they members of OPEC, but they are in fact the world's major political and military powers. In this case it is China, who edges the USA marginally to occupy the top spot, with Russia relegated to the third spot. Only two Middle Eastern countries, Saudi Arabia and Iran, both of which are all members of OPEC, feature in the top list, in the fourth and seventh spots respectively, with their combined production share is still less than that of Russia, though in this case, Saudi Arabia remains the leading OPEC producer. In addition, Indonesia, nominally still an OPEC member (though with suspended membership), also makes an appearance in the ninth place in the list. In total, the Middle East's total share is only just over 16.0%, while that of OPEC is 24.4%. Both these shares are significantly less than the corresponding Middle East and OPEC's massive shares, in terms of oil or natural gas production.

Table 12.15: Total overall fossil fuels production in the Arab countries (2008)

Country	Total Oil Production thousand bpd	Total Natural Gas Dry Production thousand boepd	Total Coal Primary Production thousand boepd	Total Fossil Fuels Production thousand boepd	Rank	Share %
Bahrain	48.52	228.56	0.00	277.08	13	0.14
Iraq	2368.91	33.99	0.00	2402.91	6	1.17
Jordan	-0.26	4.52	0.00	4.26	16	0.00
Kuwait	2741.38	229.64	0.00	2971.03	4	1.45
Lebanon	0.00	0.00	0.00	0.00		
Oman	761.00	433.97	0.00	1194.97	9	0.58
Palestine	0.00	0.00	0.00	0.00		
Qatar	1190.61	1391.99	0.00	2582.60	5	1.26
Saudi Arabia	10701.12	1454.53	0.00	12155.65	1	5.94
Syria	444.38	109.22	0.00	553.60	10	0.27
UAE	3046.47	908.45	0.00	3954.92	2	1.93
Yemen	300.12	0.00	0.00	300.12	12	0.15
Algeria	2227.33	1564.20	0.00	3791.53	3	1.85
Comoros	0.00	0.00	0.00	0.00		
Djibouti	0.00	0.00	0.00	0.00		
Egypt	630.58	873.37	0.23	1504.19	8	0.74
Libya	1854.06	287.51	0.00	2141.57	7	1.05
Mauritania	12.83	0.00	0.00	12.83	15	0.01
Morocco	-0.69	1.08	0.00	0.39	17	0.00
Somalia	0.00	0.00	0.00	0.00		
Sudan	523.25	0.00	0.00	523.25	11	0.26
Tunisia	86.93	53.70	0.00	140.63	14	0.07
ARAB WORLD	26936.56	7574.74	0.23	34511.53		16.88

Source: Calculated based on data in Chapters 9, 10 and 11.
Note 1: Totals may not add up due to rounding.

Table 12.15 shows the Arab world's overall fossil fuels production and its share of the world's total. In addition it details the overall fossil fuels production of all Arab countries. It can be seen than 17 Arab countries have active fossil fuels production; however the countries rankings changes significantly in comparison to their oil production rankings. Even though the top countries remain to be Saudi Arabia and UAE, by virtue of their large oil production, the ranking gets muddled after the top two, with Algeria promoted to third and Qatar to fifth, as a result of their significant natural gas production, while Kuwait is relegated to forth. Iraq is pushed down to sixth place, which highlights the fact that it is still lagging behind in terms of production, albeit it also highlights its potential for future production and adds weight to speculation that the Anglo-American invasion was primarily oil motivated.

Probing the above table, in conjunction with Table 12.14 shows that if taken as one entity, the Arab world will be placed at the top of the leading producers table, with its share totalling 16.9% of the world's overall production. The data reveals that the Arab world remains a major player in terms of overall fossil fuels production, though it is now only one of four major players of substantial strength. Here again, it is obvious that the Arab word's influence is significantly less than we are led to believe, where its role is far from the strong position it enjoys in terms of oil and natural gas production.

12.6 Overall Fossil Fuels Consumption
Here also, the now familiar amalgamation exercise is carried out to obtain the overall fossil fuels consumption, expressed in terms of boe, using the data previously presented in Sections 9.7, 10.6 and 11.6.

The world's overall fossil fuels consumption is shown in Table 12.16 where it can be observed that the contribution of oil, coal and natural gas is almost identical to their contribution to fossil fuels production. To expand, in terms of consumption, oil contributes the largest proportion of the overall fossil fuels consumption and accounts for approximately 41.5% of the total consumption, though, mirroring production oil is not the dominant fossil fuel any more. Coal again is still significant and remains in second place, despite all the negative publicity surrounding its carbon dioxide emissions, where it accounts for over 30.9% of the total consumption. The third and remainder consumption consistent is natural gas, with 27.6% of the total consumption. As already explained in the previous section, this highlights the current dependence of the world on oil to provide the biggest proportion of its energy needs, which is still showing no signs of diminishing.

Table 12.16: World overall fossil fuels consumption by type (2008)

Fuel Type	Consumption thousand boepd	Share %
Oil	85466.3	41.49
Natural Gas	56876.1	27.61
Coal	63673.4	30.91
TOTAL	206015.8	100.00

Source: Calculated based on data in Chapters 9, 10 and 11.
Note 1: Oil is reported in thousand bpd.
Note 2: Totals may not add up due to rounding.

Table 12.17 shows the list of the top ten countries, in terms of overall fossil fuels consumption. These countries account for almost 63.0% of the global fossil fuels consumption, which is almost identical to the share of the top ten countries, in terms of production. The table is intriguing, showing that in this instance two countries emerge as two consumption giants, namely the USA and China, with Russia and India trailing as distant third and fourth with a combined consumption barely exceeding half of the USA's consumption. Interestingly, similar to the conventional reserves situation, the world's major energy heavy-weights, in terms of consumption, are the world's major political and military powers. However in this case, the USA is placed firmly on the top, with China currently second, although it is catching up fast and is projected to overtake the USA in the near future. Only one Middle Eastern country, Iran, which is also a member of OPEC, features in the top ten list placed in the ninth spot with a share of only 1.9%. In total, the Middle East's total share is only 6.2%, while that of OPEC is 8.1%, both shares are less than half of the USA's or China's consumption. This share is significantly less than either the Middle East or OPEC oil or natural gas consumption share, due to the region's or organisation's small consumption of coal.

Table 12.17: World overall fossil fuels consumption – top ten countries (2008)

Rank	Country	Total Oil Consumption thousand bpd	Total Natural Gas Consumption thousand boepd	Total Coal Consumption thousand boepd	Total Fossil Fuels Consumption thousand boepd	Share %
1	USA	19497.96	11883.10	9867.54	41248.60	20.02
2	China	7850.00	1395.57	24890.80	34136.36	16.57
3	Russia	2900.00	8601.70	2372.38	13874.07	6.73
4	India	2940.00	777.35	5608.18	9325.53	4.53
5	Japan	4784.85	1828.89	1794.37	8408.11	4.08
6	Germany	2569.28	1732.15	2374.20	6675.63	3.24
7	Canada	2259.20	1499.59	548.72	4307.51	2.09
8	UK	1709.66	1734.81	627.86	4072.33	1.98
9	Iran	1755.41	2150.95	17.16	3923.52	1.90
10	Saudi Arabia	2380.00	1454.53	0.02	3834.55	1.86
	TOTAL	48646.37	33058.63	48101.23	129806.22	63.01
	WORLD				206015.82	

Source: Calculated based on data in Chapters 9, 10 and 11.
Note 1: Totals may not add up due to rounding.

Table 12.18 shows the Arab world's overall fossil fuels consumption and its share of the world's total. In addition it details the overall fossil fuels consumption of all Arab countries. Obviously all 22 Arab countries appear in the list, however the countries rankings are changed slightly in comparison to their oil consumption rankings. Similar to production, Saudi Arabia retains its position as the top consumer. However UAE and Algeria leapfrog Iraq and Egypt respectively. The changes in the other rankings are not as prominent since oil contributes the majority of the fossil fuels consumption in the Arab world. Almost all the rest comes from natural gas, especially in countries with significant natural gas reserves, while coal consumption is trivial and does not interfere with the rankings.

Table 12.18: Overall fossil fuels consumption in the Arab countries (2008)

Country	Total Oil Consumption thousand bpd	Total Natural Gas Consumption thousand boepd	Total Coal Consumption thousand boepd	Total Fossil Fuels Consumption thousand boepd	Rank	Share %
Bahrain	38.00	228.56	0.00	266.56	11	0.13
Iraq	637.66	33.99	0.00	671.66	5	0.33
Jordan	108.27	53.70	0.00	161.98	14	0.08
Kuwait	324.98	229.64	0.00	554.62	6	0.27
Lebanon	92.25	0.00	2.09	94.34	16	0.05
Oman	80.95	243.39	0.00	324.34	10	0.16
Palestine	23.00	0.00	0.00	23.00	18	0.01
Qatar	129.15	365.26	0.00	494.41	7	0.24
Saudi Arabia	2380.00	1454.53	0.02	3834.55	1	1.86
Syria	256.19	111.75	0.04	367.98	9	0.18
UAE	463.01	1074.50	0.00	1537.51	2	0.75
Yemen	149.19	0.00	0.00	149.19	15	0.07
Algeria	298.51	485.24	11.01	794.75	4	0.39
Comoros	0.80	0.00	0.00	0.80	22	0.00
Djibouti	12.51	0.00	0.00	12.51	20	0.01
Egypt	696.89	567.42	11.71	1276.02	3	0.62
Libya	273.32	99.45	0.00	372.77	8	0.18
Mauritania	20.95	0.00	0.00	20.95	19	0.01
Morocco	189.54	10.13	50.60	250.26	12	0.12
Somalia	5.11	0.00	0.00	5.11	21	0.00
Sudan	85.69	0.00	0.00	85.69	17	0.04
Tunisia	89.59	76.31	0.00	165.90	13	0.08
ARAB WORLD	6355.54	5033.87	75.47	11464.88		5.57

Source: Calculated based on data in Chapters 9, 10 and 11.
Note 1: Totals may not add up due to rounding.

Revisiting the above table, in conjunction with Table 12.17, shows that if taken as a single entity, the Arab world will be placed fourth in the list of the world's leading consumers, with a share of approximately 5.6% of the world's overall consumption. The data explains the reason why the Arab world can punch above its weight in the energy markets. Its fossil fuels consumption, although important, is far below that of the major two consumers, thus it possesses the much needed excess fossil fuels production to supply the other consumers.

Absolute consumption

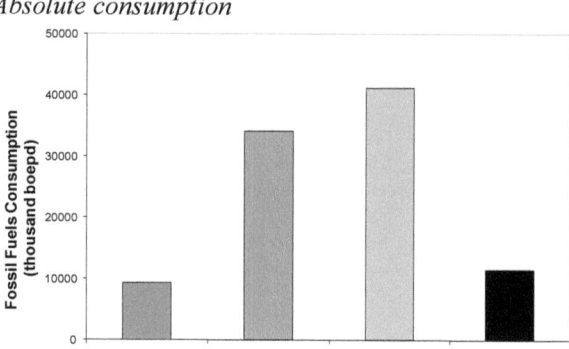

Consumption per GDP

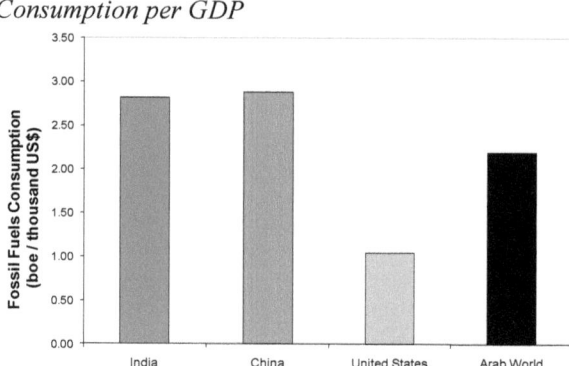

Consumption per Capita

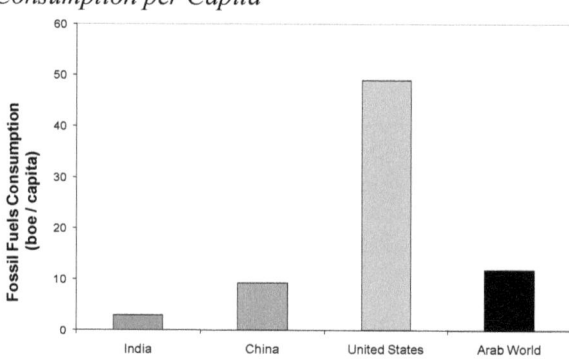

Figure 12.2: Comparison of overall fossil fuels consumption data presentation in the USA, China, India and the Arab world (2008)

Source: Calculated based on data in Chapters 9, 10 and 11.

The data in Sections 9.7, 10.6 and 11.6, Figure 12.2 shows a comparison of the Arab world fossil fuels consumption and that of the USA, China and India expressed in actual terms and per GDP or per capita. As we are already accustomed by now, the figure illustrates once more how the reporting of the data can be manipulated to make it look better, masking certain facts or presenting an unclear picture.

12.7 Overall Fossil Fuels Trade

Following the standard amalgamation methodology, we can obtain the overall fossil fuels net trade, expressed in terms of boe, using the data previously presented in Sections 9.8, 10.7 and 11.7. However caution must be practiced here, in order to obtain an actual net value for exports and imports. The correct procedure is to calculate the net exports or imports, as the difference between the sum of the net production and the net consumption of the combined overall fossil fuels production and consumption, rather than summing the net exports of net imports of the three fossil fuels types as reported in Sections 9.8, 10.7 and 11.7. This is necessary, since the latter methodology will result in errors, if a country is a net exporter, in terms of one type, but net importer, in terms of another type. This is essential to avoid ambiguity, since otherwise; we may underestimate or overestimate the net value. To illustrate, consider the UAE for example. The country is a net oil exporter, but at the same time is a net natural gas importer. Thus summing the exports only will overestimate its net exports, while summing the imports will place it incorrectly as a net importer. In reality the country is a net fossil fuel exporter but with a lower overall net value that takes the natural gas imports into account.

The world's overall fossil fuels net exports are shown in Table 12.19. It can be observed that, unlike reserves, production or consumption, oil contributes the largest proportion of the overall fossil fuels net exports, accounting for approximately 66.9% of the total exports, with natural gas placed as a distant second with a share of only 21.2% of the total net exports. Coal is the least significant in terms of net exports, where it accounts for a mere 11.9%.

Similarly, the world's overall fossil fuels net imports are shown

Table 12.19: World net overall fossil fuels exports by type (2008)

Fuel Type	Net Exports thousand boepd	Share %
Oil	44932.4	66.92
Natural Gas	14252.8	21.23
Coal	7962.7	11.86
TOTAL	67147.8	100.00

Source: Calculated based on data in Chapters 9, 10 and 11.
Note 1: Oil is reported in thousand bpd.
Note 2: Totals may not add up due to rounding.

Table 12.20: World net overall fossil fuels imports by type (2008)

Fuel Type	Net Imports thousand boepd	Share %
Oil	46578.0	67.83
Natural Gas	14418.9	21.00
Coal	7670.5	11.17
TOTAL	68667.4	100.00

Source: Calculated based on data in Chapters 9, 10 and 11.
Note 1: Oil is reported in thousand bpd.
Note 2: Totals may not add up due to rounding.

in Table 12.20, where it can be observed that the three types of fossil fuels relative contributions are almost identical to their corresponding contributions in terms of net exports, with the slight differences attributed to bunkering and storage. Quantitatively, oil is the dominant fossil fuel and contributes the largest proportion of the overall fossil fuels net imports of approximately 67.8% of the total imports. Natural gas is in the second place, with a share of 21.0%, while coal is the least significant, in terms of net imports, where it accounts for a mere 11.2%.

The tables above emphasize the dependence of the world on oil by illustrating clearly that oil is still the dominant form of energy traded on the international markets.

Table 12.21: World net overall fossil fuels exports – top ten countries (2008)

Rank	Country	Net Oil Trade (Export) thousand bpd	Net Natural Gas Trade (Export) thousand boepd	Net Coal Trade (Export) thousand boepd	Net Fossil Fuels Trade (Export) thousand boepd	Share %
1	Russia	6889.76	3372.33	760.93	11023.02	17.31
2	Saudi Arabia	8321.12	0.00	0.00	8321.10	13.07
3	Norway	2245.78	1721.97	20.71	3988.46	6.26
4	Algeria	1928.82	1078.96	0.00	2996.78	4.71
5	Canada	1093.89	1591.31	111.70	2796.89	4.39
6	Australia	0.00	199.30	2445.44	2505.97	3.94
7	Indonesia	0.00	605.74	2006.83	2417.41	3.80
8	UAE	2583.46	0.00	0.00	2416.41	3.80
9	Kuwait	2416.41	0.00	0.00	2343.87	3.68
10	Iran	2393.90	0.00	0.00	2276.42	3.58
	TOTAL	27873.15	8569.61	5345.62	41086.34	64.53
	WORLD				63665.95	

Source: Calculated based on data in Chapters 9, 10 and 11.
Note 1: Totals may not add up due to rounding.

Table 12.21 shows the list of the top ten countries in terms of overall fossil fuels net exports. These countries account for almost 64.5% of the global fossil fuels net exports. The table is different from the reserves, production and consumption tables as it is largely controlled by the net oil exporting countries, with Russia and Saudi Arabia occupying the two top places respectively. This table illustrates the strength of the Middle East and OPEC, where their countries occupy four and six places in the top ten countries respectively, with all the Middle Eastern countries in the list also being members of OPEC. In total, the Middle East's total share is only just over 32.3%, while that of OPEC is an overwhelming 51.9%, both very significant, though still less than their massive shares in terms of oil or natural gas production.

Table 12.22 shows the list of the top ten countries in terms of overall fossil fuels net imports. These countries account for almost 70.4% of the global fossil fuels net imports. As with the previous illustration, this table is different from the reserves, production and consumption tables as it is largely controlled by the net oil importing countries, with the USA and Japan occupying the two top places respectively. This table shows that, with the exception of Russia, the world's major military, industrial and economic powers are all net fossil fuels importers. This confirms the interpretation derived from Table 12.21, which demonstrates the strength of the Middle East and OPEC, where none of their countries feature in the list of leading net fossil fuels importers. In total, the Middle East's total share, in terms of net imports, is a tiny 0.9%, which is derived by summing the Middle Eastern countries separately and if taken as one region in totality, the region will have no net imports. As expected, OPEC countries are all net fossil fuels exporters and thus the organisation's share in terms on net fossil fuels imports is zero.

Table 12.22: World net overall fossil fuels imports – top ten countries (2008)

Rank	Country	Net Oil Trade (Import) thousand bpd	Net Natural Gas Trade (Import) thousand boepd	Net Coal Trade (Import) thousand boepd	Net Fossil Fuels Trade (Import) thousand boepd	Share %
1	USA	11790.72	1355.33	0.00	12708.24	19.51
2	Japan	4652.08	1731.97	1794.37	8178.42	12.56
3	Germany	2448.08	1436.30	488.59	4372.97	6.71
4	China	3876.87	20.60	0.00	3745.11	5.75
5	South Korea	2158.79	620.54	965.78	3735.01	5.73
6	Italy	1495.37	1367.52	246.40	3109.29	4.77
7	France	1915.54	874.27	185.79	2975.61	4.57
8	India	2056.49	195.11	608.73	2860.33	4.39
9	Spain	1533.92	690.12	156.72	2380.76	3.66
10	Taiwan	946.60	218.43	610.47	1775.50	2.73
	TOTAL	32874.46	8510.20	5056.84	45841.24	70.38
	WORLD				65134.42	

Source: Calculated based on data in Chapters 9, 10 and 11.
Note 1: Totals may not add up due to rounding.

The overall fossil fuels net exports of the Arab world and its share to the world's total is presented in Table 12.23. In addition the table shows the details of the overall fossil fuels net exports of all Arab countries. It can be seen than 13 Arab countries are net fossil fuels exporters. The countries rankings are a mix of the

Table 12.23: Net overall fossil fuels exports in the Arab countries (2008)

Country	Net Oil Trade (Export) thousand bpd	Net Natural Gas Trade (Export) thousand boepd	Net Coal Trade (Export) thousand boepd	Net Fossil Fuels Trade (Export) thousand boepd	Rank	Share %
Bahrain	10.53	0.00	0.00	10.53	13	0.02
Iraq	1731.25	0.00	0.00	1731.25	7	2.72
Jordan	0.00	0.00	0.00	0.00		
Kuwait	2416.41	0.00	0.00	2416.41	4	3.80
Lebanon	0.00	0.00	0.00	0.00		
Oman	680.05	190.59	0.00	870.63	8	1.37
Palestine	0.00	0.00	0.00	0.00		
Qatar	1061.47	1026.72	0.00	2088.19	5	3.28
Saudi Arabia	8321.12	0.00	0.00	8321.10	1	13.07
Syria	188.19	0.00	0.00	185.62	11	0.29
UAE	2583.46	0.00	0.00	2417.41	3	3.80
Yemen	150.93	0.00	0.00	150.93	12	0.24
Algeria	1928.82	1078.96	0.00	2996.78	2	4.71
Comoros	0.00	0.00	0.00	0.00		
Djibouti	0.00	0.00	0.00	0.00		
Egypt	0.00	305.95	0.00	228.16	10	0.36
Libya	1580.74	188.05	0.00	1768.80	6	2.78
Mauritania	0.00	0.00	0.00	0.00		
Morocco	0.00	0.00	0.00	0.00		
Somalia	0.00	0.00	0.00	0.00		
Sudan	437.56	0.00	0.00	437.56	9	0.69
Tunisia	0.00	0.00	0.00	0.00		
ARAB WORLD	21090.53	2790.28	0.00	23623.38		37.11

Source: Calculated based on data in Chapters 9, 10 and 11.
Note 1: Totals may not add up due to rounding.

rankings of oil and natural gas exporting countries, with some positions changed significantly in comparison to their oil production ranking. Thus here Saudi Arabia maintains the top ranking due to its massive oil exports, while Algeria advances to the second place due to its natural gas exports. On the other hand, Iraq is even less important occupying only the seventh place, which reflects its continuous failure to increase its production, and thus surplus to reflect its potential as a large fossil fuels exporter, rivalling Saudi Arabia. This may disappoint some who speculated that the Anglo-American invasion was primarily due to controlling the Iraqi oil.

Table 12.24: Net overall fossil fuels imports in the Arab countries (2008)

Country	Net Oil Trade (Import) thousand bpd	Net Natural Gas Trade (Import) thousand boepd	Net Coal Trade (Import) thousand boepd	Net Fossil Fuels Trade (Import) thousand boepd	Rank	Share %
Bahrain	0.00	0.00	0.00	0.00		
Iraq	0.00	0.00	0.00	0.00		
Jordan	108.53	49.18	0.00	157.72	2	0.24
Kuwait	0.00	0.00	0.00	0.00		
Lebanon	92.25	0.00	2.09	94.34	3	0.14
Oman	0.00	0.00	0.00	0.00		
Palestine	0.00	0.00	0.00	0.00		
Qatar	0.00	0.00	0.00	0.00		
Saudi Arabia	0.00	0.00	0.02	0.00		
Syria	0.00	2.53	0.04	0.00		
UAE	0.00	166.05	0.00	0.00		
Yemen	0.00	0.00	0.00	0.00		
Algeria	0.00	0.00	11.01	0.00		
Comoros	0.80	0.00	0.00	0.80	8	0.00
Djibouti	12.51	0.00	0.00	12.51	5	0.02
Egypt	66.31	0.00	11.48	0.00		
Libya	0.00	0.00	0.00	0.00		
Mauritania	8.12	0.00	0.00	8.12	6	0.01
Morocco	190.23	9.04	50.60	249.86	1	0.38
Somalia	5.11	0.00	0.00	5.11	7	0.01
Sudan	0.00	0.00	0.00	0.00		
Tunisia	2.66	22.60	0.00	25.27	4	0.04
ARAB WORLD	486.51	249.41	75.24	553.73		0.85

Source: Calculated based on data in Chapters 9, 10 and 11.
Note 1: Totals may not add up due to rounding.

The overall fossil fuels net imports of the Arab world and its share to the world's total is presented in Table 12.24. In addition the table shows the details of the overall fossil fuels net imports of all Arab countries. It can be seen than eight Arab countries are net fossil fuels importers. The countries rankings largely mirror their rankings in terms of oil imports, with the exception of Tunisia, whose ranking is due to its natural gas imports.

Note that in the above two tables, no trade numbers are reported for Palestine as its trade is still reported implicitly as part of

Israeli statistics.

Examining the above four tables shows that, if taken as a single entity, the Arab world would be placed at the top of the leading net exporters table, with its share totalling 36.6% of the world's overall fossil fuels net exports, while its share, in terms on net fossil fuels imports, is zero. The data thus reveals the real power the Arab world enjoys that allows it to remain as a major player in terms of overall net fossil fuels exports. Its share is more than double the second placed exporting country, Russia. However despite this apparent strength, the Arab world's capability in influencing the world's major powers is limited as its trading partners are not necessarily the most powerful countries. This is discussed further in Chapter 14.

12.8 The Bottom Line
Purely from a quantitative perspective, examining the results from the amalgamated numbers, one can see a very different picture emerging. This picture differs significantly from the one often described by the media and politicians, when they communicate the -so-called concrete facts about the world's energy situation and about the dependence of the western civilization on the Arab world to secure its energy supplies.

The numbers tell us that the world has plentiful resources of oil, natural gas and coal, when both conventional and unconventional deposits are taken into consideration. Furthermore, the current unstoppable trend is that more unconventional resources are being reclassified as conventional resources and thus continue to quench the world's thirst for fossil fuels.

In this 'real' picture, one can see that the Arab world is not the dominant force after all. Its influence in energy markets is similar to that of Russia or the USA, so but not as vital as we are all led to believe. So what is the true story? Well! Chapter 14 will tell you the answer!

Chapter 13
FOSSIL FUELS AND ALTERNATIVE ENERGY

We established in Chapter 12 that the world has plenty of fossil fuels. However the popular phrase worldwide is alternative energy and endless propaganda, comprising miles of pages, have been written about it with everyone giving an input from politicians to environmentalists to even religious leaders. So where do these energy types fit in this context? In this chapter, a brief discussion is presented, but is limited to answering the above question, since discussing alternative energy in detail is outside the scope of this book.

13.1 Alternative Energy Overview
Alternative energy is an umbrella term that generally encompasses several forms of energy intended to replace dominant fuel sources of the time without the undesired effects of the replaced fuels. The exact definition of alternative energy has changed through time, with the dominant fuel changing from biomass to fossil fuels. Currently the term's definition differs between various data sources, while all agree that the dominant fuel type is fossil fuels, there is disagreement on which other types are excluded from alternative energy terms, with some particularly excluding nuclear energy. In this chapter the generalised definition adopted here considers alternative energy to include all forms of energy excluding the three types of fossil fuels: oil, natural gas and coal, both conventional and unconventional.

13.1.1 Alternative Energy Types
Thus in this brief discussion, the following types of energy are considered forms of alternative energy:
- Nuclear energy, which is the energy released by

radioactive decay or by splitting or merging nuclei of atoms, with the latter two referred to as fission and fusion respectively. Nuclear fission is currently used commercially to generate electricity, using uranium as its main source of energy, with plutonium used also for military purposes. Thorium is also considered to be a suitable fuel source. Nuclear radioactive decay has also been used as an energy sources on a relatively small scale, mostly to power space missions. Nuclear fusion is undergoing extensive research to develop it as a commercial energy source albeit with no success yet, with hydrogen isotopes deuterium and tritium used as primary fuels, while helium-3 is envisaged as a second generation fuel source.

- Hydro-energy, which is the energy derived from moving water that can be harnessed for useful functions. Currently, hydro-energy is overwhelmingly used to generate electricity, with the remainder generating pure mechanical energy utilised, for example, in irrigation and operating machinery, mainly using waterwheels as practiced for millennia. At present, several forms of hydro-energy are used actively or in development. These include traditional waterwheels, hydroelectric dams, tidal energy, wave energy and water current energy.

- Solar energy, which is the energy derived directly from the sun that can be harnessed for generating electricity, or conversion into thermal, chemical or even mechanical energy. It is now widely used in many applications including heating, cooking, lighting, water treatment, and industrial processes such as evaporation and drying. At present only a fraction of the solar energy is utilised, mainly for heating purposes, with applications to generate electricity being developed rapidly, directly using photovoltaic or indirectly utilising heat engines. Numerous concepts are being developed to further harness solar energy including solar ponds and seawater thermal energy, which exploits the temperature difference between deep and shallow waters.

- Wind energy, which is the energy derived from moving wind that can be harnessed for useful functions. This energy is currently utilised to generate electricity using wind turbines or is converted to mechanical power using wind mills. In addition it has been used for millennia to propel sailing ships.
- Geothermal energy, which is the energy extracted from heat stored in the earth. Currently this energy is utilised in two main forms, either to generate electricity or for heating, which is achieved either directly or by using heat pumps.
- Bio-energy, which is a generalised term that covers a wide range of fuels which are derived from biomass. The term covers solid biomass including wood, liquid biofuels and biogases. Biofuels were discussed in Chapter 5 and interested readers are encouraged to revisit the chapter for more details.
- Chemical energy, which in this context refers to energy derived from chemical reactions, chemical solutions or retrieved from the difference in concentration between two fluids that can be harnessed usefully. Currently osmotic energy has been demonstrated successfully as an energy source derived from exploiting the difference in salt concentration using specific membranes.

13.1.2 Alternative Energy Classifications

The above types of alternative energy can be classified in several ways with the main classifications based on their origin, renewability and environmentally friendly credentials.

Based on *origin*, all types of energy are divided into three categories: solar, terrestrial and lunar. This classification is derived by tracing the origin of all energy sources. Attempting to place the alternative energy types into the above three categories results in a classification that may surprise many readers and is briefly described below:
- Solar origin includes solar energy, wind energy, the majority of hydro energy (e.g. wave energy, current

energy, part of tidal energy) and bio-energy as a result of photosynthesis. Note that fossil fuels also fall into this category, as according to the biogenic theory, they form due to the decay of plants and animals.
- Terrestrial origin includes geothermal energy, nuclear energy, part of hydro energy (due to gravitational energy) and chemical energy. Note that part of the natural gas falls into this category according to the abiogenic theory.
- Lunar origin includes part of hydro energy (majority of tidal energy).

In terms of significance, solar origin contributes the overwhelming majority of energy sources, with terrestrial a distant second and lunar an insignificant third.

Based on *renewability*, energy is either renewable or non-renewable. However, evaluating renewability is a controversial issue and is often not well defined. Different methods arrive at different conclusions and the definition of the dividing line is rather vague. There is a misconception, often promoted by popular media, that alternative energies are renewable. This is a fallacy as several types of alternative energy are strictly non-renewable and some are only renewable if their consumption is maintained at a sustainable level where they are carefully managed to avoid their depletion and maintain their consumption at a level allowing their replenishment. In broad terms, the following categorisation is outlined:
- Renewable energy includes solar energy, hydro energy, wind energy, geothermal energy and part of bio-energy if its consumption is sustainable.
- Non-renewable energy includes nuclear energy and part of bio-energy, if its consumption is unsustainable. As you already know, fossil fuels also fall in this category.

Based on *environmentally friendly credentials*, energy types are divided into three categories: friendly, neutral, and unfriendly. Here again there is a misleading notion, spread by the popular media, that alternative energies are environmentally friendly. Once again this is far from true, with many types of alternative energy being harmful for the environment. Unlike the above two classifications, this measure is totally subjective. There is no agreement on quantifying environmental friendliness and thus, several measures can be used such as carbon dioxide emissions, disruption to fauna and flora, soil salination and land erosion. Many of these factors counter each other and, while one fuel can be classified environmentally friendly based on one measure; it can be unfriendly based on the other. A fuel's favourable effects, based on one measure, can lead to adverse effects from another measure's perspective. As a result, no attempt is made here to classify alternative energy types into rigid environmental friendliness categories, but a few examples are presented below that demonstrate the confusion.

As a first example, there is almost universal agreement that fossil fuels are environmentally unfriendly in terms of carbon dioxide emissions. However various biofuels, perceived as environmentally friendly, emit more carbon dioxide, if their full life cycle is taken into account with some, like wood having higher carbon density. In addition the deforestation, resulting from commercial palm plantation to produce the biofuel palm oil, is causing additional carbon dioxide emissions, due to its adverse effects on peat bogs, amongst other factors.

A second example is that while hydro-energy is perceived to be friendly it has many adverse effects. Dams accelerate soil erosion and increase soil salinity downstream in the rivers upon which they are constructed. Plus they interfere with the marine life in the rivers and estuaries. Wave and tidal turbines also interfere with marine life. Similarly wind energy affect negatively birds' habitat.

Finally a third example is, that biofuels production is affecting food production, with many farmers opting to produce energy crops at the expense of food, due to higher profitability, leading to food shortages worldwide. In addition these 'energy crops' are also competing for the already scarce water resources in many areas.

13.2 Can Alternative Energy Replace Fossil Fuels?

For any energy source to be credible it must fulfil three main tests: reliability, availability and affordability. Nowadays fossil fuels satisfy all three tests, so if any type of alternative energy is to replace fossil fuels it has not only to satisfy the tests but also outperform them by registering a better score to justify its usage.

At the present time fossil fuels account for over 80% of the humanity's needs. See Chapter 1 (refer to Figure 1.2). As explained earlier traditional biomass is not always included in the energy production and consumption numbers, thus an exact share of all alternative energy supplies cannot be calculated accurately. However based on the most optimistic numbers presented by the IEA, which takes into account biomass usage, all alternative energy supplies contributed to just over 19% of the world's energy supply in 2007, which is distributed as follows: 10% biomass (including biofuels); 6% nuclear; 2% hydro; 1% other renewable energy sources (solar, geothermal, wind). So obviously at the moment all types of alternative energy combined cannot replace fossil fuels.

One may ask what about the future? Will alternative energy be able to replace fossil fuels? The simple answer is still a resounding no, according to the IEA[1], which forecasts that the share of all types of alternative energy combined will hardly change by 2030 and that all alternative energy types will only manage to supply approximately 20% of the world's energy by 2030, with is divided into 10% biomass (including biofuels); 6% nuclear; 2% hydro; 2% other renewables.

From the numbers above it becomes clear that despite all the hype, current alternative energy sources cannot fulfil the world's projected energy demands and that the world will still be dependent on fossil fuels for decades to come. You may be questioning the reasons for this dependence since the alternative energy sources are virtually unlimited and thus in theory should be able to fulfil the world's energy needs.

---

[1] IEA, World Energy Outlook 2009.

The answer to the above question is not straight forward. Firstly, we can correctly conclude that all types of alternative energy pass the availability criterion, especially if considered collectively. However all types of alternative energy fail at least one of the remaining two criteria identified earlier. Firstly all types are still expensive in comparison to fossil fuels and thus they fail the affordability criterion. This means that the world's population would need to accept a drop in their current living standards to be able to afford the increasing energy prices, if alternative energy is to replace fossil fuels. In terms of reliability, wind and solar energies are by their nature intermittent since they rely heavily on weather conditions and thus they fail the reliability criterion. Hydro energy also fails this criterion, but to a lesser extent since waves and tides have steadier patterns, while dams are generally reliable with the exception of seasons with severe droughts.

In addition, the public in general, while it appears to embrace alternative energy in principle, turns hostile to alternative energy projects it is to be constructed in their region. Many are strong supporters of the NIMBY (Not In My Back Yard) principle. For example, while the public opinion in Western Europe supports strongly wind energy, local communities oppose strongly wind farms being constructed in their nearby countryside or offshore their coastline, citing excuses such as protecting the scenery or the wildlife! Another example is the fierce opposition to dams by environmental organisations wherever and whenever they are proposed.

Following the preceding arguments, it is essential to put it in perspective and quantify the size of alternative energy sources required to substitute one cubic mile of oil, which happens to be approximately equal to the world's oil consumption in 2008. The comparison exercise throws staggering numbers. Thus to totally substitute one cubic mile of oil (remember we are not even including natural gas or coal) we will need 4 'Three Gorges' sized dams, or 52 nuclear power plants, or 104 coal-fired power plants, or 32850 wind turbines, or 91250000 rooftop solar photovoltaic panels developed each year for 50 years.[2]

We can conclude from the discussion above that alternative energy is certainly not an adequate substitution for the fossil fuel 'addiction', and it is not the magic solution to solve the world's energy demands. To satisfy the world's energy needs, we need to adopt a multi-conceptual strategy that is based on three principles: diversification, efficiency and legislation.

In terms of diversification, we must accept that a variety of energy resources have to be utilised together in a multi-source energy market.

In terms of efficiency, we need to develop technologies to maximise efficiency, to reduce energy usage, to improve electricity grids or to change our behaviour, attitudes and habits to encourage energy savings. The drive to improve technology is relentless, and is outside the scope of this text. Examples of behavioural change are plentiful, and if listed here will run into hundreds of lines. For example, I am sure you all agree that better insulation at our homes will decrease our energy consumption, reducing the world's energy demands and making us all richer by cutting our energy bills.

Nowadays, the drive to reduce our energy consumption is driven by legislation that stems from the concerns regarding climate change, which are strongly linked to carbon dioxide and other greenhouse gases emissions that are emitted mostly due to our usage of fossil fuels. Thus to reduce these harmful emissions, legislation has been and can be used as, a powerful tool to impose limits on emissions by various measures that range from the obvious to the innovative.

An obvious piece of legislation, for example, is to enforce stringent technology standards, forcing emitters to include carbon sequestration (also known as carbon capture and storage CCS)

---

[2] Harry Goldstein, William Sweet, Joules, BTUs, Quads--Let's Call the Whole Thing Off, January 2007, (http://spectrum.ieee.org/energy/fossil-fuels/joules-btus-quadslets-call-the-whole-thing-off).

processes not only to all future projects whether extracting fossil fuels or utilising them for power generation, but also retrospectively to existing projects, especially dirty coal power stations.

On the other hand, an innovative law would be to limit the size of car engines allowed on the road based on both its consumption and/or emissions per driven distance, which will reduce the transport energy consumption. Another law is to introduce car scrappage schemes to replace cars, based on their environmental footprints rather than to encourage their sales to stimulate the economy.

Even more innovative laws can be introduced to tax electrical appliances, based on their electrical consumption or impose a restriction on central heating thermostats to limit the temperatures at homes.

Unfortunately, the first example is being promoted enthusiastically by governments and public media despite being unproven and despite its high costs; while hardly anyone mentions the second and third examples, even though they are straightforward. A question pops up: is this due to pressure from the car industry lobby or from the energy lobby, who both need the 'gas guzzlers' to sell more oil?

As for the last two examples, the proposed measures could be very unpopular so most politicians will avoid them at any cost.

Until recently, the accepted wisdom by the majority of environmentalists, propped up by strong media propaganda and supported by main stream politicians, advocated reducing fossil fuel emissions as the only route to save our planet from the adverse effects of global warming. Despite agreeing that greenhouse emission are contributing to global warming, a second view is emerging that argues that the earth did in fact reach a tipping point in terms of carbon dioxide emissions and that replacing fossil fuels with alternative energy is not the solution to alleviate global warming anymore. An increasing number of scientists are proposing radical geo-engineering solutions as our only hope, although this view is still not taken seriously by main stream media or the hard core environmentalists. Before you laugh, these are not all fanciful science fiction projects, but some can be very simple, credible and practical solutions. For example if most buildings were to be painted white they could reflect more sun and thus could reduce some effects of the global warming.

From an Arab perspective, all countries fall in line with the global view regarding utilisation of alternative energy supplies. However, in numerical terms, the share of alternative energy in the Arab world energy mix is considerably less than in many parts of the world. Many reasons can be attributed to this low intake, with the most significant being the Arab world's massive oil and natural gas reserves, which alongside lax environmental policies, create no incentives to diversify their energy supplies. Nowadays with the exception of the High Dam on the River Nile in Egypt as a major alternative energy source, alternative energy sources are awfully under-utilised in the Arab world. This is despite the fact that the Arab world possesses a huge potential to utilise solar energy, due to its hot and arid environment, although recently plans were being drawn up by a consortium of European energy companies to tap into this resource to produce solar-generated electricity with a vast network of power plants and transmission grids across North Africa and the Middle East.[3] On the other hand the Arab world's arid environment limits its potential, in terms of both hydro and wind energy. Finally, due to political pressure, nuclear energy has not been utilised for the fear of developing military nuclear capabilities in the Middle East, although this started to change recently with the UAE and Jordan initiating nuclear energy programs.

---

[3] Desertec Foundation, (http://www.desertec.org).

Chapter 14
FOSSIL FUELS AND THE ARAB WORLD – DIFFERENTIATING FACTS FROM FICTION

In Chapter 12 we established, using all the numerical data the quantitative outline of fossil fuels both globally and from an Arab perspective. In this chapter the discussion is expanded by exposing a few famous widely believed myths and using the facts in the previous chapters, as well as touching on political insights, to challenge and debunk them. Finally we finish the chapter by raising a few questions for consideration where the readers can reflect on all the facts and try to come to the answers by themselves.

14.1 Myths Debunked

14.1.1 Myth 1: We are running out of oil, or as some depletionists recently extend it, of all fossil fuels.

This is fundamentally wrong.

The Stone Age did not finish because we ran out of stones and the hydrocarbon age will not end because we run out of hydrocarbons. The arguments regarding peak oil are exaggerated and, as demonstrated clearly in the preceding chapters, the world is awash with fossil fuels. The numbers presented earlier show that humanity consumed a mere 6.8%, 0.1% and 1.3% of oil, natural gas, and coal 'originally-in-place' respectively. So it is obvious to all, that the fossil fuel endowment is plenty and will last for centuries. Fossil fuels are abundant their resource base is likely to expand further. Furthermore, they will continue to be the primary energy sources, with the unconventional becoming increasingly important.

However, with demand projected to continue increasing, there is no argument that the era of easy and cheap oil specifically or easy and cheap fossil fuels generally, has come to an end; or that the peak of cheap oil is imminent if not already surpassed, with most of the remaining fossil fuels resources requiring more investment to extract and thus are more expensive.

An interesting observation is that the R/P for oil remained around 40 years for the last 80 years, which – despite the criticism of the ratio's relevance in future oil markets – proves that the industry has always managed to keep a considerable cushion of oil without any threat of running out, with little economic incentive to vigorously explore and develop other existing reserves.

14.1.2 Myth 2: Technology does not enhance fossil fuels availability.

This is fundamentally wrong.

Technology is playing an increasingly vital role in increasing the size of fossil fuel reserves, as it enables the industry to squeeze more oil and gas from reservoirs and even to mine coal from offshore reservoirs. While technology cannot alter the geology of the reservoirs, better understanding and better data, as well as novel new methods, are allowing us to extract more from the reservoirs. Furthermore, technology reduces costs when its usage becomes widespread and thus increases the size of the reserves.

Contrary to the blitz of publicity crying wolf, that the world has not discovered new fossil fuel fields and that the industry is depleting what is already there, the truth is that most of the world's fossil fuel production is coming from already mature fields, with the fields' life extended significantly by technical advancements. To put this in perspective the SPE[1] reported that while in 1960s 50-60% of oil and gas produced came from new fields, this share declined to 12-15% in 2009 and is projected to

---

[1] SPE, Oil, Gas and Energy: Myths and Realities, (tyumen.spe.org/images/tyumen/articles/36/OilGasEnergyMythsReality.ppt).

decline further to 7-10% in the future.

Moreover, it is an irrefutable fact is that fossil fuels recovery factors, whether oil, natural gas or coal, are improving all the time and thus continue to add reserves. In addition, technical advancement brought numerous unconventional resources into play; with the growing importance of shale gas as just one example. It is not a question of 'if' any more, but rather a question of 'when' when oil shale, gas hydrates and coal in-situ gasification will join the mix.

14.1.3 Myth 3: The Arab world is home to two thirds of the world's oil reserves, and thus its oil is indispensable to the world.

This is fundamentally wrong.

It is a known fact that the Arab world is underexplored in terms of fossil fuels and thus more resources will be identified in due course, especially unconventional resources. However, even taking this into consideration, the share of the Arab world's fossil fuels is grossly under the two thirds tag reported by the media.

As already explained in Section 9.1, the Arab world is truly dominant in terms of conventional oil reserves, where its share stands at 58%. However when unconventional reserves are taken into account, this share drops sharply to only 38% (see Section 9.3). The situation is similar in terms of natural gas reserves, though the numbers differ significantly, with the Arab world's share dropping from a significant 30% of conventional reserves to a mere 16%, when unconventional reserves are considered (see Sections 10.1 and 10.3). If we look at resources rather than reserves, the Arab world's share even drops to 47% and 21% of conventional oil and natural gas respectively (see Sections 9.1 and 10.1) and reduces further to only 30% and 10% of overall oil and natural gas respectively (see Sections 9.3 and 10.3) when unconventional resources are included. As for coal, the Arab world has no significant reserves or resources as explained in Chapter 11.

Consequently as can be seen in Chapter 12, the overall share of fossil fuels of the Arab world is a fraction of what we are lead to believe standing at only 15% and 8% of overall reserves and resources respectively.

14.1.4 Myth 4: OPEC and both national and international oil and gas companies control oil price.

This is fundamentally wrong.

The price of fossil fuels is determined by many factors, chiefly the fundamental market factors of supply and demand. However in addition, many factors play major roles in influencing the price. These include: global political instability; producing countries internal political states of affairs, which may lead to disruption in supply due to local unrest, protectionism, and nationalisation; global economic instability; market speculators; local energy subsidies; natural disasters, environmental incidents; etc. Thus it is clear that fossil fuels pricing is extremely complicated and beyond the scope of this book.

To illustrate the effect of one of the above factors that, besides market forces, plays a vital role in pricing, we note that most conventional oil and natural gas reserves are located in remote areas, mostly in politically unstable countries, where the political situation plays a key role in determining the oil and natural gas price. This far outweighs market fundamental factors. Examples of this political role include the insurgency in Iraq, the unrest in Nigeria, the nationalisation of oil and gas industry in Venezuela, the state interference in Russia, etc.

Another main factor, which is often overlooked, but may become a determining contributing factor in the pricing equation in the near future, is the lack of skilled personnel to design, construct and operate fossil fuels projects. As a result of the 1980s low fossil fuel prices, the interest in studying engineering (chemical, petroleum, mining) and geosciences, as well as related subjects reached a record low, leading to a potential acute shortage in qualified people to do the jobs which will inevitably lead to additional costs and thus affect the fossil fuel prices.

More recently the relatively high oil prices, which even having climbed down from their 2008 peak values, are still pretty high in actual terms. If the price persists at this high level and the price trend continues, there may be significant consequences for the oil industry with more oil reserves becoming feasible, being developed and coming online. This will have an adverse effect on the price of oil in the longer term.

14.1.5 Myth 5: The USA is dependent on Arab oil.

This is fundamentally wrong.

It is true that the USA is dependent on imported oil for over 60% of its oil needs. However its dependence is considerably lower in terms of natural gas and it is a net coal exporter. In addition, projections for future dependency show that the USA's reliance on foreign fossil fuels is declining, with a higher percentage coming from its domestic supply.

It is totally untrue that the USA depends on the Middle East or the Arab world to secure its oil and gas supplies. In fact, the USA does *not* depend on the Arab world for its energy needs, it imports only a small fraction of its oil from the Arab countries, and since the Iranian revolution, the USA stopped importing any Iranian oil, so the small fraction of its supplies which it imports from the Middle East is effectively from Arab countries.

The truth of the matter that many of the USA allies, especially in Asia, depend fully on Middle Eastern or Arab oil and natural gas, thus cynics may suggest that the desire of the USA to control energy resources in the Middle East is a ploy to indirectly keep a firm grip on its allies. This book does not try to analyse the politics or dig into the intentions of the American foreign and defence policies, but here we only present the numbers and ask the readers to think for themselves. Table 14.1 shows the imports of oil and natural gas from the Arab countries and Iran (if applicable) for the USA, EU, China, Japan, South Korea, India and Taiwan. Coal is excluded from the table as the Arab countries and Iran are not coal exporters.

Table 14.1: Imports of oil and natural gas from the Arab countries and Iran for selected countries

Country / Union	Oil		Natural Gas		Source
	Arab World %	Iran %	Arab World %	Iran %	
USA	19.7	0.0	4.6	0.0	1, 2
EU	21.5	4.0	28.0	0.0	3,4,5
China	46.6	11.4	15.9	0.0	6
Japan	80.0	10.1	11.4	0.0	7,8
India	45.4	16.8	71.4	0.0	9
South Korea	74.4	7.3	25.6	0.0	10
Taiwan	69.6	13.4	8.3	0.0	11

Sources:
1: EIA (http://www.eia.doe.gov/dnav/pet/pet_move_impcus_a2_nus_ep00_im0_mbbl_m.htm).
2: EIA (http://www.eia.doe.gov/dnav/ng/ng_move_impc_s1_a.htm).
3: EuroStat (http://epp.eurostat.ec.europa.eu/portal/page/portal/energy/data/database).
4: Francis Ghilès, Algeria: A Strategic Gas Partner For Europe, Journal of Energy Security. http://www.ensec.org/index.php?option=com_content&view=article&id=176:algeria-a-strategic-gas-partner-for-europe&catid=92:issuecontent&Itemid=341
5: Assessment Report of Directive 2004/67/Ec On Security Gas Supply, Commission of the European Communities, Brussels, 16.7.2009 SEC(2009) 978 final.
6: ChinaOilWeb (http://data.chinaoilweb.com).
7: Petroleum Association of Japan (http://www.paj.gr.jp/english/statis).
8: EIA (http://www.eia.doe.gov/emeu/international/LNGimp2008.html).
9: Department of Commerce, India (http://commerce.nic.in/edib).
10: Korea National Assembly Research Service, Report for Effective Official Regulation (in Korean).
11: Energy Statistics Databook, Bureau of Energy, MOEA (2009).

14.1.6 Myth 6: Energy independence is possible.

This is fundamentally wrong.

We live in global economy where trade is the essence of the new way of life. Thus, for nations to aspire to achieve energy independence runs in contrast to this new way of life, the principles of free trade and global integration. What matters now for most governments is to guarantee the security of energy supply, to optimise and reduce energy costs rather than obtain energy at any cost and to reduce the adverse environmental effects of energy production and consumption. Consequently, for most governments, the emphasis should be to drop the aim of energy independence and concentrate instead on achieving security of energy supply, strengthening international energy trade and investment, encourage energy sources diversification, curb energy usage and increase energy efficiency.

To illustrate the concepts above, consider that, for example, while it may be possible for a country to mine its coal resources at high costs and incalculable environmental costs, it may be cheaper in many instances to import its coal needs from other countries, and thus sacrificing energy independence, whilst still fulfilling its energy needs.

14.1.7 Myth 7: Alternative energy can replace fossil fuels.

This is fundamentally wrong.

As discussed already in Chapter 13, alternative energy will not be able to replace fossil fuels in the near or medium term future. No doubt it will contribute to reducing the world's reliance on fossil fuels but, without a total change in policy and attitude, the goal of reducing the dependence on fossil fuels remains elusive.

14.2 Questions For Thought

After debunking seven well publicised myths, it is now fitting to raise a few questions for consideration, where the readers can reflect on what they have read in this book and try to figure out the answers. The questions raised are:

- What is the reason of the world's obsession with the Arab world in general and Arab oil in particular?
- Why keep corrupt Arab governments in power?
- Why invade Iraq to replace a tyrant with a failed state?
- Who benefits from high fossil fuel prices? Are they the current producers or the countries with unconventional deposits?
- Why promote expensive alternative energy rather than energy efficiency and energy economising via legislation?
- Why not curb carbon emissions by restricting emission

sources rather than insisting on promotion and expansion of carbon sequestration?
- Why dismiss geoengineering?
- Who benefits from concealing the facts and promoting the myths and the fiction?

As I said earlier these questions will not be answered here, but my hint to guide readers to the answer, is the word 'Politics'.

Finally, one fact that is clear but most people still avoid acknowledging is that our real problem as human beings, is not that we are running out of fossil fuels, or our increasingly high greenhouse gas emissions leading to global warming. The real problem is that our beloved Earth is becoming exponentially over populated and that this population explosion is causing severe scarcity affecting all resources, not only fossil fuels, but also water and metals, as well as food. So my main question left here, why is this argument often ignored and overlooked?

APPENDIX A: ARAB WORLD STATISTICS

Table A1: Arab world's basic information

Country	Capital	Area km²	Rank	Population million	Rank
Bahrain	Manama	741	22	0.738	22
Iraq	Baghdad	438317	10	29.672	5
Jordan	Amman	89342	14	6.407	12
Kuwait	Kuwait City	17818	17	2.789	18
Lebanon	Beirut	10400	19	4.125	14
Oman	Muscat	309500	11	2.968	17
Palestine	East Jerusalem	6220	20	4.119	15
Qatar	Doha	11586	18	0.841	19
Saudi Arabia	Riyadh	2149690	3	29.207	6
Syria	Damascus	185180	12	22.198	8
UAE	Abu Dhabi	83600	15	4.976	13
Yemen	Sana'a	527968	9	23.495	7
Algeria	Algiers	2381741	2	34.586	3
Comoros	Moroni	2235	21	0.773	20
Djibouti	Djibouti	23200	16	0.741	21
Egypt	Cairo	1001450	6	80.472	1
Libya	Tripoli	1759540	4	6.461	11
Mauritania	Nouakchott	1030700	5	3.205	16
Morocco	Rabat	712550	7	32.119	4
Somalia	Mogadishu	637657	8	10.112	10
Sudan	Khartoum	2505813	1	41.980	2
Tunisia	Tunis	163610	13	10.589	9
ARAB WORLD		14048858		352.574	
Arab world Share (%)		9.43		5.16	

Source: CIA Factbook.
Note 1: Population as estimated in July 2010.
Note 2: Area and population of Morocco include Western Sahara
Note 3: Area and population of Palestine is for the West Bank including East Jerusalem and Gaza Strip. Population number excludes Israeli settlers.
Note 4: East Jerusalem is the proclaimed Palestinian capital. Ramallah is the current administrative centre of the Palestinian Authority, while Gaza city is the administrative centre of the Hamas-led government.

www.ingramcontent.com/pod-product-compliance
Ingram Content Group UK Ltd.
Pitfield, Milton Keynes, MK11 3LW, UK
UKHW041258180426
11947UKWH00008B/555